Advances in Hydraulics
and Hydroinformatics

Advances in Hydraulics and Hydroinformatics

Special Issue Editors

Jian Guo Zhou
Jianmin Zhang
Yong Peng
Alistair G. L. Borthwick

MDPI • Basel • Beijing • Wuhan • Barcelona • Belgrade • Manchester • Tokyo • Cluj • Tianjin

Special Issue Editors

Jian Guo Zhou
Manchester Metropolitan University
UK

Jianmin Zhang
Sichuan University
China

Yong Peng
Sichuan University
China

Alistair G. L. Borthwick
The University of Edinburgh
UK

Editorial Office
MDPI
St. Alban-Anlage 66
4052 Basel, Switzerland

This is a reprint of articles from the Special Issue published online in the open access journal *Water* (ISSN 2073-4441) (available at: https://www.mdpi.com/journal/water/special_issues/Hydraulics_Hydroinformatics).

For citation purposes, cite each article independently as indicated on the article page online and as indicated below:

LastName, A.A.; LastName, B.B.; LastName, C.C. Article Title. *Journal Name* **Year**, *Article Number*, Page Range.

Volume 1
ISBN 978-3-03936-124-3 (Hbk)
ISBN 978-3-03936-125-0 (PDF)

Volume 1–2
ISBN 978-3-03936-128-1 (Hbk)
ISBN 978-3-03936-129-8 (PDF)

Contents

About the Special Issue Editors vii

Preface to "Advances in Hydraulics and Hydroinformatics" ix

Fangfang Wang, Ang Gao, Shiqiang Wu, Senlin Zhu, Jiangyu Dai and Qian Liao
Experimental Investigation of Coherent Vortex Structures in a Backward-Facing Step Flow
Reprinted from: *Water* 2019, *11*, 2629, doi:10.3390/w11122629 1

Ruonan Bai, Dejun Zhu, Huai Chen and Danxun Li
Laboratory Study of Secondary Flow in an Open Channel Bend by Using PIV
Reprinted from: *Water* 2019, *11*, 659, doi:10.3390/w11040659 15

Ming Chen, Haijin Huang, Xingxing Zhang, Senpeng Lv and Rengmin Li
Experimental Investigation on Mean Flow Development of a Three-Dimensional Wall Jet
Confined by a Vertical Baffle
Reprinted from: *Water* 2019, *11*, 237, doi:10.3390/w11020237 31

Xuneng Tong, Xiaodong Liu, Ting Yang, Zulin Hua, Zian Wang, Jingjing Liu and Ruoshui Li
Hydraulic Features of Flow through Local Non-Submerged Rigid Vegetation in the Y-Shaped
Confluence Channel
Reprinted from: *Water* 2019, *11*, 146, doi:10.3390/w11010146 45

Kwang Seok Yoon, Seung Oh Lee and Seung Ho Hong
Time-Averaged Turbulent Velocity Flow Field through the Various Bridge Contractions during
Large Flooding
Reprinted from: *Water* 2019, *11*, 143, doi:10.3390/w11010143 63

Zhong Tian, Wei Wang, Ruidi Bai and Nan Li
Effect of Flaring Gate Piers on Discharge Coefficient for Finite Crest-Length Weirs
Reprinted from: *Water* 2018, *10*, 1349, doi:10.3390/w10101349 77

**Roman Gabl, Jeffrey Steynor, David I. M. Forehand, Thomas Davey, Tom Bruce and
David M. Ingram**
Capturing the Motion of the Free Surface of a Fluid Stored within a Floating Structure
Reprinted from: *Water* 2019, *11*, 50, doi:10.3390/w11010050 89

Qiulin Li, Lianxia Li and Huasheng Liao
Study on the Best Depth of Stilling Basin with Shallow-Water Cushion
Reprinted from: *Water* 2018, *10*, 1801, doi:10.3390/w10121801 105

Jun Deng, Wangru Wei, Zhong Tian and Faxing Zhang
Design of A Streamwise-Lateral Ski-Jump Flow Discharge Spillway
Reprinted from: *Water* 2018, *10*, 1585, doi:10.3390/w10111585 121

Mochammad Meddy Danial, Kiyosi Kawanisi and Mohamad Basel Al Sawaf
Characteristics of Tidal Discharge and Phase Difference at a Tidal Channel Junction Investigated
Using the Fluvial Acoustic Tomography System
Reprinted from: *Water* 2019, *11*, 857, doi:10.3390/w11040857 135

Ruichang Hu and Jianmin Zhang
Numerical Analysis on Hydraulic Characteristics of U-shaped Channel of Various Trapezoidal Cross-Sections
Reprinted from: *Water* **2018**, *10*, 1788, doi:10.3390/w10121788 . . . . **157**

Yongfei Qi, Yurong Wang and Jianmin Zhang
Three-Dimensional Turbulence Numerical Simulation of Flow in a Stepped Dropshaft
Reprinted from: *Water* **2019**, *11*, 30, doi:10.3390/w11010030 . . . . **181**

Arun Kamath, Gábor Fleit and Hans Bihs
Investigation of Free Surface Turbulence Damping in RANS Simulations for Complex Free Surface Flows
Reprinted from: *Water* **2019**, *11*, 456, doi:10.3390/w11030456 . . . . **199**

Yaru Ren, Min Luo and Pengzhi Lin
Consistent Particle Method Simulation of Solitary Wave Interaction with a Submerged Breakwater
Reprinted from: *Water* **2019**, *11*, 261, doi:10.3390/w11020261 . . . . **225**

Haifei Liu, Zhexian Zhu, Jingling Liu and Qiang Liu
Numerical Analysis of the Impact Factors on the Flow Fields in a Large Shallow Lake
Reprinted from: *Water* **2019**, *11*, 155, doi:10.3390/w11010155 . . . . **241**

Xin Li, Maolin Zhou, Jianmin Zhang and Weilin Xu
Numerical Study of the Velocity Decay of Offset Jet in a Narrow and Deep Pool
Reprinted from: *Water* **2019**, *11*, 59, doi:10.3390/w11010059 . . . . **255**

Shicheng Li and Jianmin Zhang
Numerical Investigation on the Hydraulic Properties of the Skimming Flow over Pooled Stepped Spillway
Reprinted from: *Water* **2018**, *10*, 1478, doi:10.3390/w10101478 . . . . **271**

Wenjun Liu, Bo Wang, Yunliang Chen, Chao Wu and Xin Liu
Assessing the Analytical Solution of One-Dimensional Gravity Wave Model Equations Using Dam-Break Experimental Measurements
Reprinted from: *Water* **2018**, *10*, 1261, doi:10.3390/w10091261 . . . . **291**

Jie Ren, Xiuping Wang, Yinjun Zhou, Bo Chen and Lili Men
An Analysis of the Factors Affecting Hyporheic Exchange based on Numerical Modeling
Reprinted from: *Water* **2019**, *11*, 665, doi:10.3390/w11040665 . . . . **307**

Anping Shu, Guosheng Duan, Matteo Rubinato, Lu Tian, Mengyao Wang and Shu Wang
An Experimental Study on Mechanisms for Sediment Transformation Due to Riverbank Collapse
Reprinted from: *Water* **2019**, *11*, 529, doi:10.3390/w11030529 . . . . **329**

About the Special Issue Editors

Jian Guo Zhou graduated from Wuhan University with a BSc in River Mechanics and Engineering and subsequently completed his MSc in Fluvial Mechanics at Tsinghua University. He received his PhD in Fluid Mechanics from the University of Leeds. He specialises in formulating mathematical models and developing numerical methods for flow problems in fluids and water engineering, such as coastal, estuary, hydraulic, river and environmental engineering.

Jianmin Zhang graduated from Shihezi University with a BSc, and subsequently completed his MSc at Xinjiang Agricultural University. In 2000, he received his PhD in Hydraulics and River Dynamics from Sichuan University. His core research focuses on engineering hydraulics, flood discharge, energy dissipation, hydraulic experiments, two-phase flows and numerical simulations of turbulence.

Yong Peng graduated from Northwest A&F University with a BSc in Hydraulic and Hydropower Engineering, and subsequently completed his MSc in Hydraulics and River Dynamics at Sichuan University. He received his PhD in Computational Hydraulics from the University of Liverpool. His main research focuses on flooding, contaminant transport, sediment transport, the lattice Boltzmann method, shallow water flow, two-phase flow, cavitation, and hydraulic experiments.

Alistair G. L. Borthwick has more than 40 years' experience in civil engineering. He holds a BEng and a PhD from the University of Liverpool, a DSc degree from Oxford University, and an honorary degree from the Budapest University of Technology and Economics. His research interests include environmental fluid mechanics, coastal and ocean engineering, and marine renewable energy. In 2019, he was awarded the Gold Medal of the UK Institution of Civil Engineers.

Preface to "Advances in Hydraulics and Hydroinformatics"

Great progress has been made in the research on hydraulics, hydrodynamics and hydroinformatics over the past few decades. This includes theoretical, experimental and numerical studies, leading to a new understanding and knowledge of water-related problems, covering a wide range of topics and applications such as hydrology, water quality, river and channel flows. For example, coherent vortex structures in a backward-facing step flow and secondary flow in an open channel bend are measured using PIV; the flow through a Y-shaped confluence channel partially covered with rigid vegetation on its inner bank is measured by ADV. Meanwhile, the velocity decay of an offset jet in a narrow and deep pool, and the skimming flow over a pooled stepped spillway, are studied numerically. In addition, mechanisms for sediment transport due to riverbank failure and a method of predicting maximum scour depth are investigated, and cavitation bubble collapse and its impact on a wall are studied. In order to accelerate knowledge transfer in resolving engineering problems, this Special Issue reports both on the state-of-the-art of the aforementioned research topics, and on advances in sediment transport dynamics, two-phase flows, flow-induced vibration, and hydropower station hydraulics. The goals of this Special Issue are to improve our understanding of the foregoing flow problems and to stimulate future research in these areas, aiming for an improved quality of life.

Jian Guo Zhou, Jianmin Zhang, Yong Peng, Alistair G. L. Borthwick
Special Issue Editors

Article

Experimental Investigation of Coherent Vortex Structures in a Backward-Facing Step Flow

Fangfang Wang [1,*], Ang Gao [1], Shiqiang Wu [1,*], Senlin Zhu [1], Jiangyu Dai [1] and Qian Liao [2]

[1] State Key Laboratory of Hydrology-Water Resources and Hydraulic Engineering, Nanjing Hydraulic Research Institute, Nanjing 210029, China; gaoang@whu.edu.cn (A.G.); slzhu@nhri.cn (S.Z.); jydai@nhri.cn (J.D.)

[2] Department of Civil Engineering and Mechanics, University of Wisconsin-Milwaukee, Milwaukee, WI 53201, USA; liao@uwm.edu

* Correspondence: ffwang@nhri.cn (F.W.); sqwu@nhri.cn (S.W.)

Received: 14 October 2019; Accepted: 10 December 2019; Published: 13 December 2019

Abstract: Coherent vortex structures (CVS) are discovered for more than half a century, and they are believed to play a significant role in turbulence especially for separated flows. An experimental study is conducted for a pressured backward-facing step flow with Reynolds number (Re_h) being 4400 and 9000. A synchronized particle image velocimetry (PIV) system is developed for measurement of a wider range of velocity fields with high resolution. The CVS are proved to exist in the separation-reattachment process. For their temporal evolution, a life cycle is proposed that vortices form in the free shear layer, develop with pairings and divisions and finally shed at the reattachment zone, and sometimes new vortical structures are restructured with recovery of flow pattern. The CVS favor the free shear layer with frequent pairings and divisions particularly at the developing stage around $x/h = 2\sim5$ (x: distance from the step in flow direction, h: step height), which may contribute to the high turbulent intensity and shear stress there. A critical distance is believed to exist among CVS, which affects their amalgamation (pairing) and division events. Statistics show that the CVS are well organized in spatial distribution and show specific local features with the flow structures distinguished. The streamwise and vertical diameters (D_x and D_y) and width to height ratio (D_x/D_y) all obey to the lognormal distribution. With increase of Re_h from 4400 to 9000, D_x decreases and D_y increases, but the mean diameter ($D=0.5 \times (D_x + D_y)$) keeps around (0.28~0.29) h. As the increase of Re_h, the vortical shape change toward a uniform condition, which may be contributed by enhancement of the shear intensity.

Keywords: coherent vortex structure; backward-facing step; synchronized PIV; separation and reattachment; free shear layer; vortical evolution

---

1. Introduction

Flow separation is common in various applications in our daily life (airfoils at large attack angle, spoiler flows, flow behind a vehicle, flow inside a combustor/condenser, flows in an enlarged pipe or a channel with steps, and flows around a boat or a building, etc.). Backward-facing step (BFS) is one basic model and fundamental in geometry and engineering design [1–5]. Flow separation behind the step changes transport of momentum, heat, and mass, and increases fluid mixing [6], and it also possibly leads to additional flow resistance and noise, and even for stalling of an aircraft under critical conditions. In a combustor or chamber, effective mixing is preferred and can be designed with detailed BFS geometric considerations [5,7–9]. In addition, the enlarged boundary in a pipe or channel sometimes generates unfavorable boundary erosion and structural vibration, and sometimes energy dissipation caused by structural changes in hydraulic engineering arouses people's curiosity. In field of scouring downstream of a structure, motions and fluctuations of the vortices will cause sediment transport and river-bed scour [10]. However, because of the irregularity and complexity of turbulence, the flow characteristics, various mechanisms, and controlling studies on this flow are still worth further efforts with developments of measurement techniques [1–5,11]. In particular, coherent vortex structures (CVS) are discovered to play an important role and may be a promising way to investigate this flow [1,3–6,11–16]. However, quantitative studies of the CVS (e.g., the conception, identification, structure, scale, and evolution, etc.,) are deficient, and it is still difficult to fully understand the hydrodynamic mechanisms and unsteady interactions in many applications (e.g., pipe cavitation, earth dam break, riverbed and bank erosion, sand dune evolution, and energy dissipation).

Figure 1 shows the typical flow structures in a two-dimensional BFS. Flow separates at the step edge (step height, h) and reattaches the bottom wall at some distance downstream, forming a recirculation region behind the step. This process consists of most physical features of a separation-reattachment flow, such as the free shear layer, the reattachment zone, and the redeveloping boundary layer [5,11,13,16–19]. The recirculation region is extensively studied in the previous researches. Length of this region (recirculation length, X_r) keeps a relative stable value of $(7 \pm 1)\,h$ for a developed turbulent flow, while it increases for a laminar flow and decreases for a transitional flow with the increase of Reynolds number [20]. The time-averaged flow structure is believed to be stable in a developed turbulent flow such as the free shear region (FSR), the corner region (CR), the redeveloping region (RR), and the reattachment zone [11]. These regions show various flow characteristics. The FSR is generated from the step top corner with strong shear intensity and finally limited by the bottom wall. The CR is located at the step corner with a secondary backflow. The RR shows a new development for the turbulent boundary layer after reattachment. The reattachment zone is defined as the locus of points, where the limiting streamline of the time-averaged flow rejoins the surface [21]. For the temporal flow, perturbation of the step generates successive CVS which promotes the fluid mixing and intensifies the unsteady interaction. The CVS shows four typical developing stages with various characteristic frequencies in the flow direction: forming, developing, shedding, and redeveloping [11].

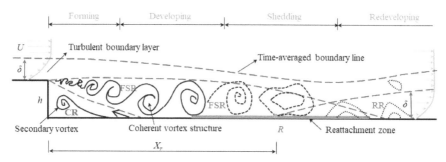

Figure 1. Simplified representation of a two-dimensional backward-facing step (BFS) flow: (FSR: free shear region; CR: corner region; RR: redeveloping region; R: mean reattachment point).

The complexities of BFS flows are summarized by Roos and Kegelman (1986) [12]: (1) development of CVS in the free shear layer; (2) substantial unsteadiness of the entire separated flow, especially near the reattachment zone and particularly at low frequencies; (3) slow relaxation to equilibrium of the reattached shear layer, and the third one is also related to the former two. Recent decades, researchers are working on the CVS and unsteady motions in the complex flows.

Flow visualization is first used to investigate this flow. Tani et al. (1961) [22] showed aluminum-power pictures of an instantaneous BFS flow and found that short exposure pictures reveal some distinct vortex structures (CVS) in the recirculation (Figure 2a), while pictures of long exposure only indicate something like that shown in the mean dividing streamlines. The CVS present formation, merging, and increasing irregularity with the increase of shear layer turbulence and persistence of the structures downstream of reattachment (Figure 2b–d) [6,14]. The CVS are believed to be rolled up in the strong shear layer, which grows with a vortex pairing mechanism until it approaches the reattachment zone [6,12,14,15,22]. In the reattachment zone, the CVS are torn roughly into two in the neighborhood of the reattachment point (*R*), where part of the flow is deflected upstream into the recirculation region to supply the entrainment and another part continues downstream [16]. McGuinness (1978) [23] believes that some CVS are swept upstream with the recirculating flow, while others proceed downstream. The fact that CVS move alternately up and downstream is also supported by the measurements of Chandrsuda (1975) [24] and Kim et al. (1978) [25]. In first part of the FSR, CVS and their evolution are usually explained by the Kelvin-Helmholtz instability that proposed in a plane mixing layer. However, BFS flow shows some differences [1,12]: (1) The free shear layer has a finite length due to the reattachment, which limits the development of the CVS; (2) The free shear layer supports an adverse pressure gradient and is curved; (3) Strong interaction between the unsteady shear layer and the reattachment wall generates substantial acoustic and convective disturbances, that feed back through the recirculation to influence the shear layer at separation. The peak value of turbulence intensity in a BFS flow is 10~20% higher than that of a plane mixing layer [1]. Additionally, local forcing with certain frequency is proved to contribute to amalgamation of the CVS and fluid mixing (Figure 2b,c), which also shortens X_r [6].

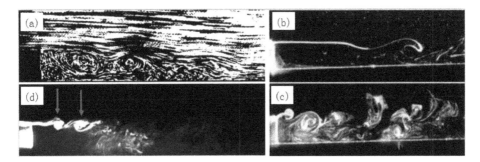

Figure 2. Visualizations of BFS flow: (**a**) Tani et al. (1961) [22]; (**b**) Chun et al. (1998), without excitation [6]; (**c**) Chun et al. (1998), with excitation [6]; (**d**) Kostas et al. (2002) [14].

Various quantitative measurements are applied to obtain velocity or pressure information such as hot-wire, x-wires, hydrogen bubble, laser doppler velocimetry, and particle image velocimetry (PIV) [11,16,20,26–30]. Many of the measurements can only measure and analyze point-by-point characteristics or time-averaged flow fields information such as X_r, turbulent shear stress, turbulent intensity, backflow, and characteristic frequency, etc. Recent developments of experimental measurements and numerical simulations allowed in-depth investigation of the flow dynamics. For example, PIV makes it possible to obtain some velocity fields and flow structures with temporal vortical evolutions. Scarano et al. (1999) [29] investigate the trajectories and sizes of CVS with $Re_h = 5000$, utilizing a pattern recognition analysis. They concluded that the length scale of the CVS ranges from 0.12 *h* to 0.44 *h*, and those with smaller diameters occur more frequently. They suggest that the clockwise CVS arise from a Kelvin–Helmholtz instability in the FSR, while the counter-clockwise CVS in the CR are due to a three-dimensional breakdown of the clockwise ones. Staggered vortex trains and pairings within the FSR are identified in the instantaneous PIV measurements by Kostas et al. (2002) [14]. They propose that the vortical interactions are most likely to contribute to the high Reynolds stresses and turbulent kinetic energy production. Furthermore, the CVS seem to be largely responsible for the persistence of the streamwise turbulent intensity (u'^2) and the shear stress ($u'v'$) in the flow downstream of reattachment, while the vertical turbulent intensity (v'^2) is predominantly by the fine scale structures. Hudy et al. (2007) [15] further propose a "wake model" that the CVS grow in place before reaching a height equivalent to *h*, and at this position the CVS shed and accelerate downstream to the ultimate convection speed. This opinion compares the evolution of this flow structure to that in the wake of bluff bodies, which is a little different with the traditional "shear-layer model."

Some other researches deal with the unsteadiness of separation flow with some low-frequency motions [6,11,13,31–35]. Four motions are summarized in separated flows including the BFS configuration that respectively correspond to flapping (FP), oscillation of X_r (OX), Kelvin-Helmholtz instability (KH), and pairings of the Kelvin-Helmholtz vortices (PKH). These motions are associated with the main instability mechanisms occurring in the shear layer and the recirculation region. The FP is the universal frequency, implying an absolute instability. While the OX appears obviously around the reattachment zone. The KH and PKH particularly concentrate at the shear layer [11]. Hu et al. (2016) [35] shows the temporal CVS in their numerical simulation and proposes a shear layer model and a shedding model. The former model is coincided with the Kelvin-Helmholtz instability mechanism with Strouhal number ($S_t = fh/U_0$, U_0 is the maximum velocity and f is the characterized frequency) of 0.2, and the later one is characterized by shedding frequency of $S_t = 0.074$. An obviously declining trend for the characteristic frequency along the streamwise behind the step is showed by Lian et al. (1993) [27] and Wang et al. (2019) [11]. They suggest this declining trend to be caused by the evolution of CVS.

Overall, some new advances have focused on the unsteady dynamics and CVS in BFS flow. However, it is still difficult to illustrate their essential developments and evolutions. Quantitative studies of the basic definition and features of the CVS are still deficient. In this study, we present the CVS particularly in the temporal streamlines and attempt to quantitatively investigate their characteristics and evolutional laws.

2. Methods

2.1. Experimental Model

A water tunnel flow system with a BFS model was built (Figure 3) and some configuration parameters were listed in Table 1. A rectangular pressured tunnel was set horizontally (top wall slope, $i = 0$). It enlarges halfway and forms a vertical step at the bottom surface with $h = 50$ mm. The expansion ratio, $E_r = H_d/H_u$, equals to 2, where H_d and H_u are respectively water depth upstream and downstream the step. The tunnel width was suggested to be no less than 10 h to keep the flow bidimensionality [16], therefore $A_r = L_z/h = 10$ was set. The channel lengths of upstream and downstream the step were $L_{xu} = 44\,h$ and $L_{xd} = 50\,h$, respectively. Honeycomb tubes were set near the inlet of the channel to obtain fully developed turbulent flow before the step.

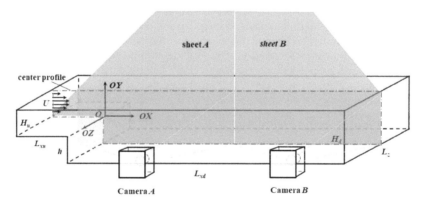

Figure 3. Schematic of the test section arrangements with a synchronous particle image velocimetry (PIV) system.

Table 1. Configuration parameters of the BFS model.

Parameters	Values	Units
h	50	mm
L_{xu}	2220	mm
L_{xd}	2500	mm
H_u	50	mm
H_d	100	mm
L_z	500	mm
i	0	°
E_r	2:1	-
A_r	10:1	-

The flow tunnel was made of transparent PMMA (polymethyl-methacrylate) to provide a clear view for the PIV system. PIV was employed to measure the velocity fields at the central profile (Figure 3).

2.2. Measurement System

According to the previous literatures, X_r behind step was about (4.2~8.6) h when flow was fully turbulent, and it was even longer for a lower Reynold number [1,11,20,30]. To study the evolution process of the CVS, a global view of velocity fields with no less than 10 h (about 60 cm) behind the step was expected. A large-scale field of view was difficult to realize with high vector density and high precision. To obtain the global velocity field, some previous studies took pictures of different subareas and matched the time-averaged velocity fields spatially [14,29]. However, this was still unavailable for the global view of the instantaneous velocity fields with CVS' evolution process. This study developed a synchronous PIV system, by which a doubled and effective view of the instantaneous velocity field was achieved (Figure 3, Sheets A and B).

The synchronous PIV system mainly consisted of laser module, image acquisition module, and control module. The laser module was a laser whose ray scanned to a light sheet optics with thickness of about 1 mm. The laser power energy was 3 W and the maximum scanning frequency reached to 40 Hz. The image acquisition module was made of a charge coupled device (CCD) camera with a resolution about 2560 pix × 2048 pix. Hollow sphere glass particles were seeded into the water with an average diameter of 15 μm. Two light sheets (A and B) and cameras (A and B) were synchronously controlled by a control module. This PIV system successfully made a global view of an instantaneous velocity field of approximate 12 h × 5 h with a resolution of 7.35 pix/mm. Interrogation area of 20 pix × 15 pix (2.722 mm × 2.042 mm) was used when computing velocity vectors. The measurement accuracy was within 1% of the maximum velocity.

2.3. Flow Conditions

The experiment was carried out with a constant water head for each case. The Reynolds number, $Re_h = U_b h/\nu$, was defined by the step height and the averaged velocity (U_b) located at 1 h upstream the step, and ν was the kinematic viscous coefficient of water at 20 °C.

Figure 4 shows the relationship between X_r and Re_h. Three typical flow regions were identified: the laminar flow (region I, $Re_h < 740$), transition flow (region II, $740 < Re_h < 4070$), and turbulent flow (region III, $Re_h > 4070$). A relative stable recirculation region with a constant X_r value (4.2~8.6 h) occurs in the turbulent flow (III) [1,20,36–38]. The flow structures in this flow region (III) were proposed to be stable, while the X_r values in another two flow patterns (I and II) were both changeable. Since most flow are with high Reynolds number in real applications, the turbulent flow region (III) were particularly favored here. For comparison, two cases with $Re_h = 4400$ and $Re_h = 9000$ were analyzed (Table 2). In addition, the good agreement of the data points in Figure 4 also verify the reliability of the measuring instrument.

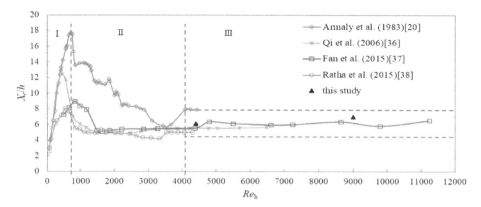

Figure 4. The reattachment length (X_r) vs. Reynolds number (Re_h).

Table 2. Flow conditions in the experiment.

h (cm)	U_b (m/s)	E_r	A_r	Re_h	X_r/h
5	0.088	2:1	10	4400	6.1
5	0.182	2:1	10	9000	7.0

3. Results and Discussions

3.1. Vortex Visualization and Definition

The vortex structures are visualized by the ghosting pictures of the inspired particles, and the pictures are also compared with the instantaneous streamlines (Figure 5). Comparison shows that the two figures present similar flow phenomena. The main flow diffuses and forms a recirculation region behind the step. The instantaneous flow is turbulent with the main flow interacting with the recirculation region.

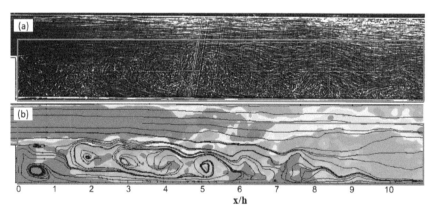

Figure 5. Coherent vortex structures (CVS) in the instantaneous flow: (**a**) visualization, (**b**) streamlines with velocity contours.

Particularly, the streamlines show some spanwise and large-scale vortex structures behind the step. They are well organized in time and space. They successively roll up in the FSR, move downstream and approach to the reattachment zone, with maximum length scale of the same order with the step height. These vortex structures present mostly an elliptical shape from a side view particularly in the early developing stage, and then break down when approaching the reattachment zone. These flow phenomena are consistent with the visualization of Tani et al. (1961) in Figure 2a [22].

One may ask if these elliptical streamlines are vortices or not. Lugt (1979) [39] defined a vortex as "multitude of material particles rotating around a common center." Robinson (1991) [40] further concluded that a vortex exists when instantaneous streamlines mapped onto a plane normal to the vortex core exhibit a roughly circular spiral pattern, when viewed from a reference frame moving with the center of the vortex core. These two definitions both propose two major features ("a common center" and "rotating"), and the later one further points to use streamlines to show a vortex. Although the concept of CVS has been proposed for centuries, its physical definition is still ambiguous. The vortex structures in the flow visualizations or by streamlines have the basic vortical features, particularly the streamlines show some elliptical patterns which rotates around a common center. Thus, they are consistent with the vortical definitions. Here, we define these vortex structures shown by streamlines to be the coherent vortex structures.

To quantitatively describe these CVS, some basic parameters such as the locations and patterns are defined (Figure 6). The common center of streamlines is regarded as the vortex core ($Z(x, y)$), representing location of a vortex. The typical shape is regarded close to an ellipse with streamwise diameter (D_x), and vertical diameter (D_y). Thus, D_x/D_y represents width to height ratio and $D = 0.5 \times (D_x + D_y)$ indicates the mean scale. Two CVS are separated by an equilibrium point, called the saddle point.

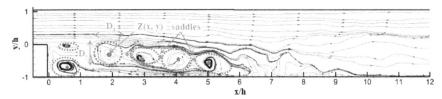

Figure 6. Parameter definition for the CVS (D_x—streamwise diameter, D_y—vertical diameter, $Z(x, y)$—vortex center, blue point—saddle point).

3.2. Vortex Motions with Developing Stages

Figure 7 shows the spatial and temporal evolution process of the CVS in streamline pictures. Totally 12 instantaneous streamline pictures with an interval time of 0.5 s are compared ($T = tU_b/h = 0.00 \sim 9.68$). These streamlines are generated with the corresponding velocity matrix by TECPLOT 2013.

In the first picture ($T = 0.00$), eight representative CVS (#1~#8) are indexed, which show various patterns and characteristics with flow structures (flow regions and developing stages see Figure 1). An analysis for these CVS' features based on the flow structures divided is conducted here.

(1) Vortex #1 is located in the forming stage (the first stage, $x = (0\sim2)\, h$ in FSR) just behind the step. Strong free shear layer is developing from the step corner with increasing thickness. Successive vortex structures generate and grow with similar size (diameter < 0.5 h) when travelling downstream (green lines). Finally, they are mixed by larger and fully developed ones at the end of this stage ($x = (1\sim2)\, h$). Their average speed toward downstream is about $(0.3\sim0.6)\, U_b$ in this case.

(2) Vortex #2~#5 occur in the developing stage (the second stage, $x = (2\sim5)\, h$ in FSR). The CVS are fully developed and energetic with larger length scales that close to h. Abundant amalgamation (pairing) and division events happen in this stage (red arrow lines). Because of the high mixing rate and flow unsteadiness, the turbulent intensity and shear stress reach a peak there [1,11,14]. This response may be a proof that the CVS is an important contributor for the Reynolds stress. Compared with the former stage, their travelling speeds reduce to approximately $(0.05\sim0.3)\, U_b$ before attaching to the reattachment zone (red lines).

(3) Vortex #6~#7 appear in the shedding stage (the third stage, $x = (5\text{~}8)\,h$ in the reattachment zone). Flow is strongly disturbed by the bottom wall in this stage. The streamlines are twisted sometimes when approaching to the wall. This may be induced by the enhancement of the three-dimensional characteristics. The CVS slow down and mostly attach on the bottom wall (blue lines), and sometimes they vanish there. Beyond R, some streamlines begin to roll up and generate some new vortex structures in a relatively smaller length scale and weaker rotating intensity. These vortex structures usually keep for a short time.

(4) Vortex #8 is usually found in the corner region ($x = (0\text{~}2)\,h$ in CR). It is induced by the recirculation, where the backflow separates from bottom wall and induces a counter-clockwise vortex [14,20,27,35]. The temporal evolution shows that the separation streamline is usually generated at the separated point of the backflow leaving the wall. The diameter of this vortex is about $0.5\,h$ [11]. This vortex has a much weak rotating intensity. And it travels upstream at the speed of about $0.1\,U_b$ in this case (yellow lines, $Re_h = 4400$).

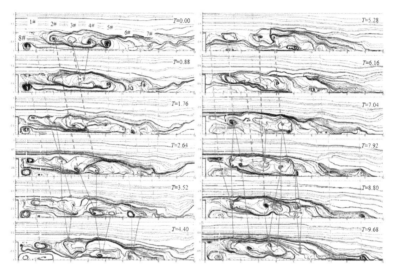

Figure 7. Motions and evolutions of the CVS with $Re_h = 4400$ (Dash lines with green, red, blue and yellow represent vortex trajectory in different developing stages; Arrow lines represent the amalgamation and division behaviors; $T = 0.00\text{~}9.68$).

There are totally dozens of CVS in each streamline picture. The evolution shows a life process of forming, developing, shedding, and redeveloping when travelling downstream. Their characteristics such as size, rotation direction and intensity, and evolution behaviors vary with the flow regions and developing stages. In particular, amalgamation (pairing) and division behaviors often occur in the second developing stage.

3.3. Vortex Trajectory and Evolution

The life process of the CVS is also reflected in the temporal evolution of an individual vortex structure. The blue line tracks the trajectory of a typical vortex center and other color lines show some break up and amalgamation (Figure 8). In this figure, the blue diamond point means the vortex center and the arrows show its moving direction. It starts nearby the step corner, and then travels toward the reattachment zone. It initially wanders in the strong shear layer, forming a relatively narrow path in the first stage (forming stage, 0~2 h). When arriving to the end of this stage, the center "jump" at about $x = 1.5\,h$, which can be explained by the amalgamation (merged by the vortex in later stage) shown in Figure 7. Then the vortex lingers and fluctuates particularly around $x = 2.2\,h$ and $x = 3.4\,h$ in the developing stage, where frequent amalgamation and division events occur there. The branches lines with other colors indicate vortex structures breaking up from the original one. When travelling to the end of this stage ($x = 4{\sim}5\,h$), the original vortex could not be found in the streamline pictures.

Figure 8. Trajectory and evolution of vortex centers ($Re_h = 4400$).

To further investigate the evolution events, a typical pairing process of two CVS is analyzed (Figure 9). To our knowledge, there is no experimental analysis for this process. A critical distance, λ_c, between two vortex patches was numerically studied by the contour dynamics method, and a distance parameter, λ, was defined when considering two scheduled circular vortex patches with the same diameters and shown by certain vorticity contour [41,42]:

$$\lambda = L/r \; (L > 2\,r) \tag{1}$$

where L is the distance between the two vortex patches, and r is the radius of each vortex patch.

When $\lambda > \lambda_c$, these two vortex patches rotate and change in shape, and their shape is close to a tear drop with the decrease of λ. When $\lambda < \lambda_c$ ($\lambda_c = 3.5$), these two vortex patches start to interact with each other. With the decrease of λ, they will get tangled up and finally they merge together.

In this experimental study, a similar parameter λ' is defined between two vortex structures with equivalent radiuses of r_1 and r_2, respectively (Equation (2)). A critical value is also expected when the CVS start to pair. It should be pointed out that these CVS are defined by streamlines instead of vorticity contours, and they are mostly elliptical but not circular in shape.

$$\lambda' = 2\,L'/(r_1 + r_2) \; (2\,L' > r_1 + r_2) \tag{2}$$

where L' is the distance between two vortex centers, and r_1 and r_2 are of the equivalent radiuses of two CVS, respectively.

As is shown in Figure 9, two CVS keep independent initially when they stay at a sufficiently far distance. As they are getting closer, they change in shape and start to interact with each other and finally come to an amalgamation when their distance is less than the critical value ($\lambda' < \lambda_c$). This indicates that critical distance still exists in current CVS. From the point view of the direction of the streamlines, these two CVS have an opposite streamline direction, one converges to the center (right) and another originates from the center (left). The former one is usually stable, while the other one is unstable. They finally merge into a stable one (streamlines converge to the vortex center).

3.4. Vortex Center and Diameter Distributions

Temporal evolution of the specific vortex structure is particularly disclosed above, their features of spatial distribution and length scales are to discuss here. Totally more than 2000 CVS were extracted from 201 instantaneous streamline pictures (picture no. 100–300), and the distributions of vortex centers ($Z(x, y)$) and diameters parameters ($D_x, D_y, D, D_x/D_y$) were statistically analyzed (Figures 10 and 11).

Figure 10 shows the distribution of the vortex centers with $Re_h = 4400$ and 9000 separately. The distribution seem to be regular, that fit well with the divided flow regions and developing stages (Figure 1). The vortex centers mainly distribute in the FSR, CR, and RR. In other words, these CVS are directly and closely related to the flow separation and reattachment characteristics. And they present specific features in different regions, going through the process of forming, developing, shedding, and redeveloping when travelling downstream. This statistical distribution also supports well the above discussion on vortex center trajectory and vortex evolution. With increase of Re_h from 4400 to 9000, the distribution of vortex centers is almost unchanged in space. It provides a good explanation for the various features with the specific flow structures.

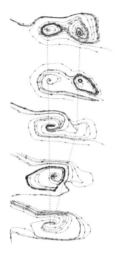

Figure 9. A vortex pairing process.

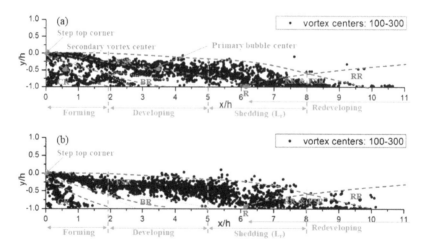

Figure 10. Distribution of the CVS with different Re_h: (**a**) Re_h = 4400 [11] and (**b**) Re_h = 9000; (FSR: free shear layer region, CR: corner region; BR: backflow region; RR: redeveloping region; R: reattachment point; L_r: length of reattachment zone).

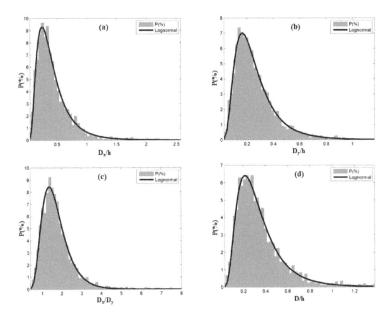

Figure 11. Distribution of length scale parameters (Re_h = 4400): (**a**) D_x; (**b**) D_y; (**c**) D_x/D_y; (**d**) D.

Besides the vortex center, the distribution of parameters D_x, D_y, D_x/D_y, and D are also well organized. Over 2000 vortex structures are analyzed. Figure 11 shows that four parameters (D_x, D_y, D_x/D_y, and D) all obey to the lognormal distribution. In other words, their logarithms follow a normal distribution. We know that lognormal distributions are positively skewed with long right tails because of low mean values and high variances in the random variables. The mean value ($\exp(\mu)$) and standard deviation (σ) were used to evaluate the length scale for the CVS, and also compare the influences of Reynolds number (Re_h = 4400 and 9000) and geometric model (BFS and turbulent boundary layer, TBL). The mean value ($\exp(\mu)$) and standard deviation (σ) were obtained by data fitting and are compared in Table 3.

Table 3. Comparison of the length scale parameters (BFS and TBL).

Literature	Configuration	Parameter	μ	Exp (μ)	σ
This study	BFS Re_h = 4400	D_x/h	−1.03	0.36	0.61
		D_y/h	−1.49	0.22	0.56
		D_x/D_y	0.41	1.51	0.38
		D/h	−1.22	0.29	0.56
	BFS Re_h = 9000	D_x/h	−1.13	0.32	0.70
		D_y/h	−1.43	0.24	0.68
		D_x/D_y	0.33	1.40	0.37
		D/h	−1.26	0.28	0.68
Wang (2009) [43]	TBL Re_b = 4677	D_x/h_w	−1.37	0.25	0.69
		D_y/h_w	−2.15	0.12	0.73
		D_x/D_y	0.78	2.18	0.64

With the increase of Re_h from 4400 to 9000, these four variables change with different trends. The streamwise diameter (D_x/h) decreases slightly from 0.36 to 0.32, while the vertical diameter (D_y/h) increases from 0.22 to 0.24. As a result, the width to height ratio (D_x/D_y) reduces from 1.51 to 1.40. This means that the long axis of the ellipse becomes shorter and the short axis grows, making the shape more circular with increasing of Re_h. However, the mean diameter (D/h) keeps a relative stable value around 0.28~0.29 for current experimental conditions. The standard deviation σ of variables (D_x/h, D_y/h, and D/h) all increase with increase of Re_h, while that of D_x/D_y has almost no response to Re_h. With the increase of Re_h, previous study showed that the shear intensity behind the step increases, and flow turbulence or unsteadiness is enhanced [20]. This will contribute to the vortex mixing and accelerate vortex evolution. We can understand that the streamwise diameter (D_x) is generally greater than the vertical diameter (D_y) as the streamwise shear direction and vertical wall constraints.

An experimental study of a turbulent boundary layer (TBL) on the distribution for vortex length scales was reported by Wang (2009) [43]. Different to the BFS configuration, there is a free water surface and no disturbance of step in the TBL model, yet the Reynold number, $Re_b = U_b R/\nu = 4677$ (R is hydraulic radius), is the same order with this study. In comparison, the CVS are more circular because of the vortical pattern $D_x/D_y = 1.40-1.51$ in the BFS flow is smaller than $D_x/D_y = 2.18$ in TBL flow. This may be explained by the different features of their shear layers, where the CVS are initially formed. The shear intensity is particularly strong near the bottom wall with a uniform mean velocity direction in TBL. While shear layer in BFS initially generates and develops from the step with a recirculation below. The extra recirculation and less impact of the wall in BFS flow both contribute to the more circular vortical shape. As for the standard deviation (σ), the BFS flow shows a smaller value than that of TBL flow. This means a more concentrated vortex scale distribution in BFS flow.

4. Conclusions and Recommendations

Velocity measurements and CVS investigation for the experiments at a BFS water flow with $Re_h = 4400$ and 9000 were reported. From a viewpoint of CVS defined in the streamline pictures, the vortical spatio-temporal coherences were verified and quantitatively studied. The following conclusions may be drawn:

(1) A synchronized PIV system is necessary for the measurement of a wider range of velocity fields with sufficiently high resolution to study the global evolution process of CVS.

(2) CVS are proved to exist in the separation and reattachment process of BFS flow. They are well organized in the recirculation regions with forming, developing, shedding, and redeveloping in the shear layer. Flow structures with various local characteristics are disclosed and explained by the motions and evolutions of CVS.

(3) The spatial distribution of CVS is almost unaffected by the Reynolds number when flow is fully turbulent. The CVS are particularly active at the developing stage with frequent pairings and divisions, which contribute to the high turbulent intensity and shear stress. And the shedding of CVS can explain the rapid decrease of the Reynolds stress at the reattachment zone.

(4) Statistics show the streamwise (D_x/h), vertical (D_y/h), and mean (D/h) diameters, and width to height ratio (D_x/D_y) of the CVS all obey to the lognormal distribution. The mean diameter (D/h) is about 0.28~0.29 for a BFS flow with $Re_h = 4400$ and 9000. Compared with the flow in a turbulent boundary layer, they have the same order of vortex length scale, while the shape of CVS in BFS is more circular. A critical distance is supposed to exist between two CVS.

Concept of CVS has been put forward for more than half a century and it is proved to play a significant role particularly in the separated flows. With the development of measurement techniques and numerical simulations, more quantitative information should be used to verify the previous hypothesis and suggestions. The quantitative vortex length scales measured in the experiments may help to choose the grid scale in certain simulations considering the large-scale vortex structures (such as the Large Eddy Simulation). In terms of the local energy dissipation at the enlarged step, the significant impact of the CVS cannot be ignored.

Author Contributions: F.W. and S.W. conceived and designed the experiment; F.W. and A.G. performed the experiment and analyzed the data and wrote the manuscript. S.Z., A.G., J.D., and Q.L. contributed to scientific advising and provided a substantial input to improve the paper.

Funding: This research is funded by the National Natural Science Foundation of China (Grant No. 51909169), the Science and Technology Support Program of Jiangsu Province (Grant No. SBK2019042181), and the financial support from Nanjing Hydraulic Research Institute (NHRI) (Grant No. Y119005). The authors gratefully acknowledge these financial supports.

Acknowledgments: The authors would also like to thank Corporation Hawksoft for their assistances in the laboratory measurements.

Conflicts of Interest: The authors declare no conflict of interest.

References

1. Eaton, J.K.; Johnston, J.P. A review of research on subsonic turbulent flow reattachment. *AIAA J.* **1981**, *19*, 1093–1100. [CrossRef]

2. Vogel, J.C.; Eaton, J.K.; Adams, E.W. Combined heat transfer and fluid dynamic measurements behind a backward-facing step. *Am. Soc. Mech. Eng.* **1985**, *107*, 922–929.

3. Jacob, M.C.; Guerrand, S.; Juvé, D.; Guerrand, S. Experimental study of sound generated by backward-facing steps under wall jet. *AIAA J.* **2001**, *39*, 1254–1260. [CrossRef]

4. Hu, R.Y.; Wang, L.; Fu, S. Review of backward-facing step flow and separation reduction. *Sci. Sin. Phys. Mech. Astro.* **2015**, *45*, 124704. (In Chinese)

5. Chen, L.; Asai, K.; Nonomura, T.; Xi, G.; Liu, T. A review of backward-facing step (BFS) flow mechanisms, heat transfer and control. *Therm. Sci. Eng. Prog.* **2018**, *6*, 194–216. [CrossRef]

6. Chun, K.B.; Sung, H.J. Visualization of a locally-forced separated flow over a backward-facing step. *Exp. Fluids* **1998**, *25*, 133–142. [CrossRef]

7. Chen, L.; Zhang, X.R.; Okajima, J.; Maruyama, S. Thermal relaxation and critical instability of near-critical fluid microchannel flow. *Phys. Rev. E* **2013**, *87*, 043016. [CrossRef]

8. Baigmohammadi, M.; Tabejamaat, S.; Farsiani, Y. Experimental study of the effects of geometrical parameters, Reynolds number, and equivalence ratio on ethaneoxygen premixed flame dynamics in non-adiabatic cylindrical meso-scale reactors with the backward facing step. *Chem. Eng. Sci.* **2015**, *132*, 215–233. [CrossRef]

9. Shahi, M.; Kok, J.B.W.; Pozarlik, A. On characteristics of a non-reacting and a reacting turbulent flow over a backward facing step (BFS). *Int. Commun. Heat Mass Transf.* **2015**, *61*, 16–25. [CrossRef]

10. Dodaro, G.; Tafarojnoruz, A.; Sciortino, G.; Adduce, C. Modified Einstein sediment transport method to simulate the local scour evolution downstream of a rigid bed. *J. Hydraul. Eng.* **2016**, *142*, 04016041. [CrossRef]

11. Wang, F.F.; Wu, S.Q.; Huang, B. Flow structure and unsteady fluctuation for separation over a two-dimensional backward-facing step. *J. Hydrodyn.* **2019**. [CrossRef]

12. Roos, F.W.; Kegelman, J.T. Control of coherent structures in reattaching laminar and turbulent shear layers. *Aiaa J.* **1986**, *24*, 1956–1963. [CrossRef]

13. Wee, D.; Yi, T.; Annaswamy, A.; Ghoniem, A.F. Self-sustained oscillations and vortex shedding in backward-facing step flows: Simulation and linear instability analysis. *Phys. Fluids* **2004**, *16*, 3361–3373. [CrossRef]

14. Kostas, J.; Soria, J.; Chong, M. Particle image velocimetry measurements of a backward-facing step flow. *Exp. Fluids* **2002**, *33*, 838–853. [CrossRef]

15. Hudy, L.M.; Naguib, A.; Humphreys, W.M. Stochastic estimation of a separated-flow field using wall-pressure-array measurements. *Phys. Fluids* **2007**, *19*, 1093. [CrossRef]

16. Bradshaw, P.; Wong, F.Y.F. The reattachment and relaxation of a turbulent shear layer. *J. Fluid Mech.* **1972**, *52*, 113–135. [CrossRef]

17. Nadge, P.M.; Govardhan, R.N. High Reynolds number flow over a backward-facing step: Structure of the mean separation bubble. *Exp. Fluids* **2014**, *55*, 1657–1678. [CrossRef]

18. Calomino, F.; Tafarojnoruz, A.; De Marchis, M.; Gaudio, R. Experimental and numerical study on the flow field and friction factor in a pressurized corrugated pipe. *J. Hydraul. Eng.* **2015**, *141*, 04015027. [CrossRef]

19. Calomino, F.; Alfonsi, G.; Gaudio, R.; D'Ippolito, A.; Lauria, A.; Tafarojnoruz, A.; Artese, S. Experimental and numerical study of free-surface flows in a corrugated pipe. *Water* **2018**, *10*, 638. [CrossRef]

20. Armaly, B.F.; Durst, F.J.; Pereira, J.C.F.; Schönung, B. Experimental and theoretical investigation of backward-facing step flow. *J. Fluid Mech.* **1983**, *127*, 473–496. [CrossRef]

21. Simpson, R.L. Review—A review of some phenomena in turbulent flow separation. *J. Fluids Eng.* **1981**, *103*, 520–533. [CrossRef]

22. Tani, I.; Iuchi, M.; Komoda, H. *Experimental Investigation of Flow Separation Associated with a Step or a Groove*; Aeronautical Research Institute, University of Tokyo: Tokyo, Japan, 1961; p. 364.

23. McGuinness, M. Flow with a Separation Bubble—Steady and Unsteady Aspects. Ph.D. Thesis, Cambridge Univ., Cambridge, UK, 1978.

24. Chandrsuda, C. A Reattaching Turbulent Shear Layer in Incompressible Flow. Ph.D. Thesis, Dept. of Aeronautics, Imperial College of Science and Technology, London, UK, 1975.

25. Kim, J.; Kline, S.J.; Johnston, J.P. *Investigation of Separation and Reattachment of a Turbulent Shear Layer: Flow over a Backward-Facing Step*; Thermosciences Div., Dept. of Mechanical Engineering, Stanford Univ., Rept. MD-37: Stanford, CA, USA, 1978.

26. Etheridge, D.W.; Kemp, P.H. Measurements of turbulent flow downstream of a rearward-facing step. *J. Fluid Mech. Digit. Arch.* **1978**, *86*, 22. [CrossRef]

27. Lian, Q.X. An experimental investigation of the coherent structure of the flow behind a backward facing step. *Acta Mech. Sin.* **1993**, *9*, 129–133. (In Chinese)

28. Stevenson, W.H.; Thompson, H.D.; Craig, R.R. Laser velocimeter measurements in highly turbulent recirculating flows. *J. Fluids Eng.* **1984**, *106*, 173–180. [CrossRef]

29. Scarano, F.; Benocci, C.; Riethmuller, M.L. Pattern recognition analysis of the turbulent flow past a backward facing step. *Phys. Fluids* **1999**, *11*, 3808–3818. [CrossRef]

30. Ma, X.; Geisler, R.; Schroder, A. Experimental investigation of separated shear flow under subharmonic perturbations over a backward-facing step. *Flow Turbul. Combust.* **2017**, *99*, 71–91. [CrossRef]

31. Driver, D.M.; Seegmiller, H.L.; Marvin, J.G. Time-dependent behavior of a reattaching shear layer. *AIAA J.* **1987**, *25*, 914–919. [CrossRef]

32. Le, H.; Moin, P.; Kim, J. Direct numerical simulation of turbulent flow over a backward-facing step. *J. Fluid Mech.* **1997**, *330*, 349–374. [CrossRef]

33. Aider, J.; Danet, A.; Lesieur, M. Large-eddy simulation applied to study the influence of upstream conditions on the time-dependant and averaged characteristics of a backward-facing step flow. *J. Turbul.* **2007**, *8*, 1–30. [CrossRef]

34. Kopera, M.A.; Kerr, R.M.; Blackburn, H.M.; Barkley, D. Direct numerical simulation of turbulent flow over a backward-facing step. *J. Fluid Mech.* **2014**, under publication.

35. Hu, R.Y.; Wang, L.; Fu, S. Investigation of the coherent structures in flow behind a backward-facing step. *Int. J. Numer. Methods Heat Fluid Flow* **2016**, *26*, 1050–1068. [CrossRef]

36. Qi, E.R.; Huang, M.H.; Li, W.; Zhang, X. An experimental study on the 2D time average flow over a backward facing step via PIV. *J. Exp. Mech.* **2006**, *21*, 225–232. (In Chinese)

37. Fan, X.J.; Wu, S.Q.; Zhou, H.; Xiao, X.; Wang, Y. Investigation on the Characteristics of Water Flow over a Backward Facing Step under High Reynolds Number with Particle Image Velocimetry. In Proceedings of the International Conference on Industrial Technology and Management Science, Tianjin, China, 27–28 March 2015.

38. Ratha, D.; Sarkar, A. Analysis of flow over backward facing step with transition. *Front. Struct. Civ. Eng.* **2015**, *9*, 71–81. [CrossRef]

39. Lugt, H.J.; Gollub, J.P. *Vortex Flow in Nature and Technology*; Wiley: Hoboken, NJ, USA, 1972.

40. Robinson, S.K. Coherent motions in the turbulent boundary layer. *Annu. Rev. Fluid Mech.* **1991**, *23*, 601–639. [CrossRef]

41. Deem, G.S.; Zabusky, N.J. Stationary "V-States," interactions, recurrence, and breaking. *Phys. Rev. Lett.* **1978**, *41*, 518–518. [CrossRef]

42. Wu, J.Z.; Ma, H.Y.; Zhou, M.D. *Introduction to Vorticity and Vortex Dynamics*; Advanced Education Press: Beijing, China, 1993; pp. 463–466.

43. Wang, L. Experimental Study on Coherent Structure in Open-Channel Flow. Ph.D. Thesis, Tsinghua University, Beijing, China, 2009. (In Chinese).

Article

Laboratory Study of Secondary Flow in an Open Channel Bend by Using PIV

Ruonan Bai [1], Dejun Zhu [1], Huai Chen [2] and Danxun Li [1],*

[1] State Key Laboratory of Hydroscience and Engineering, Tsinghua University, Beijing 100084, China; bairuonan1991@163.com (R.B.); zhudejun@tsinghua.edu.cn (D.Z.)
[2] State Key Laboratory of Hydrology, Water Resources and Hydraulic Engineering, Nanjing Hydraulic Research Institute, Nanjing 210029, China; chenhuai@nhri.cn
* Correspondence: lidx@tsinghua.edu.cn; Tel.: +86-10-6278-8532

Received: 7 March 2019; Accepted: 28 March 2019; Published: 30 March 2019

Abstract: The present paper aims to gain deeper insight into the evolution of secondary flows in open channel bend. A U-shaped open channel with long straight inflow/outflow reaches was used for experiments. Efforts were made to precisely specify flow conditions and to achieve high precision measurement of quasi-three-dimensional velocities with a multi-pass, two-dimensional PIV (Particle Image Velocimetry) method. The experimental results show that the flow begins to redistribute before entering the bend and it takes a long distance to re-establish to uniform conditions after exiting the bend. Complex secondary flow patterns were found to be present in the bend, as well as in the straight inflow and outflow reaches. A "self-breaking" (process was identified, which correlates stream-wise velocity with the intensity of flow circulation.

Keywords: open channel bend; secondary flow; velocity distribution; PIV

---

1. Introduction

Complex flow patterns exist in open channel bends and they play important roles in alluvial and ecological processes. Curvature-induced secondary flows, for instance, occur as an intrinsic balance between the centrifugal force and the pressure gradient due to water surface tilting. These secondary flows act to redistribute bulk velocity, alter sediment transport, shape bar-pool topography, enhance mass mixing, increase energy loss, and reduce conveyance capacity of the channel [1–10].

There have been abundant experimental studies on secondary flow in open channel bends. In particular, two important issues were repeatedly addressed: the strength of the secondary flow [11–15] and the interaction between the secondary flow pattern and the main flow [11,16–19].

Miao et al. [12] reported that, depending on the bend curvature, the normalized magnitude of the secondary flow grows linearly or un-linearly with the ratio of flow depth over bend radius. The Reynolds number was found to play an important role in shaping the secondary flow [11], and the maximum secondary flow strength occurred at the second half of the bend [13]. Roca et al. [14] observed that the maximum vorticity increases along a curved channel from the bend entry until the cross-sections from 40° to 60°. Ramamurthy et al. [15] found that bends with vanes exhibit a lower intensity of secondary flow.

Experiments on the interaction between main flow and secondary flow show that the transverse transport of main flow momentum by the secondary circulation is the principal cause of velocity redistribution, and the deflection of the maximum velocity toward the outer wall of the channel occurs at the cross-section where the secondary flow reaches its maximum strength [11,16–18,20]. Particularly, two parameters for the total additional secondary flow term have been defined and examined for their impact on velocity distributions [17].

An extensive literature review also reveals the lack of a reliable quantitative description of the secondary flow along the entire channel. The insufficiency of previous experimental studies can be attributed to three aspects.

Firstly, the limitation in instrumentation hinders precise measurement in some important flow areas. For instance, experiments with ADV/ADVP (Acoustic Doppler Velocity/Acoustic Doppler Velocity Profiler) fail to sample the near-surface and near-bed flow regions [6,13], making it difficult, if not impossible, to explore the number and position of secondary flow cells at each cross-section [17,21]. Other instruments, e.g., velocity meter PROPLER [20] and P-EMS velocimeter [18], are unable to provide three components of a velocity vector. The popularization of PIV (Particle Image Velocimetry) in fluid mechanics shows promise as being a better instrumentation for measurement of curved flows [22–26]. In particular, some novel multi-pass window deformation approaches have been used to improve the performance of PIV in measurement of various flows [27–29].

Secondly, the experimental design in most curved flow studies is insufficient for a closer observation of the evolution of the flow along the entire channel, including the bend and the straight inflow/outflow reaches. Some experiments either examined merely one cross-section [4,17,30] or used too long a distance between cross-sections to track the evolution of secondary patterns [11]. In the studies by Blanckaert [31], Vaghefi et al. [13], Abhari et al. [18], Bai et al. [11], Zeng et al. [32], and Booij [30], velocity measurements did not cover the straight inflow or outflow reaches at all.

Thirdly, the influence of the straight inflow/outflow reaches on the secondary flow receives little attention. The inflow/outflow reaches are often short, e.g., only 2 m long compared with a 3.77 m bend in the study by Bai et al. [11]. The curved flume used by Blanckaert [33] has no straight outflow reach at all. The influence of the inflow/outflow reaches still remains unclear, and as a result, the evolution of secondary flow in the bend involves much ambiguity. Some quantitative guidelines obtained by previous studies introduced too many assumptions [12].

To further investigate the evolution of secondary flow in open channel bend, a U-shaped bend with long straight inflow/outflow reaches has been designed and experiments have been conducted. In particular, the paper aims to illustrate the characteristics of the secondary flow at simple but precisely specified conditions to avoid possible ambiguity.

The paper has the following objectives:

(1) To set up a well-designed experimental system that facilitates accurate specification of flow conditions and detailed observation of the three components of flow velocity in a U-shaped open channel flume with long straight inflow/outflow reaches. The experimental design effectively eliminates ambiguity in both the inflow and the outflow reaches and enhances measurement reliability.

(2) To report and interpret detailed and high-quality data on the secondary flow, with focus on the distribution of the main flow, the topography of water surface, and the evolution of the circulation cells.

(3) To present high-quality data on the secondary flows that may be used for validation and calibration of numerical schemes.

2. Experimental Set-Up and Methods

2.1. U-Shaped Flume

Experiments were conducted to investigate flow properties in a U-shaped flume with a 180° bend. Focus was given to the evolution of secondary flows along the entire channel.

The flume, sketched in Figure 1a, was built with a glass bed and glass side walls. It had a width (B) of 0.3 m, a length along the centerline of approximately 26 m, and a centerline curvature radius (R_c) of 1.5 m. The bend, with a ratio of $R_c/B = 5$, falls into the category of medium bend [8]. The straight inflow and outflow reaches are both 10 m long, more than twice the length of the bend (center line 4.7 m). A honeycomb was placed at the flume entry to stabilize the flow and attenuate initial large-scale flow structures. A rolling shutter as a tail gate helps to control the water level. The overall stream-wise channel slope is adjustable, but it was fixed at S = 0.001 in the present study.

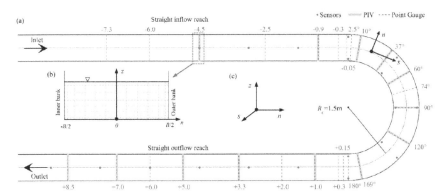

Figure 1. (**a**) Plan of the U-shaped flume with indication of positions of measurements; (**b**) Measuring grids of PIV (Particle Image Velocimetry) system (11 vertical lines and 7 horizontal lines); (**c**) the definition of the coordinate system: stream-wise (s), span-wise (n), and vertical (z).

Figure 1b shows the grids for PIV measurements. Details of the PIV technique will be given later in Figure 2.

The stream-wise, span-wise, and vertical directions are denoted by s, n, and z, respectively (Figure 1c); three components of the time-averaged velocity are U, V, and W.

Flow discharge was measured and controlled by an electromagnetic flow meter. Water surface elevations along the flume were automatically recorded with 20 ultrasonic sensors (red dots in Figure 1a) placed at 15 cross-sections. At each cross-section in the bend (0°, 45°, 90°, 135°, 180°), two sensors were used to monitor the difference in the water level along the span-wise direction. The ultrasonic sensors have a measuring range of 20–200 mm and a measuring error of ±0.2 mm. Real-time monitoring of discharge and water depth facilitates quick stabilization of the flow to steady and uniform conditions. Note that these ultrasonic sensors were used only to monitor the flow rather than to record the water level for further analysis. More precise measurements of water level were achieved with manual point gauge.

2.2. Flow

The flow in the straight inflow/outflow reaches was kept steady and uniform during the experiment. Table 1 shows the hydraulic parameters based on measurements at the cross-section −4.5, 4.5 m upstream of the bend entry. The Reynolds number and Froude number in the straight reaches were 14,917 and 0.557, respectively.

The flow in the experiment is sub-critical, thus the bend plays an important role on the secondary circulation and exerts influence on the inflow by upstream propagation. For sub-critical flows, such influence will diminish.

A water depth of H = 6 cm was used in the experiment, and the corresponding hydraulic radius was R = 4.29 cm. The straight reach of the flume is about 233 R, sufficiently long for the flow to fully develop compared with previous studies by Odgarrd and Bergs [34] and Yen [35], which recommended a straight length of 52 R and 100 R, respectively.

Table 1. Hydraulic and geometric conditions.

B (cm)	Q (L/s)	S	H (cm)	R (cm)	g (ms^{-2})	U$_f$ (cm/s)	Re	Fr	u* (cm/s)	Re*
30	7.15	0.001	6	4.29	9.8	39.7	14817	0.557	2.05	764.4297

B is flume width, Q is flow discharge, S is bed slope, H is flow depth, R = BH/(B + 2H) is hydraulic radius, g is gravitational acceleration, U$_f$ = Q/(BH) is cross-averaged stream-wise velocity, Re = UR/ν is Reynolds number, Fr = U/(gH)$^{(1/2)}$ is Froude number, Re* = u*R/ν is shear Reynolds number, u* = $\sqrt{gRJ}$ is shear velocity, J is hydraulic slope.

2.3. Measurements

Measurements of water levels and velocity distribution were conducted by means of point gauge and PIV, respectively.

Water surface was measured by manual point gauge with an accuracy of ±0.02 mm at 25 cross-sections (denoted by blue dot lines in Figure 1a) within the entire flume. Information on these 25 cross-sections is given in Table 2. The cross-sections in the straight reaches are denoted by their distance from bend entry (inflow, with minus sign) or from bend exit (outflow, with plus sign), and sections within the bends are denoted by their central angles in degree. For each cross-section, thirteen water depths in the span-wise direction were obtained, and their locations are n = 0, ±3, ±6, ±9, ±12, ±13, and ±14 cm, respectively. Note that the coordinate origin was set at the bottom of the mid-span plane of the channel (see Figure 1a) at each cross section.

Table 2. Cross-sections of water surface measurements.

Inflow		Bend		Outflow	
−7.3	−0.3	2.5°	90°	+0.15	+5.0
−6.0	−0.05	10°	120°	+0.3	+6.0
−4.5		37°	169°	+1.0	+7.0
−2.5		60°	180°	+2.0	+8.5
−0.9		74°		+3.3	

Thirteen locations within the width (B = 30 cm) of the channel have been measured for each cross-section (shown in Figure 1b).

A PIV system was employed to measure velocity distributions at 13 cross-sections (green lines in Figure 1a), which are located at the straight inflow reach (−4.5 m and −0.9 m), the bend (10°, 37°, 60°, 90°, 120°, and 169°), and the straight outflow reach (+1.0 m, +3.3 m, +5.0 m, +7.0 m, and +8.5 m).

At each cross-section, images in the s-z plane were recorded at eleven span-wise locations, i.e., n = 0, ±3, ±6, ±9, ±12, and ±14 (Figure 1b), to investigate velocity distributions of the stream-wise and vertical components. The image plane, which is normal to the channel bed and parallel to flume wall, is about 8 mm in the stream-wise direction and 6 cm in the vertical direction.

The laser and camera were placed vertically and horizontally, respectively, as shown in Figure 2a. To facilitate operation, a frame was designed with a gear system through which both the laser and the camera can be flexibly adjusted and precisely positioned.

Measurements at the s-n plane were achieved by exchanging the positions of the laser and the camera, as shown in Figure 2b. Images at seven layers were recorded, i.e., z/H = 0.05, 0.1, 0.2, 0.35, 0.5, 0.7, and 0.9, respectively (see Figure 1b). Thus, the stream-wise and span-wise velocity distributions at seven different vertical locations were obtained. A combination of measurements at both s-z and s-n planes provides three-component velocity data in a 11 × 7 grid.

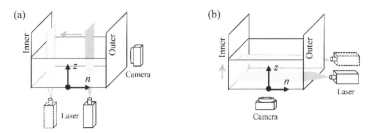

Figure 2. PIV setup with a laser and a camera: (**a**) s-z plane; (**b**) s-n plane.

The PIV system uses a continuous wave laser (8 W) at 532 nm for illumination (laser sheet 1 mm thick). The flow was seeded manually with spherical polyamide particles with a density of 1.03×10^3 kg/m^3 and a mean diameter of 5 μm. Particle images were recorded at a frame rate of 500 Hz by an 8-bit 2560 × 1920 pixel high-speed CMOS camera with a Canon EF 50 mm f/1.2 USM lens. Each measurement yields a sample of 15,000 images for analysis, corresponding to a sampling interval of 30 s. The image sizes are 96 pixels × 1120 pixels in Figure 2a, and 2192 pixel × 88 pixel in Figure 2b. Particle images were analyzed using a multi-pass, iterative multi-grid window deformation method. For grid refinement, the following sequence of the interrogation window size in terms of pixel was used: 64 × 64, 32 × 32, and 16 × 16. There is an overlap of 50% in the horizontal direction of the window. Details of the PIV algorithms can be found in Chen et al. [36] and Zhong et al. [37].

As the measurements were conducted in repeated experiments, the repeatability is essential for the validity of the measurements. This was ensured by using a sufficiently large water reservoir to minimize discharge fluctuation, carefully positioning the laser device and camera, and precisely establishing the flow with fixed depth and energy gradient. Comparison of water depth, energy slope, and velocity distribution in the vertical has indicated that the influence of environment change on measurement repeatability was negligible.

2.4. Methods for Data Analysis

2.4.1. Ensemble Average

It is worth noting that the current PIV measurements cannot be used to reconstruct the instantaneous three-dimensional velocity field (or the instantaneous vorticity field) because they refer to different instants. The quasi-3D information of the velocity field is achieved through ensemble average.

For each case, a total number of 15,000 consecutive images were captured at a frame rate of 500 Hz. Previous experience has indicated that such a sample size/interval is sufficient for statistical analysis of mean velocity [36,37]. Based on these instantaneous velocity (u, v, w) samples, the mean velocities (U, V, W) and fluctuating velocity components (u', v', w') were calculated through ensemble average. Examination of the results against the log-law velocity profile and the linear Reynolds stress profile has shown that the data are adequate to evaluate overall properties of the flow.

2.4.2. Data Interpolation

Based on water level measurements at 25 cross-sections, a linear interpolation function has been applied within the bend at an interval of 4.5°.

At each PIV measurement cross-section, based on the three-dimensional velocities at a 7 × 11 grid obtained by matching the s-z plane velocity and s-n plane velocity, an interpolation for the mean velocity has been made to get a final velocity field at a 27 × 57 grid for further analysis.

2.4.3. Calculation of Secondary Flow Strength

Secondary flows at channel bends can be conveniently visualized with velocity vectors and streamlines [13,18,26,38,39]. A more detailed description, however, necessitates the introduction of a quantitative parameter, such as vorticity [4,8,10,13,21,40,41].

The stream-wise component of the instantaneous vorticity vector can be calculated as follows:

$$w_s = \frac{\partial v}{\partial z} - \frac{\partial w}{\partial n} = \frac{\partial v^*}{\partial z} - \frac{\partial w}{\partial n}, \tag{1}$$

where v and w represent instantaneous span-wise and vertical velocity respectively, as defined in Figure 1. Obviously, $v = V + v'$, where V represents averaged span-wise velocity and v' represents fluctuating velocity. However, v can be decomposed into a translator part, v_n, called instantaneous depth-averaged span-wise velocity, and a circulatory part v^*, called the span-wise component of the secondary flow as well as $v = v_n + v^*$.

The vorticity provides a good basis for the analysis of transverse circulation.

3. Bulk Flow Distribution

Figure 3 shows contour maps of the vector representations of averaged stream-wise velocity, U, at ten cross-sections. It is obvious that the flow is symmetrical in the straight inflow reach (cross-sections −4.5 and −0.9), deflects to the inner bank in the bend (cross-section 10°) before transferring to the outer bank (cross-sections 60°, 90°, 120°, and 169°), and tends to reconstruct symmetry after exiting the bend in the outflow reach. Note that the flow resumes symmetry at cross-section 38°, a transition cross-section where the main flow achieves balance between the inner and outer banks.

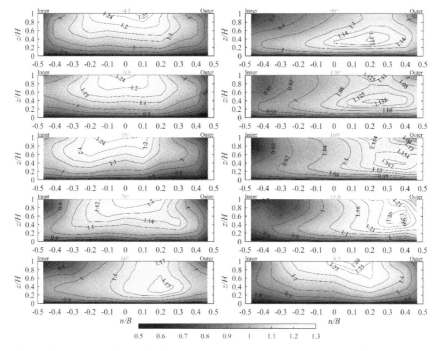

Figure 3. Stream-wise velocity distribution at ten cross-sections (−4.5, −0.9, 10°, 38°, 60°, 90°, 120°, 169°, +1.0, +8.5).

A detailed and quantitative description of the maximum stream-wise velocity, U_{max}, is given in Figure 4, in terms of both magnitude and position.

The magnitude of U_{max}, as can be seen in Figure 4a, shows a general decreasing trend when the flow enters the bend, reaches its minimum at cross-section 90°, and then turns to increase in the cross-sections well after the bend. At cross-section +1.0, however, U_{max} exhibits an unexpected rise over that in the straight inflow reach. This phenomenon may be due to the sudden change of the water level, which creates a positive stream-wise gradient near the outer bank and a negative one near the inner bank.

From the vertical position of U_{max} in Figure 4b, one can see that the maximum velocity occurs at the water surface in both the straight inflow and outflow reaches. In the channel bend, however, the flow sinks, resulting in a submersion of the maximum velocity well below the water surface. This is particularly noticeable in the central reach of the bend (60°–169°). After the bend, cross-section +1.0 witnesses a rise of water surface followed by a gradual decrease.

The change in the span-wise position of U_{max}, in Figure 4c, is consistent with established knowledge of curved flows. The flow starts to go toward the inner bank at cross-sections −0.9 and 10°, turns to the outer bank in the rest of the bend, and keeps left-deviated even in the straight outflow reach. It is interesting that the maximum deviation occurs at cross-section +1.0, i.e., 1.0 m away from the bend exit. This indicates that the flow needs a sufficiently long way to re-establish itself to uniform conditions.

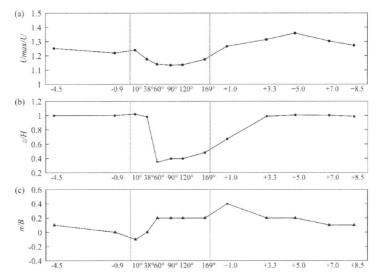

Figure 4. (**a**) Stream-wise evolution (U_{max}/U) of the magnitude of the stream-wise velocity; (**b**) Evolution of the vertical location of $U_{max}(z/H)$; (**c**) Evolution of the span-wise location (n/B) of U_{max}.

To further illustrate the asymmetry of the flow, we introduced another parameter, m, as the horizontal position where the flow discharge is split fifty–fifty. The result is shown in Figure 5. A comparison of Figures 4c and 5 indicates that the change of m in the stream-wise direction is similar to that of n/B, except in the reach from cross-section +3.3 to cross-section +5.0, where the increase of magnitude (Figure 4a) plays a more important role rather than the deflection of position (Figure 5).

Both Figures 4 and 5 reveal that the flow needs a long distance after exiting the bend to re-establish itself, and the curvature has a significant influence not only on the deflection, but also on the magnitude of U_{max}. For field engineering and numerical simulation of curved flows, such complexity needs in-depth consideration.

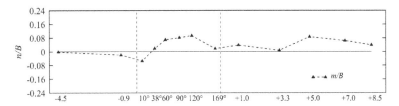

Figure 5. Stream-wise evolution of m along the flume (at 13 cross-sections).

4. Secondary Flow Characteristics

4.1. Surface Topography

Based on the measured water depth data from 25 cross-sections, the topography of the water surface in the entire channel was obtained, as shown in Figure 6. Note that the contour represents a subtraction of 6 cm from actual measurement data.

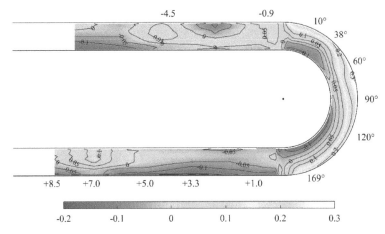

Figure 6. Topography of water surface in the channel derived from point gauge measurement (values are in centimeters and 6 cm defines '0' reference).

The presence of tilting water surface is a typical feature at channel bends. As expected, a transverse slope was found in the curved flow, i.e., it maintains a negligibly small presence in the straight inflow reach, becomes evident when the flow enters the bend, and reduces to nearly zero immediately after the bend. Within the bend, the outer banks witness a higher transverse slope than the inner bank. The difference in magnitude of transverse slope between the outer and inner banks diminishes in the last third of the bend as the water surface turns to be linear.

The transverse slope depends on a combination of factors, e.g., the incoming flow, the geometry of the bend, and the tail gate. When these factors vary, the transverse slope may display seemingly inconsistent characteristics. For instance, in their experimental and numerical study of a 90° bend with a short exit, Gholami et al. [21] reported a higher degree of surface tilting closer to the inner bank rather than the outer bank, a finding that is contradictory to the present result. In the straight reach of the curved open channel, Blanckaert [7] revealed a significantly larger transverse slope than the present study. Such differences indicate the dependence of flow characteristics on the bend geometry as well as the length of inflow/outflow reaches. If the inflow/outflow characteristics cannot be exactly specified, flow in the bend may exhibit significant variations which hinder verification and calibration of numerical schemes.

Figure 7 shows the change of the transverse slope along the channel. Note that for each cross-section the transverse slope (in degree) was calculated based on the water depth difference between the outer and inner banks.

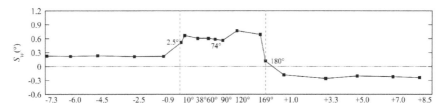

Figure 7. Evolution of the transverse slope along the flume.

In the bend, as we can see, the transverse slope increases quickly and remains at about 0.6 degrees until the end of the bend. At cross-section 120°, the transverse slope reaches its maximum, about 0.8 degrees.

In the straight reaches of the channel, as has been generally acknowledged, the transverse slopes are distinctly smaller. Interestingly, S_w remains positive in the inflow reach and becomes negative in the outflow reach. A positive S_w in the inflow reach is inconsistent with the water tilting in the bend. A negative S_w in the outflow reach is possibly due to the redistribution of flux and transfer of momentum.

4.2. Secondary Flow and Vice Cells

Vice cells of secondary circulation are visible in the experiments. This is critical in the investigation of the number and position of secondary flow.

The cross-section-averaged magnitude of vorticity along the flow is shown in Figure 8. Note that the vorticity is normalized by water depth H and cross-section-averaged stream-wise velocity U_f.

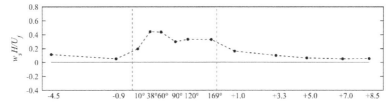

Figure 8. Evolution of dimensionless secondary flow strength.

In the bend, a gradual increase of transverse circulation is recognized, followed by a decrease. The maximum value occurs at cross-section 60°. The decrease is not distinct after cross-section +3.3. Similar findings have been reported by Rovovikii [42], Blanckaert [31], and Roca et al. [14].

It is interesting that the vorticity reaches its peak at cross-section 60° rather than at 90°, and it keeps roughly constant until 169°. Force analysis below helps to explain this phenomenon.

Within a curved channel, secondary circulation occurs as a result of local imbalance between the inward-directed pressure gradient force, $-1/(\rho \cdot \partial p / \partial n)$, and the outward-directed centrifugal force, u^2/r, as shown in Figure 9. The blue areas in Figure 9 represent the local force imbalance which corresponds to the last term in the simplified transverse momentum equation [12].

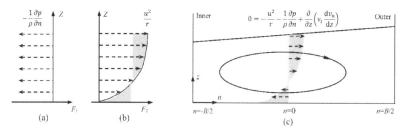

Figure 9. Schematic representation of curvature-induced secondary flow in a cross-section of a curved river reach; F1 represents the inward pressure difference; F2 represents the outward centrifugal force.

Obviously, the secondary flow is basically proportional to the local imbalance, which depends on the profile of the stream-wise velocity distribution along the vertical direction and the transverse pressure gradient [4]. The advective momentum transport by the cross-circulation cell brings down the stream-wise velocities in the upper part of the water column and enhances them in the lower part [4,16]. When the stream-wise velocity becomes more evenly distributed along the vertical, the outward centrifugal force will decrease in the upper part and increase in the lower side, as can be seen in Figure 9b. As a result, the intensity of circulation is reduced. There occurs, as it was, a "self-breaking" of circulation.

Figure 10 shows the pattern of the stream-wise vorticity and the cross-stream motions in ten cross-sections. The color map represents the vorticity in two senses of rotation, i.e., positive for anticlockwise and negative for clockwise. It is worth noting that the flow takes a relatively long distance for the circulation to decay after exiting the bend.

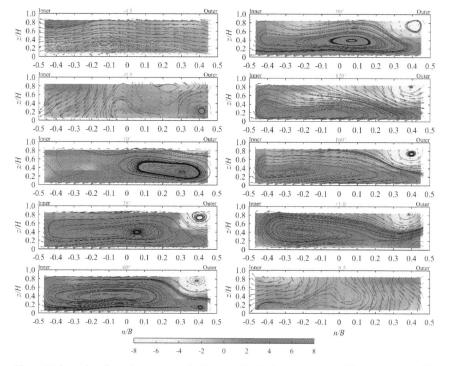

Figure 10. Secondary flow phenomena: velocity projection in the cross-sections (blue vectors); the 2D streamline patterns of the cross-flow (the black line); the dimensionless values of the vorticity (color).

In cross-section −4.5, one can see that the radial velocity components of the transverse circulation are all oriented to the inner bank. Similar results were also obtained by Gholami et al. [20] and Rozovskii [42]. This one-way radial flow leads to asymmetric velocity distributions, which are consistent with the inner-bank deflection of the stream-wise velocity (Figure 4). Accordingly, the high-velocity zone deviates toward the inner bank at the beginning of the bend.

The changing patterns of the secondary flow are illustrated by the streamlines in Figure 10. As can be seen, circulation appears before the flow enters the bend. The commencement of circulation can be recognized at cross-section −0.9 where the velocity vectors are generally oriented to the inner bank, but a visible circulation cell occurs in the outer-bank region near the bed. This finding is contradictory to the conclusion made by Gholami et al. [20], that no secondary flow exists at the entry of channel bends.

At cross-sections 38°, 60°, 90°, 120°, 169°, and +1.0, a large secondary cell presents across each cross-section and a counter-rotating weaker cell is also identified. These results are consistent with previous findings in both laboratory and field experiments, e.g., Einstein and Harder [43], Rozovskii [42], de Vriend [16], Bridge [44], and Blanckaert et al. [45]. When it rotates, the large secondary cell will transport sediment from the outer bank (where erosion occurs) to the inner bank (where deposition develops) of the bend. The presence of the small counter-rotating cell, in contrast, will attenuate such transportation process and help to protect the bank.

A comparison of cross-sections 10° and 38° reveals a dramatic change in the pattern of the rotating cell. This indicates a fast development of secondary flow in the entry of the bend due to the centrifugal acceleration. At cross-section 10°, the circulation is unique, with a large outer bank cell in the middle of the water depth. This is closely related to the fact that cross-section 10° is the only cross-section whose U_{max} is close to the inner bank and the core area of the contour (U) deflects to the inner bank (see Figure 3).

A closer look at cross-section 60° shows that near the outer bank, two small circulations occur with different rotation senses, one below the water surface (denoted the third cell) and the other close to the channel bed. This is probably due to the strong dip and deflection of the core of stream-wise velocity (Figures 3 and 4). The mechanism underlying the circulation is complicated by the effect of turbulence anisotropy near the outer bank. Actually, the U_{max} at cross-section 60° is located near the bed at $n/B = 0.3$ and $z/H = 0.4$ (Figure 3), where the big central region separates from the two small outer bank circulation cells.

The third cell occurring at cross-section 60° seems to decay in the following cross-sections and totally disappears at about 90°. This anisotropy generates a stream-wise vorticity of the additional secondary circulation moment, and thus enhances the tendency of the secondary circulation to split up and recur at cross-section +1.0. As the identification of such a small cell requires measurement with high spatial resolution, few previous studies have succeeded in providing detailed information.

The presence of an additional small cell at cross-sections 60° and +1.0 is closely related to the role of the kinetic energy fluxes from turbulence to the mean flow on the generation of the outer bank cell. The results in this paper may help to gain more insight into the kinetic energy transportation.

The outer-bank cell provides a buffer layer that protects the outer bank from any influence of the center-region cell and keeps the core of U_{max} a certain distance from the bank (Figure 3). At cross-section +1.0, the outer bank cell is very weak, failing to provide a good protection. This phenomenon agrees well with the stream-wise velocity distribution, in that the bulk flow deflects to the outer bank significantly (Figure 4c).

As general knowledge, the outer bank of a bend is more susceptible to erosion. However, the present results, consistent with those of Rovovikii [42], indicate that the most erosion-prone area is the concave bank near the exit from the bend at cross-section +1.0, where a sudden increase in U_{max} and an attenuation of the counter-rotating cell significantly enhances the erosional force of the flow.

It is surprising that at cross-section +8.5, some weak circulations still exist. This, again, indicates the importance of the length of the straight outflow reach for the flow to re-establish itself.

5. Conclusions

The study reported velocity measurements for the experiments at 25 cross-sections in a U-shaped open channel, based on PIV technique. The PIV system was designed as one which is able to obtain quasi-three-dimensional velocities. The observations focused not only on the bend, but also on the straight inflow/outflow reaches which were designed to be sufficiently long. The velocity distribution, the maximum velocity path, the flow surface topography, and the secondary flow have been evaluated and analyzed. The following conclusions may be drawn:

(1) Streamlines in cross-sections show secondary flows of different patterns along the entire channel. Both the curvature-induced secondary flow and weaker secondary flows, including vice cells, were recognized in the straight inflow/outflow reaches.

(2) The maximum secondary flow occurs at the first half of the bend (cross-section 60°), and a "self-breaking" process was observed, which is consistent with previous studies.

(3) The flow near the entry and exit of the bend is complex due to the sudden change of water surface. Flow redistribution also occurs in the straight inflow and outflow reaches, except the channel bend, indicating that the interaction between the main flow and secondary flow exists along the entire channel. If the inflow/outflow cannot be exactly specified, the flow in the bend may exhibit significant variations.

Author Contributions: R.B. and D.L. conceived and designed the experiments; R.B. collected the data; R.B. and H.C. designed the framework and analyzed the data of this study; D.L. and D.Z provided significant suggestions on the methodology and structure of the manuscript; R.B wrote the paper with the contribution of all co-authors.

Funding: The research was funded by National Key R&D Program of China (2016YFC0402308) and the National Natural Science Foundation of China (Grant no.91647107).

Acknowledgments: The authors are grateful to Wang Xingkui for his help in design of the curved channel.

Conflicts of Interest: The authors declare no conflict of interest.

Abbreviations

ADV	Acoustic Doppler Velocity
ADVP	Acoustic Doppler Velocity Profiler
B	channel width (cm)
R	hydraulic radius (m)
R_c	radius of curvature of channel centerline (m)
S	stream-wise channel slope (−)
Q	water discharge (m³/s)
H	averaged water flow depth along the flume (cm)
U_f	cross-averaged stream-wise velocity in the flume (m/s)
u*	shear velocity (cm/s)
Re	UR/ν is the Reynolds number
Re*	u*R/ν is the shear Reynolds number
Fr	the Froude number
g	gravitational acceleration (ms^{-2})
s	stream-wise reference coordinate (−)
n	span-wise reference coordinate; −15 cm represents the inner bank and 15 cm represents the outer one (−)
N	sample size (−)
z	vertical reference coordinate (−)
u	instantaneous stream-wise velocity (m/s)
v	instantaneous span-wise velocity (m/s)
w	instantaneous vertical velocity (m/s)

u′	instantaneous fluctuating stream-wise velocity (m/s)
v′	instantaneous fluctuating span-wise velocity (m/s)
w′	instantaneous fluctuating vertical velocity (m/s)
U	averaged stream-wise velocity (m/s)
V	averaged span-wise velocity (m/s)
W	averaged vertical velocity (m/s)
v_n	instantaneous depth-averaged span-wise velocity (m/s)
V_n	time-averaged value of v_n (m/s)
v*	span-wise component of the secondary circulation, v* = v − V (m/s)
V*	time averaged value of v* (m/s)
Inner	Inner bank (−)
Outer	Outer bank (−)
θ	cross-section angle (°)
w_s	stream-wise vorticity, $w_s = \partial w/\partial n - \partial v/\partial z$ (s^{-1})
ρ	density of water (kg/m^3)
h	local water flow depth (cm)
U_{max}	the highest value of U in a cross-section (m/s)
m	the deviation of equipartition position (m)
N	sample size (−)
P	pressure (Pa)
S_w	transverse water surface slope (°)

References

1. Shiono, K.; Muto, Y.; Knight, D.W.; Hyde, A.F.L. Energy losses due to secondary flow and turbulence in meandering channels with overbank flows. *J. Hydraul. Res.* **1999**, *37*, 641–664. [CrossRef]
2. Edwards, B.F.; Smith, D.H. Critical wavelength for river meandering. *Phys. Rev. E* **2001**, *63*, 45304. [CrossRef] [PubMed]
3. Lancaster, S.T.; Bras, R.L. A simple model of river meandering and its comparison to natural channels. *Hydrol. Process.* **2002**, *16*, 1–26. [CrossRef]
4. Blanckaert, K.; Graf, W.H. Momentum Transport in Sharp Open-Channel Bends. *J. Hydraul. Eng.* **2004**, *130*, 186–198. [CrossRef]
5. Sin, K. *Methodology for Calculating Shear Stress in a Meandering Channel*; Colorado State University: Fort Collins, CO, USA, 2010.
6. Roca, M.; Martín-Vide, J.P.; Blanckaert, K. Reduction of Bend Scour by an Outer Bank Footing: Footing Design and Bed Topography. *J. Hydraul. Eng.* **2007**, *133*, 139–147. [CrossRef]
7. Blanckaert, K. Flow separation at convex banks in open channels. *J. Fluid Mech.* **2015**, *779*, 432–467. [CrossRef]
8. Kashyap, S.; Constantinescu, G.; Rennie, C.D.; Post, G.; Townsend, R. Influence of Channel Aspect Ratio and Curvature on Flow, Secondary Circulation, and Bed Shear Stress in a Rectangular Channel Bend. *J. Hydraul. Eng.* **2012**, *138*, 1045–1059. [CrossRef]
9. Silva, A.M.F.D.; Yalin, M.S. *Fluvial Processes*, 2nd ed.; CRC Press/Balkema: London, UK, 2001.
10. Roca, M.; Martín-Vide, J.P.; Blanckaert, K. Bend scour reduction and flow pattern modification by an outer bank footing. In Proceedings of the International Conference on Fluvial Hydraulic, Lisbon, Portugal, 6–8 September 2012; pp. 1793–1799.
11. Bai, Y.; Song, X.; Gao, S. Efficient investigation on fully developed flow in a mildly curved 180° open-channel. *J. Hydroinform.* **2014**, *16*, 1250–1264. [CrossRef]
12. Wei, M.; Blanckaert, K.; Heyman, J.; Li, D.; Schleiss, A.J. A parametrical study on secondary flow in sharp open-channel bends: Experiments and theoretical modelling. *J. Hydro-Environ. Res.* **2016**, *13*, 1–13. [CrossRef]
13. Vaghefi, M.; Akbari, M.; Fiouz, A.R. An experimental study of mean and turbulent flow in a 180 degree sharp open channel bend: Secondary flow and bed shear stress. *KSCE J. Civ. Eng.* **2015**, *20*, 1582–1593. [CrossRef]
14. Roca, M.; Blanckaert, K.; Martín-Vide, J.P. Reduction of Bend Scour by an Outer Bank Footing: Flow Field and Turbulence. *J. Hydraul. Eng.* **2009**, *135*, 361–368. [CrossRef]
15. Ramamurthy, A.S.; Han, S.S.; Biron, P.M. Characteristics of Flow around Open Channel 90° Bends with Vanes. *J. Irrig. Drain. Eng.* **2011**, *137*, 668–676.

16. De Vriend, H.J. Velocity redistribution in curved rectangular channels. *J. Fluid Mech.* **1981**, *107*, 423. [CrossRef]

17. Tang, X.; Knight, D.W. The lateral distribution of depth-averaged velocity in a channel flow bend. *J. Hydro-Environ. Res.* **2015**, *9*, 532–541. [CrossRef]

18. Abhari, M.N.; Ghodsian, M.; Vaghefi, M.; Panahpur, N. Experimental and numerical simulation of flow in a 90° bend. *Flow Meas. Instrum.* **2010**, *21*, 292–298. [CrossRef]

19. Termini, D.; Piraino, M. Experimental analysis of cross-sectional flow motion in a large amplitude meandering bend. *Earth Surf. Process. Landf.* **2011**, *36*, 244–256. [CrossRef]

20. Gholami, A.; Akhtari, A.A.; Minatour, Y.; Bonakdari, H.; Javadi, A.A. Experimental and numerical study on velocity fields and water surface profile in a strongly-curved 90° open channel bend. *Eng. Appl. Comput. Fluid Mech.* **2014**, *8*, 447–461. [CrossRef]

21. Farhadi, A.; Sindelar, C.; Tritthart, M.; Glas, M.; Blanckaert, K.; Habersack, H. An investigation on the outer bank cell of secondary flow in channel bends. *J. Hydro-Environ. Res.* **2018**, *18*, 1–11. [CrossRef]

22. Van der Kindere, J.W.; Laskari, A.; Ganapathisubramani, B.; de Kat, R. Pressure from 2D snapshot PIV. *Exp. Fluids* **2019**, *60*, 32. [CrossRef]

23. Scharnowski, S.; Bross, M.; Kähler, C.J. Accurate turbulence level estimations using PIV/PTV. *Exp. Fluids* **2019**, *60*, 1. [CrossRef]

24. Kalpakli Vester, A.; Sattarzadeh, S.S.; Örlü, R. Combined hot-wire and PIV measurements of a swirling turbulent flow at the exit of a 90° pipe bend. *J. Vis.* **2016**, *19*, 261–273. [CrossRef]

25. Nishi, Y.; Sato, G.; Shiohara, D.; Inagaki, T.; Kikuchi, N. A study of the flow field of an axial flow hydraulic turbine with a collection device in an open channel. *Renew. Energy* **2019**, *130*, 1036–1048. [CrossRef]

26. Lindken, R.; Merzkirch, W. A novel PIV technique for measurements in multiphase flows and its application to two-phase bubbly flows. *Exp. Fluids* **2002**, *33*, 814–825. [CrossRef]

27. Termini, D.; Di Leonardo, A. Efficiency of a Digital Particle Image Velocity (DPIV) Method for Monitoring the Surface Velocity of Hyper-Concentrated Flows. *Geosciences* **2018**, *8*, 383. [CrossRef]

28. Sarno, L.; Carleo, L.; Papa, M.N.; Villani, P. Experimental Investigation on the Effects of the Fixed Boundaries in Channelized Dry Granular Flows. *Rock Mech. Rock Eng.* **2018**, *51*, 203–225. [CrossRef]

29. Sarno, L.; Carravetta, A.; Tai, Y.C.; Martino, R.; Papa, M.N.; Kuo, C.Y. Measuring the velocity fields of granular flows—Employment of a multi-pass two-dimensional partical image velocimetry (2D-PIV) approach. *Adv. Power Technol.* **2018**, *29*, 3107–3123. [CrossRef]

30. Booij, R. Measurements and large eddy simulations of the flows in some curved flumes. *J. Turbul.* **2003**, *4*, 1–17. [CrossRef]

31. Blanckaert, K. *Flow and Turbulence in Sharp Open-Channel Bends*; EPFL: Lausanne, Switzerland, 2002.

32. Zeng, J.; Constantinescu, G.; Blanckaert, K.; Weber, L. Flow and bathymetry in sharp open-channel bends: Experiments and predictions. *Water Resour. Res.* **2008**, *44*. [CrossRef]

33. Blanckaert, K.; Graf, W.H. Mean flow and turbulence in open-channel bend. *J. Hydraul. Eng.* **2001**, *127*, 835–847. [CrossRef]

34. Odgaard, A.J.; Bergs, M.A. Flow processes in a curved alluvial channel. *Water Resour. Res.* **1988**, *24*, 45–56. [CrossRef]

35. Yen, B.C. On establishing uniform channel flow with tail gate. *Proc. Inst. Civ. Eng.-Water Marit. Eng.* **2003**, *156*, 281–283. [CrossRef]

36. Chen, H.; Zhong, Q.; Wang, X.; Li, D. Reynolds number dependence of flow past a shallow open cavity. *Sci. China Technol. Sci.* **2014**, *57*, 2161–2171. [CrossRef]

37. Zhong, Q.; Li, D.; Chen, Q.; Wang, X. Coherent structures and their interactions in smooth open channel flows. *Environ. Fluid Mech.* **2015**, *15*, 653–672. [CrossRef]

38. Moser, R.D.A.P. *Direct Numerical Simulation of Curved Turbulent Channel Flow*; National Aeronautics and Space Administration: Moffett Field, CA, USA, 1984.

39. Nezu, I.; Onitsuka, K. Turbulent structures in partly vegetated open-channel flows with LDA and PI V measurements. *J. Hydraul. Res.* **2001**, *39*, 629–642. [CrossRef]

40. Bradshaw, P. Turbulence secondary flows. *Fluid Mech.* **1987**, *19*, 53–74. [CrossRef]

41. Blanckaert, K.; De Vriend, H.J. Secondary flow in sharp open-channel bends. *J. Fluid Mech.* **1999**, *498*, 353–380. [CrossRef]

42. Rozovskii, I.L. *Flow of Water in Bends of Open Channels*, 2nd ed.; Academy of Sciences of the Ukrainian SSR: Jerusalem, Israel, 1957.

43. Einstein, H.A.; Harder, J.A. Velocity distribution and the boundary layer at channel bends. *Eos Trans. Am. Geophys. Union* **1954**, *35*, 114–120. [CrossRef]

44. Bridge, J.S. *Rivers and Floodplains*; Blackwell Publishing: Hoboken, NJ, USA, 2009.

45. Blanckaert, K.; Duarte, A.; Chen, Q.; Schleiss, A.J. Flow processes near smooth and rough (concave) outer banks in curved open channels. *J. Geophys. Res. Earth Surf.* **2012**, *117*. [CrossRef]

Article

Experimental Investigation on Mean Flow Development of a Three-Dimensional Wall Jet Confined by a Vertical Baffle

Ming Chen [1,2,*], Haijin Huang [1], Xingxing Zhang [1], Senpeng Lv [1] and Rengmin Li [1]

[1]   Key Laboratory of Hydraulic and Waterway Engineering of the Ministry of Education, Chongqing Jiaotong University, Chongqing 400074, China; hhj910219@163.com (H.H.); zhangddy@126.com (X.Z.); sinplor@126.com (S.L.); lrengmin@126.com (R.L.)

[2]   Key Laboratory of Navigation Structure Construction Technology, Ministry of Transport, Nanjing 210029, China

*   Correspondence: chenmingjy@126.com; Tel.: +86-023-62652714

Received: 26 December 2018; Accepted: 27 January 2019; Published: 30 January 2019

Abstract: Three-dimensional (3D) confined wall jets have various engineering applications related to efficient energy dissipation. This paper presents experimental measurements of mean flow development for a 3D rectangular wall jet confined by a vertical baffle with a fixed distance (400 mm) from its surface to the nozzle. Experiments were performed at three different Reynolds numbers of 8333, 10,000 and 11,666 based on jet exit velocity and square root of jet exit area (named as B), with water depth of 100 mm. Detailed measurements of current jet were taken using a particle image velocimetry technique. The results indicate that the confined jet seems to behave like an undisturbed jet until $16B$ downstream. Beyond this position, however, the mean flow development starts to be gradually affected by the baffle confinement. The baffle increases the decay and spreading of the mean flow from $16B$ to $23B$. The decay rate of 1.11 as well as vertical and lateral growth rates of 0.04 and 0.19, respectively, were obtained for the present study, and also fell well within the range of values which correspond to the results in the radial decay region for the unconfined case. In addition, the measurements of the velocity profiles, spreading rates and velocity decay were also found to be independent of Reynolds number. Therefore, the flow field in this region appears to have fully developed at least $4B$ earlier than the unconfined case. Further downstream (after $23B$), the confinement becomes more pronounced. The vertical spreading of current jet shows a distinct increase, while the lateral growth was found to be decreased significantly. It can be also observed that the maximum mean velocity decreases sharply close to the baffle.

Keywords: experiment; particle image velocimetry; 3D confined wall jet; mean flow

1. Introduction

It is well understood that three-dimensional (3D) wall jets are typically characterized by the interaction between a turbulent boundary layer and a free jet. Because of their diverse practical engineering applications (e.g., film cooling [1], heat transfer [2], and energy dissipation in hydraulic structure [3–8]), a number of experimental studies on the mean flow development in the jet have been conducted in the past few decades. The experiments of 3D wall jet issuing from rectangular orifices have been firstly performed by Sforza and Herbst [9]. From their results, it is well known that the flow field of jet can be divided into three regions: the potential core region (PC), the characteristic decay region (CD) and the radial decay region (RD). Subsequent to their reviews, considerable investigations were carried out for the unconfined case. For example, Padmanabham and Gowda [10] measured the mean flow characteristics of 3D wall jets using a technique of the total pressure probe and determined the influence of the geometry on the characteristic decay region. Law and Herlina [11] investigated the velocity and concentration characteristics of 3D turbulent circular wall jets using a combined PIV and Planar Laser Induced Fluorescence approach. The results showed that velocity profiles collapsed well in both the longitudinal and lateral directions after 20 nozzle diameters and 25 nozzle diameters, respectively. Additionally, the spreading of the jet and decay of local maximum velocity were critically presented. Agelin-Chaab and Tachie [12] performed PIV (particle image velocimetry) measurements for the 3D wall jet issuing from a square nozzle. The main goal of their measurements was to examine the effect of the rough surface on the flow development of the jet. Later, they presented more detailed laboratory investigations on the jet, including the mean flow in the developing and self-similar regions [13]. Recently, detailed flow structures of a 3D curved wall jet have been reported by Kim et al. [14]. The results indicated that, due to the Coanda effect [15], the jet developed on the cylinder surface after the impingement of the circular jet and self-preserving wall jet profile did not clearly occur in such jet.

In the past, there have been only a few studies on the 3D confined wall jet. In order to determine whether the large lateral growth of the jet is induced by the secondary flow, Després and Hall [16] measured the flow field in a 3D wall jet with and without the grid using hot-wire anemometry and PIV. They found that the grid delayed the lateral growth of the jet and increased its vertical growth. Meanwhile, the grid also decreased the mass entrainment and mixing performance. The grid, however, was placed at the nozzle exit in their studies; thus, the confinement condition differs significantly from that in the present study. Onyshko et al. [17] provided the PIV data for a deflected wall jet. The results showed that the baffle placed on the bed had dramatic impact upon the flow feature in the jet: a wall jet-like flow was observed before reaching the baffle; after the baffle, a plane jet-like was formed and then the jet flow was deflected toward the water surface along a curvilinear trajectory. Successively, experimental studies for a wall jet impinging onto a forward-facing step in a cross-flow were comprehensively carried out by Langer et al. [18] using planar laser induced fluorescence (PLIF). They presented the jet flow regime after the initial impingement and found that the perimeter and aspect ratio of the jet were dependent of jet-step distance, height of the step and Reynolds numbers. Additionally, predictive correlations for the shape and size of the jet after impingement were discussed. In these two investigations, although the baffle or step was positioned away from the jet exit, their heights were relatively low and submerged in water. Consequently, the jet flow development was not highly confined by the baffle.

The present investigation is to focus mainly on the mean flow development of the 3D confined wall jet. In fact, the 3D confined wall jet plays a significantly important role in hydraulic engineering related to its powerful efficacy for enhancing energy dissipation. The filling and emptying system of navigation lock is a typical case of hydraulic structure producing 3D confined wall jet, which issues from rectangular nozzles (side ports) located in longitudinal culvert. In general, the large amount of water energy is dissipated by the interaction among the jet, bounded ambient fluid in the lock chamber and chamber wall so as to provide better mooring conditions for vessels [19,20]. A schematic sketch of the composite flow in current jet is shown in Figure 1, where the deflected streamlines in the lateral and wall-normal directions are clearly illustrated due to the confinement effect. In particular, after the normal impingement of the jet onto the baffle, corner wall jets in the lateral directions [21,22] and upward wall jet are easily generated in the vicinity of the baffle due to the Coanda effect. The figure also servers to define the coordinate system; x, y and z represent the longitudinal, wall-normal and lateral directions, respectively; U_m denotes the local maximum mean velocity; y_m is the wall-normal location where U_m occurs; $y_{m/2}$ is the distance from the bottom wall to the point in the outer layer where the velocity is half of U_m (called half-height); $z_{m/2}$ is the lateral location where the velocity has a half value of U_m (called half-width). It should be noted that the lateral confinement is neglected.

Water **2019**, 11, 237

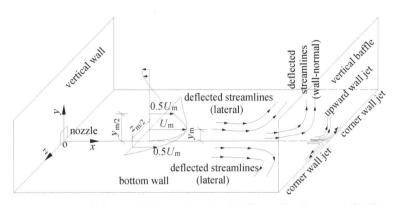

Figure 1. Schematic diagram of a three-dimensional wall jet confined by a vertical baffle.

2. Experimental Setup

The experiments were performed in a test section which is 800 mm long, 400 mm wide and 400 mm deep. Figure 2 shows schematically the experimental arrangement. The side walls and bottom of the test section were made of clear glass to facilitate the PIV measurements. The wall jet was formed by water passing through a long rectangular pipe, which allowed the flow to fully develop. The pipe has a 14 mm × 16 mm (width × height) cross section and was placed to flush the test section floor. The jet exit velocity was conditioned by a flow control valve and an electromagnetic flowmeter. In this experimental facility, two settling basins attached directly to both ends of the test section were specially designed to condition the flow and obtain the various water depths. Moreover, four constant-head skimming weirs, which were used to stabilize the water surface and ensure the overflow to return back to the supply tank, were constructed and installed at each corner of the settling basins. The Cartesian coordinate system, as shown in Figures 1 and 2, was employed. Note that $x = 0$ is at the jet exit plane, $y = 0$ is on the test section floor, and $z = 0$ is at the symmetry plane of the nozzle.

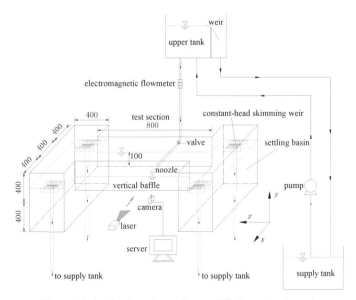

Figure 2. A sketch of experimental setup. (All dimensions in mm).

3. PIV System and Data Analysis

The velocity field for the jet was acquired using a PIV technique. For the present investigation, the water was seeded with 10 μm hollow glass spheres which have a specific gravity of 1.4. This size and density can ensure that the particles follow the flow synchronously [23,24]. A laser with a continuous energy of 10 W was used to illuminate the flow field. The light sheet formed by the laser was set around 1 mm thick and included the central axis of the jet. All the illuminated images were captured by an 8-bit high-resolution digital camera (NX5-S2 series) with a 2560 × 1920 pixels charge-coupled device (CCD). A 50 mm lens (Canon 50 mm f/1.2) was fitted to the camera. All the image pairs were captured at each position with a sampling rate of 1 Hz, a value that is low enough for the images to be uncorrelated [24]. The arrangements of the laser and camera required to be adjusted depending on the plane of measurements. For the *x-z* plane measurements, the laser and camera were positioned at the side and bottom of the test section, respectively. In case of the *x-y* plane measurements, the laser was positioned at the bottom of the test section while the camera was positioned at the side of the settling basin. It should be noted that the location of *x-z* plane measurement depends on the position of U_m, and thus varies with the streamwise distance due to the baffle confinement. It is anticipated that the offset *x-z* plane measurements at different *x*-axial locations can describe the vertical variations for mean velocities such as *x*- and *z*-axial velocity, *U* and *W*, respectively. Regarding *x-y* plane measurements, Law and Herlian [11] conducted the offset tests and found that the self-similar velocity profiles still occurred at various sections. Therefore, the offset *x-y* plane measurements from the jet centerline were not performed in the present study.

In the particle image processing, each velocity field involved two consecutive frames, and a time interval between the two frames was critically determined to be 1250 μs such that the maximum particle displacement satisfied the one-quarter rule for PIV correlation analysis [25]. The exposure time for each frame was fixed at 400 μs as a compromise between minimizing image streak and maximizing brightness [26]. The frames presented were divided into numerous small interrogation regions, and the cross-correlation method was used to determine the displacement of the particles in the interrogation window through the peak of the cross-correlation. Subsequently, the local velocity vector of each pair of images was calculated by the displacement and time interval mentioned previously. Detailed information of PIV algorithms is available in the investigation reported by Westerweel et al. [27]. Meanwhile, to improve the computation accuracy for measurements, particle images were processed with the iterative multigrid image deformation method [28]. A three-point Gaussian curve fit was used to determine the peak of displacement with subpixel accuracy. Spurious vectors were removed by the normalized median test method recommended by Westerweel et al. [29] and new vectors were filled by a weighted interpolation approach. The minimum size of the interrogation window of 16 × 16 pixels with 50% overlap was used to process the data. The instantaneous image processing program was developed by Beijing Jiang Yi technology co., LTD (Beijing, China). The mean velocity field was calculated using a MATLAB script developed in our laboratory. Considering the effect of the number of instantaneous image pairs on the calculation accuracy of mean velocity, Hu et al. [30] measured the vertical velocity profiles in a high-precision flume, and obtained mean velocities at various vertical locations based on different sample size of image pairs. The mean velocities obtained were compared with that averaged by expected sample size, thereby gaining the standard deviations between the two mean velocities at various vertical locations. In terms of this analysis method, the convergence test for the experimental data, including three mean velocity components at a typical gauge point location, is shown in Figure 3, where 5000 image pairs were selected as the expected sample size. Small enough deviation for *x*-axial mean velocity, *U*, was found after the sample size *N* = 2500, while the corresponding small deviations for *y*- and *z*-axial mean velocities, *V* and *W*, respectively, were obtained at least after *N* = 4000. To ensure faithful mean flow quantities, 5000 PIV instantaneous image pairs were chosen in the present study. In terms of the curve-fitting algorithm for instantaneous vectors, the size of interrogation window, and the required number of instantaneous image pairs for the calculation of the mean velocity, uncertainties in the mean velocities were estimated to be ±3.4% and ±2.5% for the local velocity close to and away from the wall, respectively.

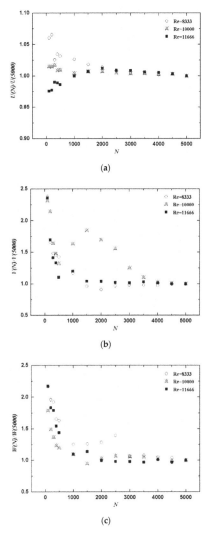

(a)

(b)

(c)

Figure 3. Convergence test for the experimental data for three different Reynolds numbers (Re) of 8333, 10,000, and 11,666: (**a**) x-axial mean velocity (U) varies with the number of instantaneous image pairs (N), (**b**) y-axial mean velocity (V) varies with the number of instantaneous image pairs (N), and (**c**) z-axial mean velocity (W) varies with the number of instantaneous image pairs (N).

4. Results and Discussion

In the present study, several important mean flow quantities, including velocity profiles and their similarity, growth rate of the half-width and half-height, and decay rate of local maximum velocity, are discussed. Experiments were performed at three Reynolds numbers of 8333, 10,000, and 11,666 (Re = U_0B/v, where U_0 is jet exit velocity, B is square root of jet exit area, and v is kinetic viscosity of water) The jet exit velocities of 0.5 m/s, 0.6 m/s and 0.7 m/s, which correspond to the three Reynolds numbers, were determined from the PIV measurements. The corresponding flow rates (Q) of 0.089 l/s,

0.109 l/s and 0.131 l/s were measured by the electromagnetic flowmeter. A water depth of 100 mm was set for the test section. All measurements in the lateral (*x-z*) and symmetry (*x-y*) planes are presented at least for the range of $10 \leq x/B \leq 24$. It should be noted that the measurement locations have not been extended to both the wall and water surface considering the effect of reflection of the laser light on the accuracy of PIV data. The corresponding distances from the measurement edge to the wall and water surface are 1 mm and 5 mm, respectively.

4.1. Spreading Rates

Figure 4 shows the variations of the velocity half-height $y_{m/2}$ and half-width $z_{m/2}$ with downstream distance. More specifically, $y_{m/2}$ and $z_{m/2}$ are the wall-normal and lateral locations where $0.5U_m$ occurs, respectively. In this figure, they were normalized by the square root of jet exit area (*B*) which is an appropriate scaling parameter as suggested by Padmanabham and Gowda [10] and Agelin-Chaab and Tachie [13]. The results for unconfined case obtained by Law and Herlian [11] are also included for comparison. The half-height increases approximately linearly in the region $16 < x/B < 23$, but after $x/B = 23$ the value of $y_{m/2}$ grows dramatically (Figure 4a). This behavior may be closely related to a clockwise vortex formed in the region $y/y_{m/2} < 0.25$ as mentioned in Section 4.2. Therefore, the position of U_m tends to be deflected away from the bottom wall. The half-width starts to spread after $x/B = 6$ and varies nearly linearly with downstream distance in the region $6 < x/B < 23$ (Figure 4b). However, the $z_{m/2}$ in the region $16 < x/B < 23$ develops more rapidly compared to the early region $6 < x/B < 16$. Beyond $x/B = 23$, although the confinement of the baffle can enhance the development of the jet flow field, the spreading of $z_{m/2}$ tends to significantly decrease, which is contrary to the variation of $y_{m/2}$ in the corresponding region. This is because the value of U close to the baffle gets considerably dropped due to most of the impinged jet fluid moving in both the lateral directions. In general, the variations of $y_{m/2}$ and $z_{m/2}$ are independent of Reynolds number within the present range.

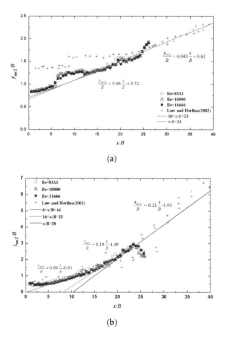

(a)

(b)

Figure 4. Mean flow developments of the 3D confined wall jet with Re = 8333, 10,000 and 11,666: (**a**) velocity half-height $y_{m/2}$, and (**b**) velocity half-width $z_{m/2}$.

To estimate the growth rate $(dy_{m/2}/dx)$, a linear fit is applied to the data in the region $16 < x/B < 23$ in Figure 4a. The variation of $y_{m/2}$ can be well described by the equation:

$$\frac{y_{m/2}}{B} = 0.04\frac{x}{B} + 0.72 \tag{1}$$

Similarly, the two linear relationships for $z_{m/2}$ (Figure 4b) in the regions $6 < x/B < 16$ and $16 < x/B < 23$ can be well fitted, respectively. These two equations are written as:

$$\frac{z_{m/2}}{B} = 0.09\frac{x}{B} - 0.01 \tag{2}$$

$$\frac{z_{m/2}}{B} = 0.19\frac{x}{B} - 1.49 \tag{3}$$

Therefore, the corresponding spreading rate $dy_{m/2}/dx$ is 0.04. The $dz_{m/2}/dx$ value of 0.09 in the region $6 < x/B < 16$ is comparable to the circular free jet. Further downstream ($16 < x/B < 23$), the lateral spreading becomes to diverge more rapidly and its slope ($dz_{m/2}/dx$) is 0.19. For comparison, some of previous investigations for 3D circular wall jets are illustrated in Table 1. It is interesting to see that, in the region $16 < x/B < 23$, the spreading rates of $y_{m/2}$ and $z_{m/2}$ for the confined wall jet in the present study, respectively, fall well within the ranges of 0.036 [31]–0.045 [10] and 0.17 [32]–0.33 [31] corresponding to the values in RD region for undisturbed jet reported in the literature. It should be pointed out that reaching RD region for the unconfined wall jet requires streamwise distance at least $x/B = 20$, as summarized in Table 1. These results imply that the baffle starts to alter the jet flow development after $x/B = 16$ and the fully developed region for the confined case appears to occur at least $4B$ earlier than the unconfined case.

Table 1. Illustration of previous studies on three-dimensional wall jet.

Authors	Measuring Technique	Re	RD Region	$dz_{m/2}/dx$	$dy_{m/2}/dx$	n
Padmanabham and Gowda [10]	HWA	95,400	>20B	0.216	0.045	1.15
Law and Herlina [11]	PIV	5500, 12,200, 13,700	>23B	0.21	0.042	1.07
Agelin-Chaab and Tachie [12,13]	PIV	5000, 10,000, 20,000	>60B	0.255	0.054	1.15
Després and Hall [16]	PIV	108,000	>45B	0.25	0.047	-
Present data	PIV	8333, 10,000, 11,666	16B–23B	0.19	0.040	1.11

Note: B = square root of jet exit area; RD is the radial decay region.

4.2. Mean Velocity Profiles

The time-averaged velocity profiles at selected x/B locations in both the vertical and lateral planes are summarized in this section and compared with the previous results, including the free jet [33], two-dimensional [34] and three-dimensional [11,13] wall jets. All the velocities were normalized by the local maximum mean velocity, U_m.

Figure 5 shows the streamwise development of the axial (U) and vertical (V) mean velocity profiles measured in the symmetry plane for Re = 11,666. The y coordinate was normalized by the velocity half-height $y_{m/2}$. The profiles of U collapse reasonably well in the region $10 < x/B < 23$, while the quality of collapse at $x/B = 10$ is relatively poor because the exit flow could not fully develop in the vertical direction, as shown in Figure 5a. Additionally, in the region $y/y_{m/2} < 1.5$, the present data are comparable to the unconfined 2D and 3D wall jet results from Verhoff [34] and Agelin-Chaab and Tachie [13], respectively. Some slight fluctuations of the profiles are observed as the jet evolves downstream to $x/B = 24$ near the baffle. This behavior is attributed to the confinement of the baffle. Further downstream, the confinement becomes more noticeable and negative values of U are observed in the region $y/y_{m/2} < 0.25$. This occurs because most of the impinged jet fluid moves in both the lateral directions and resulting low momentum in the vertical direction could not overcome the adverse pressure gradient. As a result, a clockwise vortex is formed in the corner. It also can be seen from Figure 5a, near the baffle ($x/B \geq 24$), there are some significant deviations which occur just approximately from $y/y_{m/2} = 1.5$ to 3 due to the resulting reverse flow in the vicinity of the water surface. The deviations could be supported by profiles of wall-normal mean velocity (V) in the symmetry plane shown in Figure 5b. After the normal impingement, flow separation occurs close to the baffle and the flow is divided into corner jets [21,22] in both the lateral directions and upward wall jet along the baffle surface due to the Coanda effect, followed by most positive values of V beyond $x/B = 23$ shown in Figure 5b. As expected, the reverse flow is formed after impingement of the upward jet onto the water surface. For the region $21 \leq x/B \leq 24$, some negative values of V are observed in the region $y/y_{m/2} > 3.4$ due to the reverse flow. In the range of $x/B \leq 23$, the magnitudes of V are negative over most of the water depth ($y/y_{m/2} < 3$), indicating that the ambient fluid is being drawn towards the bottom wall, owing to the presence of a secondary mean vortex presented by Launder and Rodi [1]. In addition, compared to the measurements made by Law and Herlian [11] and Agelin-Chaab and Tachie [13], similar variations of V with water depth are observed but the values are slightly lower as illustrated in Figure 5b. However, their measurements reported were selected at least after $x/B = 28$ where the jet flow has been fully developed. It should be noted that considerable scattered points are shown in this figure due to the low accuracy of PIV in the wall-normal direction as described by Law and Herlian [11].

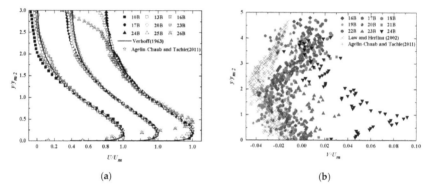

Figure 5. Mean velocity profiles measured in *x-y* plane for Re = 11,666: (**a**) U profiles, and (**b**) V profiles.

To further compare the present observations with analytical and previous results, a quantitative evaluation method with mean absolute relative error (MARE [35]) was used in the present study. The MARE is written as:

$$\text{MARE} = \frac{1}{M} \sum_{i=1}^{M} \left| \frac{U - U_p}{U_p} \right| \times 100 \qquad (4)$$

where M is total number of data on one velocity profile in given region, U is present *x*-axial mean velocities, U_p is analytical or previous *x*-axial mean velocities. The results of MARE for U profiles in *x-y* plane at different locations are summarized in Table 2. In the region ($y/y_{m/2} < 1.5$) unconfined by the water surface, the maximum error is substantially within the range of MARE < 5% before $x/B = 24$, while the maximum error increases to 18.7% after $x/B = 24$. The relatively large error between the confined and unconfined cases indicates that the baffle confinement has noticeable impact upon the mean velocity distribution.

The profiles of axial (U) and lateral (W) mean velocities at selected locations in the lateral plane are shown in Figure 6, where the *z* coordinate was normalized by the half-width ($z_{m/2}$). Being similar to the U velocity distribution in the symmetry plane, measurements ($10 < x/B < 23$) in the lateral plane show reasonable collapse in the region $z/z_{m/2} < 1.2$. However, when the flow evolves downstream ($x/B > 16$), there are some slight differences between the experimental data and previous observations

(Figure 6a). This is critically because the profiles of current jet start to be affected by the confinement of the vertical baffle after $x/B = 16$. Far downstream ($x/B \geq 24$), the confinement increases with increasing longitudinal distance and the velocity profiles across the entire sections seem to be unstable. For example, some significant fluctuations can be observed especially after $x/B = 26$ due to the presence of the baffle. For comparison, the results obtained by Law and Herlian [11] generally agree better with the present data. Similarly, Table 3 gives the MARE values for U profiles in x-z plane at different locations. Except for the region very close to the baffle ($x/B \geq 26$), the maximum error between the present data and analytical and previous results ($z/z_{m/2} < 1.2$) does not exceed the range of MARE < 5%. Figure 6b shows the W distribution at typical x/B locations. The lateral mean velocity (W) increases from zero at the symmetry plane to a peak value which occurs approximately at $z/z_{m/2} = 1.2$ within the region $x/B \leq 24$. Beyond $x/B = 24$, the location of the peak value is gradually delayed (i.e., $z/z_{m/2} = 1.6$ for $x/B = 25$). From this figure, the W profiles in the region $x/B \leq 21$ are relatively lower than those of Law and Herlian [11]. However, as the jet leaves the nozzle, the corresponding W values continuously increase. For example, the present observation at $x/B = 22$ is comparable to those of Law and Herlian [11]. When the jet develops downstream ($x/B = 24$), larger values of W can be observed compared to the results reported by Law and Herlian [11]. Further downstream ($x/B > 24$), the W profiles increase dramatically and are significantly higher as compared to previous results. This indicates that most of the fluid is deflected away from the centerline of the jet in the lateral plane due to the confinement of the vertical baffle.

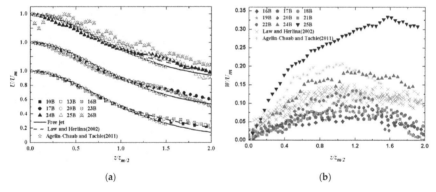

(a) (b)

Figure 6. Mean velocity profiles measured in x-z plane for Re = 11,666: (**a**) U profiles, and (**b**) W profiles.

In order to examine the effect of the baffle on the mean velocity profiles more closely, detailed velocity distribution with three Reynolds numbers are plotted in Figure 7. Collapsed curves also can be observed both in the lateral and vertical directions especially within the range of $10 \leq x/B \leq 23$. This indicates that the profiles become independent of Reynolds numbers except the regions near the baffle and water surface.

Table 2. MARE (%) values for U profiles in x-y plane at different locations ($y/y_{m/2} < 1.5$).

Location	10B	13B	16B	17B	20B	23B	24B	25B	26B
Analytical solution [34]	4.13	1.45	2.63	1.30	0.92	1.40	2.75	4.65	18.70
Agelin-Chaab and Tachie [13]	4.43	2.54	3.08	1.60	2.24	2.59	3.87	5.12	18.44

Table 3. MARE (%) values for U profiles in x-z plane at different locations ($z/z_{m/2} < 1.2$).

Location	10B	13B	16B	17B	20B	23B	24B	25B	26B
Analytical solution [33]	1.19	1.24	2.85	3.32	3.94	4.72	5.56	6.54	16.22
Law and Herlian [11]	2.06	1.83	2.73	3.11	3.05	3.74	5.47	5.58	15.19
Agelin-Chaab and Tachie [13]	1.44	1.04	2.44	3.00	3.50	4.14	5.38	6.34	15.86

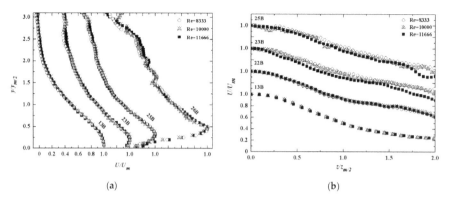

(a)   (b)

Figure 7. Variations of U profiles at typical locations with Re = 8333, 10,000 and 11,666: (**a**) *x-y* plane, and (**b**) *x-z* plane.

4.3. Decay of Local Maximum Velocity

Further insight into the development of mean velocity can be made by examining how the maximum mean velocity varies with downstream distance, as shown in Figure 8. The value of U_m virtually remains constant in the PC region ($x/B < 3.75$), while it starts to decrease gradually after $x/B = 6$. The U_m decay in both the regions $6 < x/B < 16$ and $16 < x/B < 23$, respectively, can be expressed in power-law forms:

$$\frac{U_m}{U_0} = 2.74\left(\frac{x}{B}\right)^{-0.58} \tag{5}$$

$$\frac{U_m}{U_0} = 11.67\left(\frac{x}{B}\right)^{-1.11} \tag{6}$$

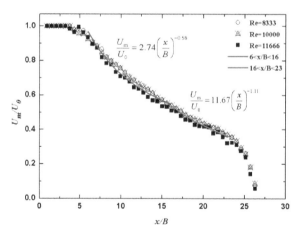

Figure 8. Decay of local maximum velocity U_m.

The decay rate of 0.58 represents the region $6 < x/B < 16$, which is comparable to the values in the CD region for unconfined 3D wall jets given by Padmanabham and Gowda [10]. In the region $16 < x/B < 23$, significant decay of U_m can be clearly seen from Figure 8 and the corresponding decay exponent is 1.11, a value that is compared very well with those of unconfined cases in the RD region summarized in Table 1. These results are consistent with the previous observations mentioned above.

However, it is anticipated that a sharp reduction of U_m can be observed by the strong confinement of the baffle as the jet develops further downstream ($x/B > 24$).

5. Conclusions

An experimental study on a three-dimensional confined wall jet was conducted to examine the effect of a fixed vertical baffle on the mean flow development. Measurements of the flow field were performed using a particle image velocimetry method. The results showed that the confined jet remained unaffected by the baffle until $x/B = 16$, while the other regions of the mean flow development are significantly characterized by the baffle confinement.

In the region ($16 < x/B < 23$), the relatively larger growth rates of 0.04 and 0.19 in the wall-normal and lateral directions, respectively, were obtained for the confined case. These spreading rates fall well within the range of values which correspond to the results in the radial decay region for the undisturbed case. Both the vertical and lateral spreading rates of current jet were found to be independent of Reynolds number. The decay rate was estimated to be 1.11, a value that is consistent with those reported for undisturbed cases in the literature. Similarly, the decay of the maximum mean velocity is independent of Reynolds number. The measurements of mean velocity profiles also exhibit self-similarity and are strongly independent of Reynolds number. Therefore, the fully developed flow of current jet appears to form at least $4B$ earlier than the unconfined case.

As the jet evolves further downstream ($x/B \geq 24$), the confinement becomes more noticeable, and corner jets and upward wall jet are generated close to the baffle. It was found that the jet flow developed more rapidly in the wall-normal direction due to the presence of the baffle. On the contrary, the lateral spreading is significantly reduced. In general, the present mean velocity profiles show reasonable collapse but some differences between the confined and undisturbed jets in the regions very close to the baffle and near the water surface.

It can be concluded that the baffle helps the jet to develop especially beyond $x/B = 16$. This study contributes to a better understanding of the energy dissipation mechanism of current jet. Given the limited measurement cases in the present study, additional tests over considerable distances from nozzle to baffle are needed to fully investigate the mean flow development of the three-dimensional confined wall jet.

Author Contributions: M.C. led the work performance and wrote the article; H.H., X.Z., S.L. and R.L. conducted the experiments and collected data through review of papers.

Funding: This research was supported by the National Key R&D Program of China, Grant number 2016YFC0402001, National Natural Science Foundation of China, Grant number 51509027, and Key Laboratory of Navigation Structure Construction Technology of China, Grant number Yt918002.

Conflicts of Interest: The authors declare no conflicts of interest.

Notation

B	square root of jet exit area
N	number of instantaneous image pairs
M	total number of data on one velocity profile in given region
n	exponent describing the decay of U_m
Q	flow rate through the jet pipe
Re	jet exit Reynolds number based on jet exit velocity and square root of jet exit area
U	x-axial mean velocity
V	y-axial mean velocity
W	z-axial mean velocity
U_0	jet exit velocity
U_m	local maximum mean velocity
U_p	analytical or previous x-axial mean velocity
x	longitudinal direction in the coordinate system
y	wall-normal direction in the coordinate system
y_m	wall-normal location where U_m occurs

Notation

$y_{m/2}$	wall-normal location where $0.5U_m$ occurs
z	lateral direction in the coordinate system
$z_{m/2}$	lateral location where $0.5U_m$ occurs
v	kinetic viscosity of water

References

1. Launder, B.E.; Rodi, W. The turbulent wall jet-measurements and modeling. *Annu. Rev. Fluid Mech.* **1983**, *15*, 429–459. [CrossRef]
2. Caggese, O.; Gnaegi, G.; Hannema, G.; Terzis, A.; Ott, P. Experimental and numerical investigation of a fully confined impingement round jet. *Int. J. Heat Mass Transf.* **2013**, *65*, 873–882. [CrossRef]
3. Chen, J.G.; Zhang, J.M.; Xu, W.L.; Peng, Y. Characteristics of the velocity distribution in a hydraulic jump stilling basin with five parallel offset jets in a twin-layer configuration. *J. Hydraul. Eng. ASCE* **2014**, *140*, 208–217. [CrossRef]
4. Yang, Y.H.; Chen, M.; Zhang, X.X.; Duan, L.M.; Miao, J.K. 3D numerical simulation of hydraulic characteristics of ditches designed for a navigation lock with high-head and large scale. *Port Waterw. Eng.* **2018**, *4*, 84–90. (In Chinese)
5. Gabl, R.; Righetti, M. Design criteria for a type of asymmetric orifice in a surge tank using CFD. *Eng. Appl. Comput. Fluid Mech.* **2018**, *12*, 397–410. [CrossRef]
6. Adam, N.J.; De Cesare, G.; Schleiss, A.J. Influence of geometrical parameters of chamfered or rounded orifices on head losses. *J. Hydraul. Res.* **2018**. [CrossRef]
7. Wu, J.H.; Ai, W.Z.; Zhou, Q. Head loss coefficient of orifice plate energy dissipator. *J. Hydraul. Res.* **2010**, *48*, 526–530.
8. Zhang, Q.Y.; Chai, B.Q. Hydraulic characteristics of multistage orifice tunnels. *J. Hydraul. Eng. ASCE* **2001**, *127*, 663–668. [CrossRef]
9. Sforza, P.M.; Herbst, G. A Study of Three-Dimensional Incompressible, Turbulent Wall Jets. *AIAA J.* **1970**, *8*, 276–283.
10. Padmanabham, G.; Gowda, B.H.L. Mean and turbulence characteristics of a class of three-dimensional wall jets—Part 1: Mean flow characteristics. *J. Fluids Eng. ASME* **1991**, *113*, 620–628. [CrossRef]
11. Law, A.W.K.; Herlina. An experimental study on turbulent circular wall jets. *J. Hydraul. Eng. ASCE* **2002**, *128*, 161–174. [CrossRef]
12. Agelinchaab, M.; Tachie, M.F. PIV study of three-dimensional wall jet over smooth and rough surfaces. In Proceedings of the FEDSM2007 5th Joint ASME/JSME Fluids Engineering Conference, San Diego, CA, USA, 30 July–2 August 2007.
13. Agelin-Chaab, M.; Tachie, M.F. Characteristics of turbulent three-dimensional wall jets. *J. Fluids Eng. ASME* **2011**, *133*, 021201-1–021201-12. [CrossRef]
14. Kim, M.; Kim, H.D.; Yeom, E.; Kim, K.C. Flow characteristics of three-dimensional curved wall jets on a cylinder. *J. Fluids Eng. ASME* **2018**, *140*, 041201-1–041201-7. [CrossRef]
15. Cîrciu, I.; Boşcoianu, M. An analysis of the efficiency of Coanda-NOTAR anti-torque systems for small helicopters. *INCAS Bull.* **2010**, *2*, 81–88.
16. Després, S.; Hall, J.W. The development of the turbulent three-dimensional wall jet with and without a grid placed over the outlet. In Proceedings of the ASME 2014 International Mechanical Engineering Congress and Exposition, Montreal, QC, Canada, 14–20 November 2014.
17. Onyshko, P.R.; Loewen, M.R.; Rajaratnam, N. Particle image velocimetry applied to a deflected wall jet. In Proceedings of the Hydraulic Measurements and Experimental Methods, Edmonton, AB, Canada, 28 July–1 August 2002.
18. Langer, D.C.; Fleck, B.A.; Wilson, D.J. Measurements of a wall jet impinging onto a forward facing step. *J. Fluids Eng. ASME* **2009**, *131*, 091103-1–091103-9. [CrossRef]
19. Stockstill, R.L. Modeling Hydrodynamic Forces on Vessels during Navigation Lock Operations. ASCE, 2002. Available online: https://doi.org/10.1061/40655(2002)83 (accessed on 8 September 2011).
20. Chen, M.; Liang, Y.C.; Xuan, G.X.; Chen, M.D. Numerical simulation of vessel hawser forces within chamber during navigation lock operations. *J. Ship Mech.* **2015**, *19*, 78–85. (In Chinese)
21. Hogg, S.I.; Launder, B.E. Three dimensional turbulent corner wall jet. *Aeronaut. J.* **1985**, *89*, 167–171.

22. Poole, B.; Hall, J.W. Turbulence measurements in a corner wall jet. *J. Fluids Eng. ASME* **2016**, *138*, 081204-1–081204-8. [CrossRef]

23. Shinneeb, A.M. Confinement Effects in Shallow Water Jets. Ph.D. Thesis, University of Saskatchewan, Saskatoon, SK, Canada, 2006.

24. Shinneeb, A.M.; Balachandar, R.; Bugg, J.D. Confinement Effects in Shallow-Water Jets. *J. Hydraul. Eng. ASCE* **2011**, *137*, 300–314. [CrossRef]

25. Adrian, R.J. Particle-imaging techniques for experimental fluid mechanics. *Annu. Rev. Fluid Mech.* **1991**, *23*, 261–304. [CrossRef]

26. Wang, H.; Zhong, Q.; Wang, X.K.; Li, D.X. Quantitative characterization of streaky structures in open-channel flows. *J. Hydraul. Eng. ASCE* **2017**, *143*, 04017040-1–04017040-10. [CrossRef]

27. Westerweel, J.; Elsinga, G.E.; Adrian, R.J. Particle image velocimetry for complex and turbulent flows. *Annu. Rev. Fluid Mech.* **2013**, *45*, 409–436. [CrossRef]

28. Scarano, F. Iterative image deformation methods in PIV. *Meas. Sci. Technol.* **2002**, *13*, R1–R19. [CrossRef]

29. Westerweel, J.; Scarano, F. Universal outlier detection for PIV data. *Exp. Fluids* **2005**, *39*, 1096–1100. [CrossRef]

30. Hu, J.; Yang, S.F.; Wang, X.K.; Lan, Y.P. Adaptability of PIV based on different working principle to turbulence measurements. *J. Hydroelectr. Eng.* **2013**, *32*, 181–186. (In Chinese)

31. Davis, M.; Winarto, H. Jet diffusion from a circular nozzle above a solid plane. *J. Fluid Mech.* **1980**, *101*, 201–221. [CrossRef]

32. Swamy, N.C.; Bandyopadhyay, P. Mean and turbulence characteristics of three-dimensional wall jets. *J. Fluid Mech.* **1975**, *71*, 541–562. [CrossRef]

33. Schlichting, H. *Boundary-Layer Theory*; Kestin, J., Translator; McGraw-Hill: New York, NY, USA, 1979.

34. Verhoff, A. *The Two-Dimensional Turbulent Wall Jet with and without an External Stream*; Princeton University: Princeton, NJ, USA, 1963.

35. Gumus, V.; Simsek, O.; Soydan, N.G.; Akoz, M.S.; Kirkgoz, M.S. Numerical modeling of submerged hydraulic jump from a sluice gate. *J. Irrig. Drain. Eng. ASCE* **2016**, *142*, 04015037-1–04015037-11. [CrossRef]

Article

Hydraulic Features of Flow through Local Non-Submerged Rigid Vegetation in the Y-Shaped Confluence Channel

Xuneng Tong, Xiaodong Liu *, Ting Yang, Zulin Hua, Zian Wang, Jingjing Liu and Ruoshui Li

Key Laboratory of Integrated Regulation and Resource Development on Shallow Lake of Ministry of Education, 1 Xikang Road, Nanjing 210098, China; txn@hhu.edu.cn (X.T.); yang92005006@163.com (T.Y.); Zulinhua@hhu.edu.cn (Z.H.); wangzian1996@hotmail.com (Z.W.); 15205168389@163.com (J.L.); 1514050118@hhu.edu.cn (R.L.)

* Correspondence: xdliu@hhu.edu.cn

Received: 5 December 2018; Accepted: 11 January 2019; Published: 15 January 2019

Abstract: A laboratory measurement with acoustic Doppler velocimeter (ADV) was used to investigate the flow through a Y-shaped confluence channel partially covered with rigid vegetation on its inner bank. In this study, the flow velocities in cases with and without vegetation were measured by the ADV in a Y-shaped confluence channel. The results clearly showed that the existence of non-submerged rigid plants has changed the internal flow structure. The velocity in the non-vegetated area is greater than in the vegetated area. There is a large exchange of mass and momentum between the vegetated and non-vegetated areas. In addition, due to the presence of vegetation, the high-velocity area moved rapidly to the middle of the non-vegetated area in the vicinity of tributaries, and the secondary flow phenomenon disappeared. The presence of vegetation made the flow in non-vegetated areas more intense. The turbulent kinetic energy of the non-vegetated area was smaller than that of the vegetated area.

Keywords: Y-shaped confluence channel; non-submerged rigid vegetation; longitudinal velocity; secondary flow; turbulent kinetic energy

---

1. Introduction

Channel confluences are a common occurrence in fluvial networks, where significant changes can occur in hydraulics, sediment transport, water environment, and ecology. Many previous studies have concentrated on hydraulic characteristics in channel confluence areas [1–6]. Lan [7] divided the confluences into two types: (1) the confluence where post-confluence channel forms a linear extension of the upstream main channel (also named asymmetrical confluences), (2) the Y-shaped junction. According to Lan's data the quantities of two types of confluences are about equal in nature. Most previous work has paid more attention to the asymmetrical river confluences, but few studies have been conducted on the Y-shaped confluences [8–13]. Many studies confirmed that there are many differences between the two types of confluence channel. Nadia el al. [14], with their experimental work, found that the higher the junction angle, the wider and longer the retardation zone at the upstream junction corner and the separation zone, and the greater the flow deflection at the entrance of the tributary into the post-confluence channel.

The aquatic vegetation is ubiquitous in natural rivers. The vegetation has many ecological, aesthetic and economic benefits, such as providing a terrestrial wildlife habitat; improving water quality; stabilizing streambanks and floodplains; supplying energy subsidies for aquatic and terrestrial ecosystems [15–17]. However, from an engineering point of view, aquatic vegetation practices have not been frequently encouraged because of increased flow resistance, sediment transport effects and decreased flood discharge efficiency compared to unvegetated regions [18]. Thus, a better understanding of the physical processes governing flow resistances in vegetated areas could help resolve these conflicting engineering and ecological considerations [19].

The studies conducted so far in this field consider straight flumes with different types of vegetation on the bed. Nepf [20] developed a physically based model to predict turbulence intensity and diffusion within rigid and emergent vegetation. Ghisalberti and Nepf [21] emphasized the importance of vegetation in open channels and used a model of vegetation to study the flow features in a straight flume. This model showed that the coherent vortices of vegetated shear layer dominate the vertical transport. Since then, many laboratory and numerical studies have been conducted to discuss the effects of vegetation on moving fluids [22–25]. For example, Nezu and Sanjou [26] investigated the turbulence characteristics and the evolution of coherent structures in a flume with rigid and submerged vegetation. They highlighted that the development of sweep and/or ejection events near the top of vegetation affects the exchange processes between vegetation and the overflow. Zhao and Cheng [27] used an array of rigid cylindrical rods to simulate emergent vegetation stems that were subject to unidirectional open channel flows. Until now, straight open channel flows with vegetation have received enough research attention. To address the new objectives of river restoration and environmental flood management, a better understanding of hydraulic features of flow through local non-submerged rigid vegetation in Y-Shaped confluences channel is required.

In this study, we focus on how local non-submerged vegetation influence hydraulic features in a Y-shaped confluence channel. A physical model was used to simulate changes of flow state, velocity distribution, turbulence structures, and turbulent kinetic energy caused by vegetation by comparing the results with non-vegetated conditions under the same flow regime. The primary objective is to investigate the features of distribution of velocity, secondary flow structure and turbulent kinetic energy with the existence of vegetation in Y-shaped confluence channel.

2. Laboratory Experiments

2.1. Experimental Setup and Measurement Technique

Experiments were conducted in a flat-bottom Plexiglas flume (Figure 1) in the Hydraulics Laboratory of Hohai University. Water was pumped into a stilling cistern and then flowed into the flume. The length of the upstream tributaries was 3 m, and the widths were 0.26 and 0.22 m, respectively. The length of the downstream main stream was 6 m, and the width was 0.4 m. In order to make this study have a certain practical significance, the design of the flume model was combined with the analysis of the morphological characteristics of the Xitiaoxi River Basin ($30°23'$–$31°11'$ N, $119°14'$–$120°29'$ E) to select the parameters (Table 1). The average width before and after the intersection of rivers and the intersection angle at the intersection were counted respectively. Based on the analysis of the morphological characteristics of river network and the actual conditions, the convergence angle between the tributaries (the angle between the geometric axes) was $60°$. A large number of non-submerged rigid vegetation, such as reeds, existed on both sides of these Y-shaped intersections. Polyvinyl chloride(PVC) baseboards ($1 \times 0.1 \times 0.01$ m) were used to cover the entire bottom of the flume. Rigid cylinders were used to simulate rigid vegetation (8 mm diameter, 0.2 m height). The flexural rigidity of simulated rigid vegetation was calculated using the approach of Łoboda et al. [28], based on Niklas [29] and ASTM [30]. The flexural rigidity of simulated rigid vegetation is $40,960$ N· mm^2.

The distance measured between the plants (rigid cylinders) was 0.025 m, and the linear spacing was 0.1 m. Vegetation was distributed on both sides of the flume in 2-m-long bands along the two tributaries, and 5-m-long bands along the main stream were planted perpendicularly with the artificial vegetation. In addition, two transition segments and a tailgate were installed to prevent large-scale disturbances from the inlet, thus enabling the development of a quasi-constant water flow by depth.

The outlet and the inlet, which were both connected to the tank, enabled continuous recirculation of the steady-state discharges.

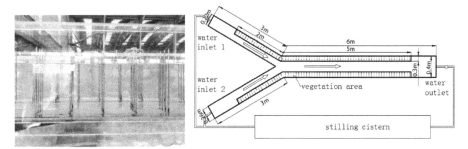

Figure 1. The sketch of the experiment.

Table 1. The morphological characteristics of the Xitiaoxi River Basin.

Y-Shaped Confluence Channel	Average Width of Left River before Confluence (m)	Average Width of Right River before Confluence (m)	Average Width of River after Confluence (m)	The Convergence Angle (°)
1	100.21	107.81	198.81	58.59
2	69.06	80.15	106.15	48.35
3	27.81	30.83	54.65	30.81
4	98.23	100.88	163.15	60.98
5	54.14	73.18	100.88	69.27
6	33.31	48.02	54.15	75.025
7	32.61	38.17	44.63	92.73
8	29.56	40.3	74.14	45.57
Average	55.616	64.918	99.570	60.166

This experiment used two flowmeters to measure the discharge of both tributaries. 3D velocities and velocity fluctuations were measured using a 3D sideways-looking acoustic Doppler velocimeter (ADV) manufactured by SonTek, Inc. (San Diego, CA, USA). The ADV technology is based on the pulse-to-pulse coherent measurement method. The instrument consists of three modules: a measuring probe, a conditioning module, and a processing module [31]. In this study, the ADV was used to sample each measurement point at a frequency of 50 Hz for 30 s. The WinADV-program, a post-processing program, was used to filter and post-process the sampled data. Data with average correlations less than or equal to 80% were filtered out. The result was that each point had a total of 1500 instantaneous data points, which ensured the adequacy and accuracy of the dataset. To achieve three-dimensional movement control of the ADV, a special regulating device was used to realize three-dimensional free positioning movement of the ADV probe in the test flume. The probe could be easily moved between measurement lines and sections. The ADV was mounted in a wood frame across the center section of the test segment and could be easily moved upstream or downstream, so that all sampling points were vertically aligned. The probe could also channel real-time data to the user's computer through a data acquisition program.

2.2. Test Series Description

To conduct comparative experiments and analyses, both the flume with vegetation and one without vegetation were measured while keeping other conditions unchanged. In this experiment, 18 sections were set up (two sections each on the tributaries and 14 sections on the main stream after the intersection). The coordinate origin was set at the bottom of the flume, directly below the intersection point of the two tributaries.

Before the intersection, five perpendiculars were established on each section of the tributaries, with each perpendicular having eight measurement points. In other words, there were 40 points in each section. After the intersection, eight perpendiculars were established on each section of the main stream, with each perpendicular having eight measurement points. This means that there were 64 points for each section. There were five survey lines on the two tributaries and seven survey lines on the mainstream. Figure 2a,b show the sections and the measuring lines.

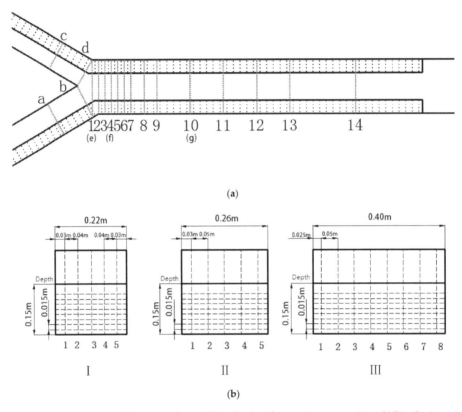

(a)

(b)

Figure 2. The sections and measuring lines; (**a**) Distribution of measuring cross sections; (**b**) Distribution of measuring lines.

The 3-D Doppler ultrasound anemometer was used to measure the flow velocities on 18 typical cross-sections. By comparing date, seven of those typical cross-sections were chosen to study. The seven sections were the initial section and intersection of both tributaries and sections 1, 4, and 10 of the main streams, as seen in Figure 2a. These seven cross sections were defined as cross Section a, cross Section b, cross Section c, cross Section d, cross Section e, cross Section f, and cross Section g. Section a is 2 m upstream of the right confluence apex, Section b is the right confluence apex, Section c is 2 m upstream of the left confluence apex, Section d is the left confluence apex, Sections e, f, g are 0, 0.2, 0.6 m downstream of the confluence apex, respectively. Vertical measuring lines (Figure 2b) were arranged on each cross-section. All the experimental data were derived from detailed high-resolution measurements on each measuring line.

In the flow characteristics test, the flow rate of the left tributary was $Q_1 = 20$ m^3/h, and the flow rate of the right tributary was $Q_2 = 20$ m^3/h. Therefore, the confluence ratio = $Q_1/Q_2 = 1.0$, and the intersection angle = $60°$. Some parameters of the experiment are shown in Table 2, Q denotes to the flow rate; b and h means the width and water section depth, respectively; V points to the velocity of the water section; R_e and F_r is the Reynolds number and Froude number of the flow, respectively.

Table 2. Parameters of the experiment.

Element	Q (m³/h)	b (m)	h (m)	V(m/s)	R_e (10 °C)	F_r
Left tributary	20	0.22	0.15	0.17	8168	0.14
Right tributary	20	0.26	0.15	0.14	7584	0.12
Main stream	40	0.4	0.15	0.19	12135	0.16

3. Analytical Methods

3.1. Velocity Distribution

Li et al. [32] proposed a method to calculate the velocity of a curved open channel with partially non-submerged rigid vegetation by comparing the integral of the longitudinal depth averaged velocity between vegetated and non-vegetated conditions to analyze the changes in the velocity distribution. The streamwise velocities at different water depths in different sections under vegetated conditions were depth averaged along the depth to obtain the corresponding average streamwise velocity. Due to the limited number of measurement points, the area method of experimental points can be used to circumscribe the area and to calculate the experimental measurement points. By comparing the integral of the streamwise depth velocity between vegetated and non-vegetated conditions, the changes in velocity distribution could be analyzed.

3.2. Secondary Flow Structure

Secondary motions are commonly present in open channel flows. In contrast to the facility in identifying longitudinal bedforms, secondary flows are much more difficult to measure because of their relatively small magnitude [33]. Around the 1950s, the existence of cellular secondary flows in rivers was only inferred from laterally periodic distribution of primary flow and sediment concentration, rather than being confirm through direct velocity measurements, by geologists and river engineers [34]. The precise filed measurement of cellular secondary flows was possible only after the ascent of the electromagnetic current meter (EM) and acoustic Doppler velocimeter (ADV) [35]. The water mixed energetically in the Y-shaped channels, and three-dimensional flow characteristics were prominent. A single vector or contour map was not enough to explain these complex hydrodynamic phenomena. For ease of explanation, this paper provides and explains a longitudinal velocity contour slice map. Each specific section of the diagram is presented and explained. The study presents a comparative analysis of the flow characteristics of the Y-shaped channels with and without vegetation. The velocity data obtained through the experiments were processed using the Tecplot software.

3.3. Turbulent Kinetic Energy

The quantitative analysis of turbulent flow is based on measurements of velocity fluctuations at a single point with local non-submerged rigid vegetation and without any vegetation in a Y-shaped confluence channel. In this study, the mean flow velocity components (u, v, and w) and the velocity fluctuation components in turbulent flow (u', v', and w') correspond to the streamwise, lateral, and vertical directions, respectively. Velocity fluctuations can be defined as deviations from the mean velocity. The turbulent kinetic energy was calculated as noise free [36].

In general, turbulent kinetic energy can be considered a measure of turbulence intensity [37]. In this study, the turbulent kinetic energy E_k was calculated using Equation (1), which was proposed by Li [38] and defines E_k as follows in terms of streamwise, lateral, and vertical directions:

$$E_k = \rho \left(\overline{u'^2} + \overline{v'^2} + \overline{w'^2} \right) / 2 \qquad (1)$$

where ρ is the density of water and u', v', and w' are the velocity fluctuations of streamwise, lateral, and vertical flow, respectively, and the overbars denote mean values.

4. Results

4.1. Distribution of Stream-Wise Velocity

The depth-averaged streamwise velocity distribution of the seven typical cross sections (Figure 3) showed that under the retarding effect of vegetation, the depth-averaged streamwise velocities (U_{avg}) in the vegetated region were much lower than in the non-vegetated region. A large velocity gradient was apparent near the junction of the vegetated and non-vegetated areas, we can see all the figures show a large velocity gradient was apparent near the junction of the veg and non-veg areas, it not only can be seen from the tributaries, but also from the mainstreams, which showed typical values as reported by Huai et al. [39].

There was a difference in the streamwise velocity distribution between non-vegetated and vegetated cases (Figure 3). The velocities in the vegetated case have a higher gradient than in the non-vegetated case, which can be ascribed to the fact that the retardation caused by vegetation makes the transverse distribution slightly more non-uniform. Along the streamwise direction, the velocities in the vegetated area decreased and those in the non-vegetated area increased when heading downstream, which is consistent with results for open channel flows found by Nikora [40] and Caroppi [41].

Three-dimensional flow characteristics were prominent after the intersection. The depth-averaged streamwise velocity distribution was not enough to explain the complex hydrodynamic phenomena. Clearly, the flow rate gradient was larger with vegetation than without vegetation, which indicates that the presence of vegetation caused drastic changes in the flow structure of the river (Figure 4). With vegetation (Figure 4a), the flow velocity in vegetated zones was smaller than in non-vegetated zones. It can also be observed that in the absence of vegetation (Figure 4b), the streamwise velocity distribution was polarized in the longitudinal direction, with the branch on the left side going downstream having a large flow area, whereas the area close to the right-side tributaries near the bottom was in a low-velocity zone. Downstream of the confluence in non-vegetated cases, a small velocity separation zone appeared on the lateral surface of the left tributary. This is consistent with Wang's results [42]. After vegetation planting (Figure 4a), the polarized character of the streamwise velocity distribution in the longitudinal direction disappeared, the downstream flow velocities in the vegetated areas in the left and right branches obviously diminished, and the flow velocity distribution with depth became insignificant.

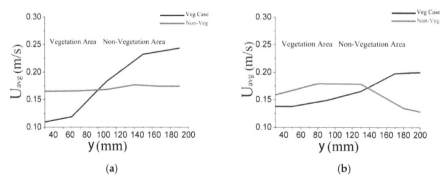

Figure 3. *Cont.*

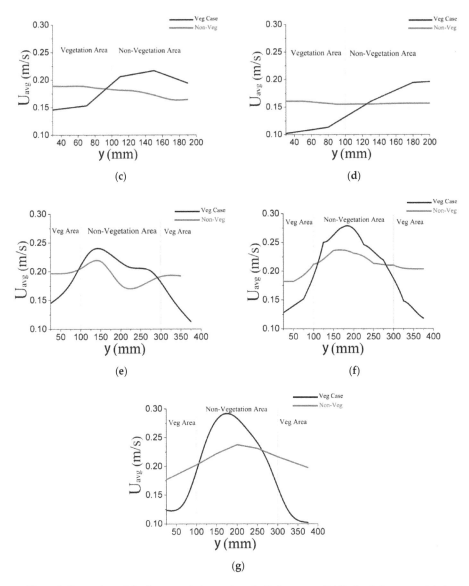

Figure 3. Comparison of depth-averaged streamwise velocity transverse distribution between vegetated and non-vegetated cases, where U_{avg} is the depth-averaged streamwise velocity and y is the flume width. (**a**) Cross Section a; (**b**) Cross Section b; (**c**) Cross Section c; (**d**) Cross Section d; (**e**) Cross Section e; (**f**) Cross Section f; (**g**) Cross Section g.

4.2. Secondary Flow Structure

There was a difference in Secondary Flow Structure by comparing the experimental results for the confluence of the Y-shaped channels with vegetation (Figure 5) and without vegetation (Figure 6). For the mainstream Sections (e), (f), and (g), when vegetation was present, the velocity difference in the confluence was more remarkable. The Section e is the confluence cross section, without vegetation (Figure 5), the left velocity was higher, and the right velocity was lower. A small clockwise circulation appeared at the bottom of the left side. But there was no circulation in the vegetated case (Figure 6). In addition, with the presence of vegetation, the velocity rapidly became partitioned, with the velocity on both sides decreasing with resistance from vegetation and increasing in the middle. Cross sections e and g showed that the higher-velocity area extended to the center in the non-vegetated case, the bottom of the separation zone became larger, the secondary flow phenomena along the left bank slowly disappeared, and a slight counterclockwise motion appeared in the lower right corner in the mainstream cross Section f. This result is similar to the report by Vaghefi et al. [43]. Under the influence of vegetation resistance, the low-velocity zone gradually became wider, the high-speed zone gradually narrowed, the low-speed area on the left changed faster than the low-speed area on the right, and there was no circulation.

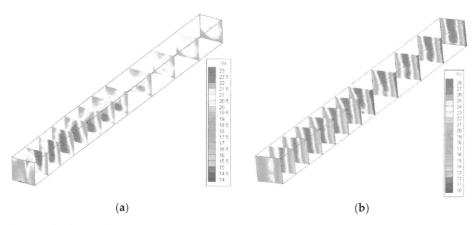

(**a**)   (**b**)

Figure 4. Slice diagram of streamwise velocity contour after intersection: (**a**) vegetated case; (**b**) non-vegetated case.

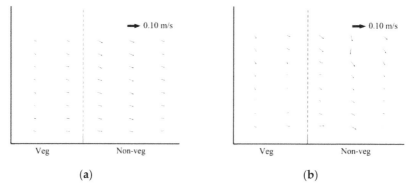

(**a**)   (**b**)

Figure 5. *Cont.*

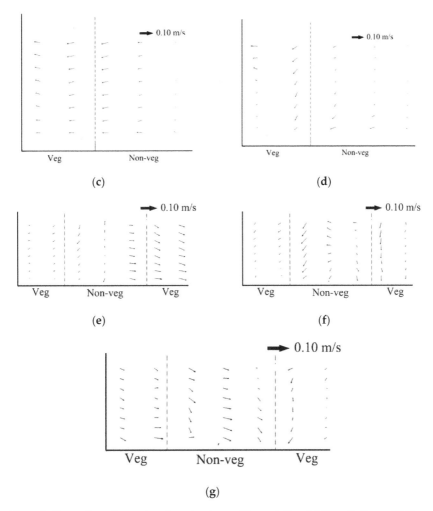

Figure 5. Comparison of secondary flow structures with vegetation. (**a**) Cross Section a; (**b**) Cross Section b; (**c**) Cross Section c; (**d**) Cross Section d; (**e**) Cross Section e; (**f**) Cross Section f; (**g**) Cross Section g.

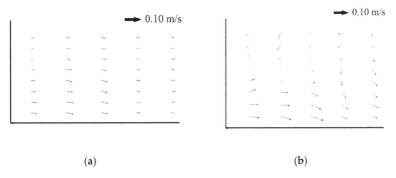

Figure 6. *Cont.*

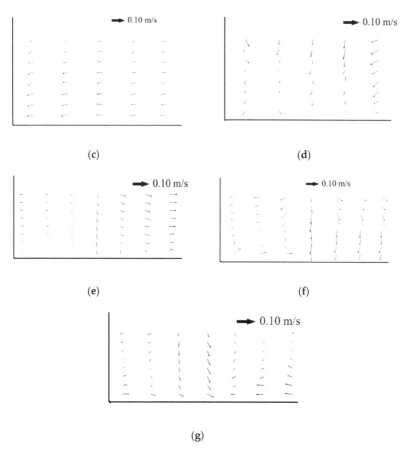

(c)

(d)

(e)

(f)

(g)

Figure 6. Comparison of secondary flow structures without vegetation. (**a**) Cross Section a; (**b**) Cross Section b; (**c**) Cross Section c; (**d**) Cross Section d; (**e**) Cross Section e; (**f**) Cross Section f; (**g**) Cross Section g.

4.3. Turbulent Kinetic Energy

The relational graphs of the change of turbulent kinetic energy with depth were shown in Figures 7 and 8, respectively. According to Figures 7 and 8, the results showed that the turbulent kinetic energy in the tributaries was much larger with vegetation than without it. The maximum value with vegetation was greater than 20 Pa (Figure 7a–d), compared to a maximum of less than 15 Pa (Figure 8a–d) without vegetation. The turbulent kinetic energy of the vegetated area was less than that of the non-vegetated area, and the turbulent kinetic energy showed a remarkable increase near the junction of the vegetated and non-vegetated areas. Although the numerical values of turbulent kinetic energy for each section at different depths were different, the distribution trends in the tributaries were consistent. After convergence, as shown in Figure 7e–g, the turbulent kinetic energy of the non-vegetated area was smaller than that of the vegetated area.

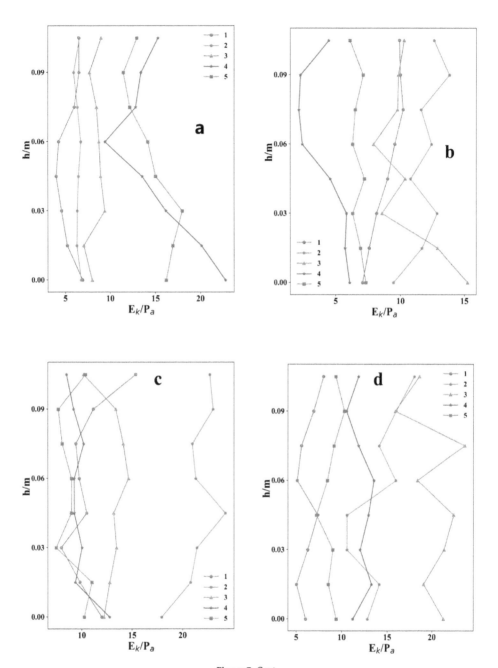

Figure 7. *Cont.*

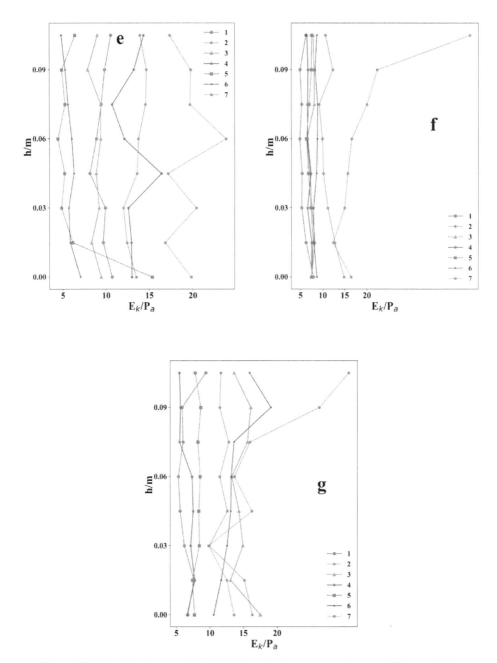

Figure 7. Turbulent kinetic energy (E_k/Pa); (**a**) Cross Section a at vegetation cases; (**b**) Cross Section b at vegetation cases; (**c**) Cross Section c at vegetation cases; (**d**) Cross Section d at vegetation cases; (**e**) Cross section e at vegetation cases; (**f**) Cross section f at vegetation cases; (**g**) Cross section g at vegetation cases.

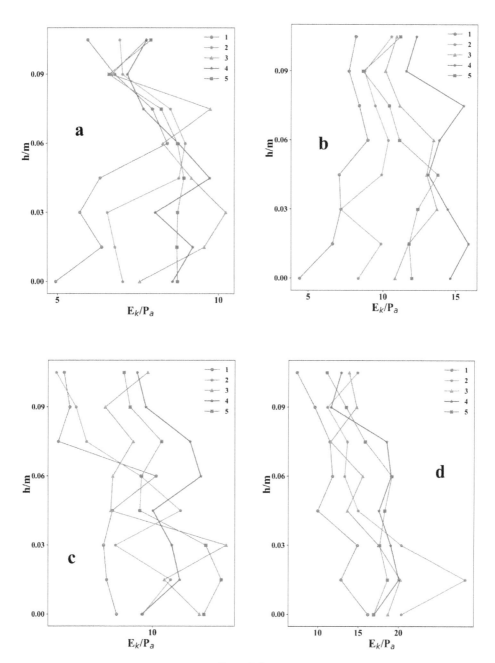

Figure 8. *Cont.*

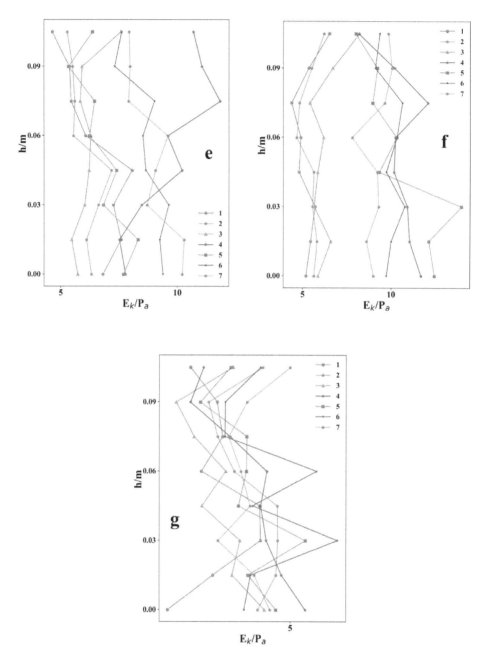

Figure 8. Turbulent kinetic energy (E_k/Pa); versus relative depth (h/m) with vegetation, versus relative depth (h/m) without vegetation. (**a**) Cross Section a at non-veg cases; (**b**) Cross Section b at non-veg cases; (**c**) Cross Section c at no-veg cases; (**d**) Cross Section d at no-veg cases; (**e**) Cross Section e at no-veg cases; (**f**) Cross Section f at no-veg cases; (**g**) Cross Section g at no-veg cases.

5. Discussion

In this study, with the effect of vegetation, the streamwise velocities in the vegetated area were much lower than in the non-vegetated area (Figures 3 and 4), which is consistent with the field study conducted by Chambers et al. [44]. They showed that in natural rivers with slower currents and therefore in the vegetated region, the water current was expected to be lower than in the main stream. The velocities in the vegetated case have a higher gradient than in the non-vegetated case, which can be ascribed to the fact that the retardation caused by vegetation makes the transverse distribution slightly more non-uniform, consistent with the conclusion drawn in the study by Mossa et al. [45]. Along the streamwise direction, the velocities in the vegetated area decreased and those in the non-vegetated area increased when heading downstream (Figures 3 and 4). The reason for this may have been that the main circulation in the Y-shaped confluence changed the internal flow structures, and the velocity distribution varied accordingly, which is consistent with the results from Wang et al. [42]. The vegetation prevented vertical mixing of fluid, and because of resistance, the velocity near the vegetated side was significantly lowered.

A large velocity gradient was apparent near the junction of the vegetated and non-vegetated areas. We can see that all the figures show that a large velocity gradient was apparent near the junction of the veg and non-veg areas, indicating there are must be remarkable mass and momentum exchanges at the junction of these two areas. The secondary flow is occurred for the momentum exchange [46]. The experimental results for the confluence of the Y-shaped channels with vegetation (Figure 5) and without vegetation (Figure 6) were compared. In the vegetation area, no clear circulation is found in the whole area.

The presence of vegetation caused a great change in the internal flow structure and made the flow in non-vegetated areas more intense. When water flows through vegetation, the flow resistance increases and the corresponding velocity decreases. At the same time, the flow will generate turbulence, consuming the flow energy. The turbulent kinetic energy of the non-vegetated area was smaller than that of the vegetated area. This is similar to Liu et al.'s results [47].

Hydraulic properties of riverine vegetation have been widely researched. Many laboratory, numerical, analytical and field studies have been conducted to account for the complex dynamic interactions between vegetation and moving fluids. The channel characteristics of natural rivers are seen to constitute an interdependent system which can be described by a series of graphs having simple geometric forms [48]. There are many Y-shaped confluence channels in natural river networks. This study can be helpful in providing some physiographic implications for practical use.

6. Conclusion and Future Works

Due to the presence of vegetation, the flow velocities in the Y-shaped confluence channel were redistributed. Velocities in the vegetation area were much smaller than those in non-vegetation area due to the presence of vegetation. Compared with the flow without vegetation, the streamwise velocity in the vegetated area was much lower than in the non-vegetated area because of the effect of vegetation. A large velocity gradient is generated between the vegetated and non-vegetated areas, indicating a remarkable mass and momentum exchange at the junction of these areas. Under the combined effect of vegetation and the Y-shaped confluence, the streamwise velocities in vegetated areas were much lower than those in non-vegetated areas, and the streamwise velocity of non-vegetated areas in the mainstream in the case with vegetation increased significantly compared with the case without vegetation.

In the tributaries, due to the presence of vegetation, the high-velocity area moved rapidly to the middle of the channel in non-vegetated areas, and the secondary flow phenomenon disappeared. In the mainstream, when vegetation was introduced, circulation disappeared, and the degree of lateral mixing decreased.

The presence of vegetation brought about great changes in internal flow structure. It caused the flow in non-vegetated areas to become more intense, and the turbulent kinetic energy of the tributaries in the cases with vegetation was significantly lower than in the cases without vegetation.

At the junction of vegetated and non-vegetated areas, the turbulent energy increased significantly. The turbulent energy of the non-vegetated areas was significantly greater than in the same position in the absence of vegetation.

This work focused on a single type of aquatic plant in the Y-shaped confluence channel. However, rivers actually include several types of plants. Further studies could investigate different vegetation types and different discharges in the Y-shaped channel. This study focused on flume experiments, but further investigations could combine this approach with numerical modeling.

Author Contributions: Conceptualization, X.T. and X.L.; Methodology, T.Y.; Software, R.L.; Validation, J.L., Z.W. and X.T.; Formal Analysis, T.Y.; Investigation, Z.W.; Resources, Z.W.; Data Curation, T.Y.; Writing-Original Draft Preparation, X.T.; Writing-Review & Editing, X.L.; Visualization, X.L.; Supervision, Z.H. and X.L.; Project Administration, X.L.; Funding Acquisition, X.L.

Funding: This research was funded by the National Natural Science Foundation of China (NSFC) (Grant No. 51479064, No. 51739002, No. 51479010, No. 51779016), the National College Students' Innovative Training Project China (Grant No. 201710294031), the National Key Project R&D of China (2016YFC0401702), Jiangsu Province Discipline Construction Funded Projects, Project Funded by the Priority Academic Program Development of Jiangsu Higher Education Institutions (PAPD) and PPZY2015A051.

Acknowledgments: We are grateful to the Hydraulic Laboratory, Hohai University, for the use of laboratory facilities and to Xu Jingzhao and Mei Shengchen for their thoughtful comments on measurements during the laboratory research.

Conflicts of Interest: The authors declare no conflict of interest.

References

1. Ashmore, P.; Parker, G. Confluence scour in coarse braided streams. *Water Resour. Res.* **1983**, *19*, 392–402. [CrossRef]
2. Best, J.L. Sediment transport and bed morphology at river channel confluences. *Sedimentology* **1988**, *35*, 481–498. [CrossRef]
3. Best, J.L.; Reid, I. Separation zone at open-channel junctions. *J. Hydraul. Eng.* **1984**, *110*, 1588–1594. [CrossRef]
4. Best, J.L.; Roy, A.G. Mixing-layer distortion at the confluence of channels of different depth. *Nature* **1991**, *350*, 411–413. [CrossRef]
5. Biron, P.M.; Lane, S.N. Modelling hydraulics and sediment transport at river confluences. In *River Confluences, Tributaries and the Fluvial Network*; Rice, S.P., Roy, A.G., Rhoads, B.L., Eds.; John Wiley: Hoboken, NJ, USA, 2008; pp. 17–43.
6. Biron, P.; De Serres, B.; Roy, A.G.; Best, J.L. Shear layer turbulence at an unequal depth channel confluence. In *Turbulence: Perspectives on Flow and Sediment Transfer*; Clifford, N., Ed.; Wiley: New York, NY, USA, 1993.
7. Bo, L.A.N. The comprehensive analysis of the special property at the tributary junction of Mountain river. *J. Chongqing Jiaotong Inst.* **1998**, *17*, 91–96. (In Chinese)
8. Best, J.L. Flow dynamics at river channel confluences: Implications for sediment transport and bed morphology. In *Recent Developments in Fluvial Sedimentology*; Ethridge, F.G., Flores, R.M., Harvey, M.D., Eds.; SEPM Society for Sedimentary Geology: Tulsa, OK, USA, 1987; pp. 27–35.
9. Taylor, E.H. Flow characteristics at rectangular open-channel junctions. *J. Trans.* **1944**, *109*, 893–902.
10. Rhoads, B.L.; Kenworthy, S.T. Flow structure at an asymmetrical stream confluence. *Geomorphology* **1995**, *11*, 273–293. [CrossRef]
11. Rhoads, B.L.; Sukhodolov, A.N. Field investigation of three-dimensional flow structure at stream confluences: 1. Thermal mixing and time-averaged velocities. *Water Resour. Res.* **2001**, *37*, 2393–2410.
12. Riley, J.D.; Rhoads, B.L.; Parsons, D.R.; Johnson, K.K. Influence of junction angle on three-dimensional flow structure and bed morphology at confluent meander bends during different hydrological conditions. *Earth Surf. Process. Landf.* **2015**, *40*, 252–271. [CrossRef]
13. Gualtieri, C.; Filizola, N.; de Oliveira, M.; Santos, A.M.; Ianniruberto, M. A field study of the confluence between Negro and Solimões Rivers. Part 1: Hydrodynamics and sediment transport. *C. R. Geosci.* **2018**, *350*, 31–42.
14. Penna, N.; De Marchis, M.; Canelas, O.B.; Napoli, E.; Cardoso, A.H.; Gaudio, R. Cardoso and Roberto Gaudio. Effect of the Junction Angle on Turbulent Flow at a Hydraulic Confluence. *Water* **2018**, *10*, 469. [CrossRef]

15. Schnauder, I.; Sukhodolov, A.N. Flow in a tightly curving meander bend: Effects of seasonal changes in aquatic macrophyte cover. *Earth Surf. Process. Landf.* **2012**, *37*, 1142–1157. [CrossRef]

16. Weber, A.; Zhang, J.; Nardin, A.; Sukhodolov, A.; Wolter, C. Modelling the Influence of Aquatic Vegetation on the Hydrodynamics of an Alternative Bank Protection Measure in a Navigable Waterway. *River Res. Appl.* **2016**, *32*, 2071–2080. [CrossRef]

17. Łoboda, A.M.; Bialik, R.J.; Karpiński, M.; Przyborowski, Ł. Two simultaneously occurring Potamogeton species: Similarities and differences in seasonal changes of biomechanical properties. *Pol. J. Environ. Stud.* **2019**, *28*, 1–17. [CrossRef]

18. Aberle, J.; Järvelä, J. Flow resistance of emergent rigid and flexible floodplain vegetation. *J. Hydraul. Res.* **2013**, *51*, 33–45. [CrossRef]

19. Järvelä, J.; Aberle, J.; Dittrich, A.; Schnauder, I.; Rauch, H.P. Flow–Vegetation–Sediment Interaction: Research Challenges. In Proceedings of the International Conference River Flow, London, UK, 6–8 September 2006; Ferreira, R.M.L., Alves, E.C.T.L., Leal, J.G.A.B., Cardoso, A.H., Eds.; Taylor & Francis: London, UK, 2006; Volume 2, pp. 2017–2026.

20. Nepf, H.M. Drag, turbulence, and diffusion in flow through emergent vegetation. *Water Resour. Res.* **1999**, *35*, 479–489. [CrossRef]

21. Ghisalberti, M.; Nepf, H. Mass Transport in Vegetated Shear Flows. *Environ. Fluid Mech.* **2005**, *5*, 527–551. [CrossRef]

22. Nikora, V. Hydrodynamics of aquatic ecosystems: An interface between ecology, biomechanics and environmental fluid mechanics. *River Res. Appl.* **2010**, *26*, 367–384. [CrossRef]

23. Nikora, V.; Lamed, S.; Nikora, N.; Debnath, K.; Cooper, G.; Reid, M. Hydraulic resistance due to aquatic vegetation in small streams: field study. *J. Hydraul.* **2008**, *134*, 1326–1332. [CrossRef]

24. Termini, D. Effect of vegetation on fluvial erosion processes: Experimental analysis in a laboratory flume. *Procedia Environ. Sci.* **2013**, *19*, 904–911. [CrossRef]

25. Termini, D. Flexible Vegetation Behavior and Effects on Flow Conveyance: Experimental Observations. *Int. J. River Basin Manag.* **2015**, *13*, 401–411. [CrossRef]

26. Nezu, I.; Sanjou, M. Turbulence structure and coherent motion in vegetated canopy open-channel flows. *J. Hydro-Environ. Res.* **2008**, *2*, 62–90. [CrossRef]

27. Zhao, K.; Cheng, N.-S.; Wang, X.; Tan, S.K. Measurements of Fluctuation in Drag Acting on Rigid Cylinder Array in Open Channel Flow. *J. Hydraul. Eng.* **2014**, *140*, 48–55. [CrossRef]

28. Łoboda, A.M.; Przyborowski, Ł.; Karpiński, M.; Bialik, R.J.; Nikora, V.I. Biomechanical properties of aquatic plants: The effect of test conditions. *Limnol. Oceanogr. Methods* **2018**, *16*, 222–236. [CrossRef]

29. Niklas, K.J. Plant Biomechanics. In *An Engineering Approach to Plant Form and Function*; University of Chicago Press: Chicago, IL, USA, 1992; ISBN 0-226-58641-6.

30. American Society for Testing and Materials (ASTM) D790-03. *Standard Test Methods for Flexural Properties of Unreinforced and Reinforced Plastics and Electrical Insulating Materials*; ASTM International: West Conshohocken, PA, USA, 2003.

31. Carollo, F.G.; Ferro, V.; Termini, D. Flow velocity measurements in vegetated channels. *J. Hydraul. Eng.* **2002**, *128*, 664–673. [CrossRef]

32. Li, C.G.; Xue, W.Y.; Huai, W.X. Effect of vegetation on flow structure and dispersion in strongly curved channels. *J. Hydrodyn.* **2015**, *27*, 286–291. [CrossRef]

33. Nezu, I.; Nakagawa, H.; Tominaga, A. Secondary Currents in a Straight Channel Flow and the Relation to Its Aspect Ratio. In *Turbulent Shear Flows 4*; Springer: Berlin/Heidelberg, Germany, 1985.

34. Karcz, I. Secondary currents and the configuration of a natural stream bed. *J. Geophys. Res.* **1966**, *71*, 3109–3112. [CrossRef]

35. Wang, Z.Q.; Cheng, N.S. Time-mean structure of secondary flows in open channel with longitudinal bedforms. *Adv. Water Resour.* **2006**, *29*, 1634–1649. [CrossRef]

36. Przyborowski, Ł.; Łoboda, A.; Bialik, R. Experimental Investigations of Interactions between Sand Wave Movements, Flow Structure, and Individual Aquatic Plants in Natural Rivers: A Case Study of *Potamogeton Pectinatus* L. *Water* **2018**, *10*, 1166. [CrossRef]

37. Blanckaert, K.; Vriend, H.J. Turbulence characteristics in sharp open-channel bends. *Phys. Fluids* **2005**, *17*, 717–731. [CrossRef]

38. Li, K.-F.; Wang, D.; Yang, K.; Zhu, M. Distribution of Turbulent Energy in Open Channel with Vegetation Patch. *Water Resour. Power.* **2017**, *2*, 122–124. (In Chinese)

39. Huai, W.; Li, C.; Zeng, Y. Curved open channel flow on vegetation roughened inner bank. *J. Hydrodyn.* **2012**, *24*, 124–129. [CrossRef]

40. Nikora, N.; Nikora, V.; O'Donoghue, T. Erratum for "Velocity Profiles in Vegetated Open-Channel Flows: Combined Effects of Multiple Mechanisms" by Nina Nikora, Vladimir Nikora, and Tom O'Donoghue. *J. Hydraul. Eng.* **2014**, *140*, 08014003. [CrossRef]

41. Caroppi, G.; Västilä, K.; Järvelä, J.; Rowin'ski, P.M. Experimental investigation of turbulent flow structure in a partially vegetated channel with natural-like flexible plant stands. In *New Challenges in Hydraulic Reseearch and Engineering, Proceedings of the 5th IAHR Europe Congress, Trento, Italy, 12–14 June 2018*; Armanini, A., Nucci, E., Eds.; Taylor & Francis: New York, NY, USA, 2018; pp. 615–616.

42. Wang, X.; Yan, Z.; Guo, W. Three-dimensional simulation for effects of bed discordance on flow dynamics at Y-shaped open channel confluences. *J. Hydrodyn.* **2007**, *19*, 587–593. [CrossRef]

43. Vaghefi, M.; Akbari, M.; Fiouz, A.R. An experimental study of mean and turbulent flow in a 180 degree sharp open channel bend: Secondary flow and bed shear stress. *KSCE J. Civ. Eng.* **2016**, *20*, 1582–1593. [CrossRef]

44. Chambers, P.A.; Prepas, E.E.; Hamilton, H.R.; Bothwell, M.L. Current velocity and its effect on aquatic macrophytes in flowing waters. *Ecol. Appl.* **1991**, *1*, 249–257. [CrossRef]

45. Mossa, M.; Ben Meftah, M.; De Serio, F.; Nepf, H.M. How vegetation in flows modifies the turbulent mixing and spreading of jets. *Sci. Rep.* **2017**, *7*, 6587. [CrossRef] [PubMed]

46. Tominaga, A.; Nezu, L.; Ezaki, K.; Nakagawa, H. Three-dimensional turbulent structure in straight open channel flows. *J. Hydraul. Res.* **1989**, *27*, 149–173. [CrossRef]

47. Liu, C.; Luo, X.; Liu, X.; Yang, K. Modeling depth-averaged velocity and bed shear stress in compound channels with emergent and submerged vegetation. *Adv. Water Resour.* **2013**, *60*, 148–159. [CrossRef]

48. Leopold, L.B.; Maddock, T. *The Hydraulic Geometry of Stream Channel and Some Physiographic Implications*; US Government Printing Office: Washington, DC, USA, 1953.

Article

Time-Averaged Turbulent Velocity Flow Field through the Various Bridge Contractions during Large Flooding

Kwang Seok Yoon [1], Seung Oh Lee [2] and Seung Ho Hong [3],*

[1] Korea Institute of Civil Engineering and Building Technology, Goyang 10223, Korea; ksyoon@kict.re.kr

[2] School of Urban and Civil Engineering, Hongik University, 94 Wausan-ro, Mapo-gu, Seoul 04066, Korea; seungoh.lee@hongik.ac.kr

[3] Department of Civil and Environmental Engineering, West Virginia University, 1306 Evansdale Drive, Morgantown, WV 26506, USA

* Correspondence: sehong@mail.wvu.edu; Tel.: +01-304-293-9926

Received: 18 December 2018; Accepted: 11 January 2019; Published: 15 January 2019

Abstract: Extreme rainfall events, larger than 500-year floods, have produced a large number of flooding events in the land and also close to the shore, and have resulted in massive destruction of hydraulic infrastructures because of scour. In light of climate change, this trend is likely to continue in the future and thus, resilience, security and sustainability of hydraulic infrastructures has become an interesting topic for hydraulic engineering stakeholders. In this study, a physical model experiment with a geometric similarity of the bridge embankments, abutments, and bridge deck as well as river bathymetry was conducted in a laboratory flume. Flow conditions were utilized to get submerged orifice flow and overtopping flow in the bridge section in order to simulate extreme hydrologic flow conditions. Point velocities of the bridge section were measured in sufficient details and the time-averaged velocity flow field were plotted to obtain better understandings of scour and sediment transport under high flow conditions. The laboratory study concluded that existing lateral flow contraction as well as vertical flow contraction resulted in a unique flow field through the bridge and the shape of velocity profile being "fuller", thereby increasing the velocity gradients close to the bed and subsequently resulting in a higher rate of bed sediment transport. The relationships between the velocity gradients measured close to the bed and the degree of flow contraction through the bridge are suggested. Furthermore, based on the location of maximum scour corresponding to the measured velocity flow field, the classification of scour conditions, long setback abutment scour and short setback abutment scour, are also suggested.

Keywords: abutment; overtopping flow; pressure flow; physical hydraulic modeling; scour and velocity field

1. Introduction

Usually, bridge foundation failure occurs due to the processes of (1) contraction scour-scouring across the entire channel due to the flow contraction caused by the bridge opening and deflection of floodplain flow into the main channel and (2) local scour at the base of piers and abutments caused by local flow contraction, down flow, and formation of a horseshoe vortex that wraps around the obstructions. Thus, the contraction scour and the local scour (pier scour and abutment scour) have been considered as two separate types of scour caused by different processes.

However, recent studies [1,2] show that abutment scour can be expressed as one type of contraction scour, not as a fundamental mechanism of local scour. When flow area is reduced by the bridge opening, velocity and bed shear stress are increased in order to satisfy continuity and momentum equations. The higher velocity resulted in increased erosive force, so more bed material is removed from the contracted section. In addition to the higher velocity due to the flow contraction, local flow structures associated with the base of the abutment result in additional erosion around the abutment. Due to the local flow structures, the scour depth near the upstream edge or corner of the abutment is usually deeper than that near the center of the channel. Thus, the combined effect of flow contraction as well as the local flow structure around an abutment made the problem of abutment scour more challenging. Although some researchers have suggested the need for more research on the subject of local flow structures and velocity distribution, past studies [3–10] have provided acceptable insight for the mechanisms of local flow structure around single and a group of bridge foundations under free flow conditions. However, for high flow conditions such as in an extremely hydrologic event, there is no widely applicable abutment scour formula, and the term has not been distinctly defined because of difficulties in understanding the complicated flow and scouring mechanism combined with complex geometries of bridge and various flow conditions.

Furthermore, a lot of research on bridge scour has focused only on simple and idealized situations where the bridge is placed in straight rectangular channels, even though many bridges are sited in non-rectangular channels whose geometry and hydraulic characteristics are site-specific in the real world. Thus, in this study, scour experiments were carried out in a compound channel cross-section using various lengths of bridge under free-surface flow (*F*) as well as in submerged orifice flow (*SO*), and overtopping flow (*OT*) cases. To investigate the effect of bridge submergences during large flooding scenarios, a bridge deck model was constructed based on the bridge design and dimensions commonly used in a rural region in the United States (USA). To understand the complex flow physics through the bridge, three components of velocities were measured by acoustic Doppler velocimeters (ADV).

The experimental results show new insights into the velocity flow field through the scour-critical bridges subject to the submerged orifice flow and overtopping flow during extreme flooding events. It is shown that vertical flow contraction by submerged flow resulted in a unique flow field through which the bridge and the shape of vertical velocity profile are "fuller" and thereby push the location of the maximum velocity closer to the bed, increasing the shear stress and erosion. Because the unique shape of velocity profile under the submerged flow causes higher erosion rate on the bed, examination of the relationships between the velocity gradient measured close to the bed and the non-dimensional parameters commonly used for the contraction scour is suggested to understand the effect of local flow variables on the scour. Furthermore, with the locations of the maximum scour corresponding to the measured velocity flow field, the scour conditions, long setback abutment scour and short setback abutment scour were classified for the practitioners to use in bridge design.

2. Methodology

2.1. Experimental Setup

In a previous study [11–13], laboratory experiments were carried out using a 1:60 scaled hydraulic model of the Towaliga River bridge at Macon, Georgia, USA including the full river bathymetry. The field data including measured discharge, bed elevation of cross sections, and gage height from the United State Geology Service (USGS) were reproduced by Froude number similarity. The previous experimental results showed that the hydraulic model can reproduce field scour data.

For the current set of experiments, the cross section shape and river geometry used in the previous experiments were slightly simplified and modified for the experiments to address more general features of the velocity flow field. The shape of the floodplain was horizontal on both sides of the main channel cross-section while preserving the original parabolic shape of the main channel. Also, the channel was constructed to have a straight alignment rather than meandering and all the piers were removed to be able to focus on the abutment scour. Figure 1 shows the modified laboratory model for the experiment in the flume and initial contour before the scouring experiment.

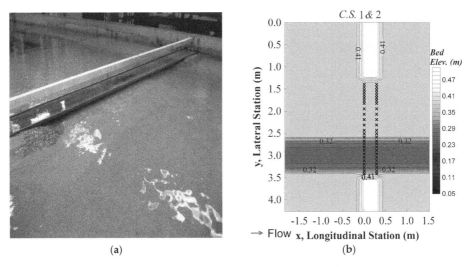

Figure 1. Laboratory model in this study in (**a**) and initial contour with velocity measurements cross sections in (**b**).

The approach channel upstream of the bridge was 11 m long followed by a working mobile bed section with a length of approximately 5.2 m in which the bridge model was placed. The fixed bed approach section was filled with 3.3 mm of gravel, and the 5.2 m moveable bed working section was filled with sand with d_{50} = 1.1 mm and σ_g = 1.3, where d_{50} and σ_g is the median diameter of sand and the geometric standard deviation often to describe the grain size distribution. To ensure a fully developed boundary layer at the bridge approach section consistent with the moveable bed sediment size, the fixed-bed approach channel consisted of a surface layer of fixed 1.1 mm sand having a depth of 3 cm. Based on the results, the approach channel has enough length to obtain a fully developed turbulent flow before reaching the moveable bed section. The range of approach section length/flow depth is 80 to 150, and the velocity measurements in the approach of the bridge, measured 3 m upstream of the bridge, shows perfect logarithmic velocity profiles for all the experimental cases (R^2 = 0.99) [14]. At the downstream of moveable bed section, there is 1.6 m long sediment trap section; this section trapped the sediment transported out of the working moveable bed section.

The model of the embankment and abutment was constructed as an erodible fill with rock riprap protection [1,2,14]. Three different lengths of erodible-embankment/abutment were modeled to simulate wide range of flow contractions on the left floodplain, but on the right floodplain, the toe of the abutment was maintained at the bankline for all experiments. This arrangement allowed the study of bankline abutment as well as setback abutments in the floodplain simultaneously under realistic geometric conditions similar to in the field. Based on the above modifications, abutment and embankment lengths and river bathymetry to be modeled in the laboratory were constructed as shown in Figure 2a. The ratio between abutment lengths (L_a) to the floodplain width (B_f), L_a/B_f, varied from 0.53 to 0.88 in the left floodplain. One of the main purposes of this study is to find the effect of velocity flow field on scour through the bridge for extreme hydrologic conditions. In those extreme conditions, overtopping or submerged orifice flow is likely to occur at the bridge. To simulate those extreme cases, the following dimensions normally used in a bridge design in rural region were used for the model bridge deck.

(a) Width of bridge deck 12.2 m, in accordance with standard two-lane roads;
(b) Bridge barrier 0.61 m high with 0.46 m top without sidewalks on non-bicycle routes;
(c) Slab depth of 0.46 m including the pavement;
(d) Girders 0.43 m wide and 0.46 m deep with 2.74 m spacing.

These design dimensions are commonly used for rural region two-lane bridges. Based on the prototype dimension, the 1:45 length scale bridge deck was constructed as shown in Figure 2b and a solid bridge deck model was supported and leveled with respect to an upper support beam as shown in Figure 1a.

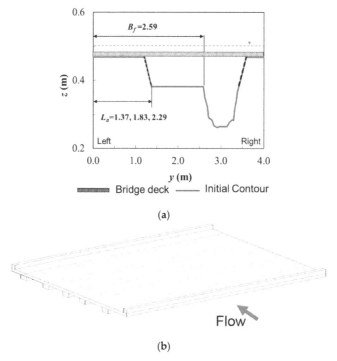

Figure 2. Geometries for (**a**) compound channel and abutment before scouring and (**b**) shape of model bridge deck.

2.2. Experimental Procedure

After completion of laboratory setup, the flume was filled with water slowly enough to saturate sand in working moveable-bed section without changing the original contours. After complete saturation, the initial bed elevations were measured in detail throughout the test section using an ADV and a point gage. The ADV gives distance from the sampling volume to the bed, which can be converted into elevation relative to the fixed datum by reading the point gage. After that, the required discharge and a flow depth larger than the target value were set by using a magnetic flow meter and by adjusting a tailgate, respectively, to prevent un-expected scour (erosion) while the experimental condition was set. Then, the tailgate was lowered slowly enough to achieve the target value of flow depth without abrupt water depth change. During this time, the point gage and/or wave gage mounted on the instrument carriage were used to measure the flow depth. Once the target flowrate and flow depth had been reached, the experiment was continued for five to six days until equilibrium scour was achieved (change in scour depth less than 2% within 24 hour). At the end of experiment, the equilibrium bed elevations were measured in the same way as in the initial bed-elevations measurements using the ADV and the point gage. The end of the experiment was defined when the scour depth reached the equilibrium state at which there were negligible changes in bed elevation with time. Once each experimental study was completed, the original river model was re-constructed, and scour experiment was re-started under same procedure as explained above using different flow variables. Total eighteen experimental conditions were used for the scour experiments to satisfy the purpose of the research.

Then, after the moveable bed experiments, the complete river bathymetry was modeled with a fixed-bed channel by spraying it with polyurethane. In the fixed-bed experiments, water surface profiles and velocities were measured to address the initial hydraulic conditions. Figure 1b shows the velocity measuring cross section (C.S. 1 and C.S. 2) along the bridge. Velocities were taken every 15 cm laterally in both the floodplain and the main channel. A minimum of six measuring points in each vertical profile and as many as 15 points were measured at both C.S. 1 and C.S. 2. With a similar way as in the bridge section, velocities were also measured at 3 m from the upstream face of the bridge for the approach flow velocity. During the velocity measurements, correlation values in these experiments were greater than 80% and the Signal Noise Ratio (SNR) was greater than 15. The sampling frequency of the ADV was chosen to be 25 Hz with a sampling duration of two minutes and perhaps as much as five minutes depending on the turbulence at each measuring location [14–19]. The phase-space despiking algorithm of Goring and Nikora [20] was also employed to remove any spikes in the time record caused by aliasing of the Doppler signal, which sometimes occurs near a boundary.

3. Results and Discussion

Initial hydraulic parameters measured in fixed bed have been summarized in Table 1: Q is the total discharge; V_{f1}/V_{fc1} and V_{m1}/V_{mc1} is approach flow intensity in the floodplain and in the main channel, respectively, where V_{f1} and V_{m1} is approach flow velocity and V_{fc1} and V_{mc1} is approach flow critical velocity calculated by Keulegan's equation; y_{f1} and y_{m1} is the water depth of floodplain and main channel in the approach section, respectively; q_{f2}/q_{f1} and q_{m2}/q_{m1} is unit discharge contraction ratio in the floodplain and main channel, respectively; W is the setback distance; L_m is the traverse distance from the toe of the abutment to the maximum scour hole depth. The definition sketch for the variables are shown in Figure 3.

Table 1. Initial experimental parameters and results for location of maximum scour depth.

Run	Flow Type	$\frac{Q}{(m^3/s)}$	$\frac{L_a}{B_f}$	$\frac{V_{f1}}{V_{fc1}}$	$\frac{V_{m1}}{V_{mc1}}$	$\frac{Y_{f1}}{(m)}$	$\frac{Y_{m1}}{(m)}$	$\frac{q_{f2}}{q_{f1}}$	$\frac{q_{m2}}{q_{m1}}$	$\frac{L_m}{y_{f1}}$	$\frac{W}{y_{f1}}$	Cond-Itions
1	F	0.093		0.61	0.83	0.074	0.152	1.818	1.543	3.70	16.46	A
2	SO	0.116		0.60	0.73	0.106	0.184	1.875	1.483	3.75	11.53	A
3	OT	0.164	0.53	0.61	0.72	0.149	0.227	1.148	1.146	2.05	8.20	A
4	F	0.085		0.58	0.77	0.075	0.153	1.755	1.441	4.05	16.19	A
5	SO	0.110		0.57	0.68	0.108	0.186	1.781	1.416	3.38	11.27	A
6	OT	0.150		0.56	0.64	0.148	0.226	1.25	1.100	1.86	8.26	A
7	F	0.085		0.54	0.74	0.076	0.154	2.236	1.756	5.24	10.08	A
8	SO	0.103		0.53	0.71	0.103	0.181	2.257	1.602	5.05	7.42	A
9	OT	0.150	0.71	0.56	0.66	0.150	0.228	1.176	1.243	1.63	5.09	A
10	F	0.074		0.49	0.72	0.073	0.151	2.208	1.539	4.15	10.37	A
11	SO	0.091		0.49	0.61	0.105	0.183	2.223	1.647	4.08	7.29	A
12	OT	0.130		0.50	0.57	0.147	0.225	1.278	1.262	1.45	5.18	A
13	F	0.074		0.44	0.69	0.076	0.156	*	1.904	5.84	4.03	C
14	SO	0.088		0.43	0.63	0.103	0.180	*	1.951	4.76	2.97	C
15	OT	0.130	0.88	0.45	0.57	0.150	0.227	*	1.422	2.46	2.04	C
16	F	0.062		0.38	0.55	0.073	0.155	*	1.976	4.72	4.15	C
17	SO	0.074		0.37	0.52	0.105	0.181	*	1.902	4.44	2.92	C
18	OT	0.110		0.40	0.50	0.147	0.224	*	1.481	2.09	2.07	C

(Subscript 1 and 2 refers to the approach section and bridge section, respectively; Condition A = long setback and Condition C = short setback abutment; symbol *: discharge per unit width in the bridge section were only measured in the main channel for short setback abutment).

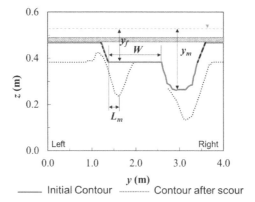

Figure 3. Definition sketch for classification of scour conditions.

As shown in Table 1, in the first three runs, the value of V_{f1}/V_{fc1} (approach flow intensity parameter) was similar. The experimental plan was to reproduce increasing tailwater with increasing discharge according to a tailwater rating curve, just as would occur in the field moving from *F* to *SO* to *OT* flow. The experiments were conducted in such a fashion that V_{f1}/V_{fc1} was held constant in the floodplain for a series of runs encompassing the three flow types for a given abutment length. This arrangement allowed us to find the effect of different flow types in scour because three flow types were encountered for different value of discharge and water depth, but their approach flow intensity parameter remained nearly constant for the three runs. The same conditions were also applied to other sets of the experiment. As explained in Hong et al. [2], the magnitude of maximum scour depth increased as the flow type changed from free flow (*F*) to submerged orifice flow (*SO*) and then decreased again for overtopping flow (*OT*) while holding the flow intensity (V_{f1}/V_{fc1}) constant. This pattern can be found in Figure 4. The scour depth is decreased in *OT* compared to *SO* even if the discharge is higher. This can be explained by the flow relief over the deck in *OT* rather than under the bridge. In most cases, the value of maximum scour depth was greater for *OT* flow than for *F* flow depending on the fraction of the total flow going under the bridge.

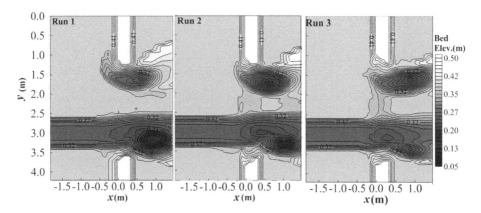

Figure 4. *Cont.*

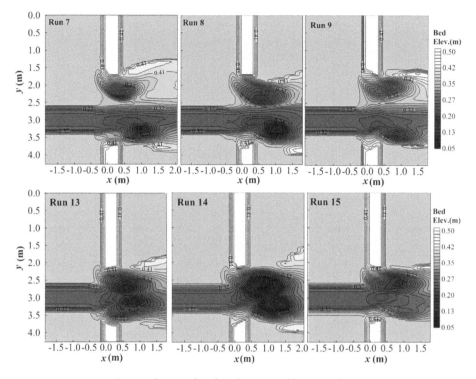

Figure 4. Contours for selected run in equilibrium condition.

3.1. Classification of Abutment Scour

Figure 4 shows the selected bed contours after each run. Based on the findings from Hong [14], the scour hole moved downstream from the abutment over time and the region of the deepest scour hole was located near the downstream region of the abutment. Because the scour process differed depending on the flow type over time, the detailed location of the deepest scour depth was case dependent. However, a similar trend can be deduced with respect to the shape of the scour hole and locations of the maximum scour depth in the experiments conducted with the same length of abutment. As shown in Figure 4, for runs 1, 2, and 3, the scour hole shape in the left floodplain showed curvature around the abutment, and the resulting point of the maximum scour was located downstream of the abutment. Similar findings can be applied for runs 7, 8, and 9. However, for runs 13, 14, and 15 with a longer abutment, the scour hole initially developed around the upstream corner of the abutment, the same as in the other runs, and then moved along the toe of the abutment in the left floodplain. However, as the scour hole elongated diagonally from the face of the abutment over time, the resulting point of maximum scour hole depth terminated at the bankline of the main channel, thus, the maximum scour depth in equilibrium was located inside the main channel on the bank side slope. For the bankline abutment in the right floodplain when looking downstream, the maximum scour depth can be found around the main channel side slope downstream of the bridge.

As explained in the previous paragraph, the location of maximum scour depth around the abutment varies in accordance with the length of abutment because the velocity flow field developed by the geometrical characteristics of an abutment is a leading factor to define the location of maximum scour depth. Thus, Melville and Coleman [4] classified the abutment to account for the effect of abutment/embankment length on the scour depth. In their classification, Case A applies to the abutment sited in a rectangular channel, while Case B represents the abutment that is sited on a floodplain and extended into the main channel. Case C is an abutment set well back from the main channel such that all scour takes place on the floodplain only. Case D is the limit of Case B and C where the abutment protrudes out to the edge of the main channel. Also, Chang and Davis [21,22] classified the abutment by the three categories as short, intermediate and long depending on the value of the setback distance and assumed that converging flow under the bridge with the abutment near the channel bank (long abutment) is mixed with the flow in the main channel and distributed uniformly. On the other hand, if the abutment is set well back from the channel bank, it is assumed that the overbank flow and the main channel flow remain separated from each other and do not mix as the flow passes under the bridge. Similar as in the previous research, scour conditions are classified as three cases, Condition A, Condition B, and Condition C in this study, but instead of using the length of the abutment, the scour conditions were classified in accordance with the location of maximum scour hole with respect to geometric ratio between length of abutment and width of the floodplain.

To find whether the maximum scour hole occurs in the floodplain or in the main channel, the non-dimensional value of W/y_{f1} was compared with L_m/y_{f1}. When the value of L_m/y_{f1} is larger than that of W/y_{f1}, the location of maximum scour hole is in the main channel. The maximum value of L_m/y_{f1} in all experiments is 5.84 in Run 13, as shown in Table 1. Thus, if the value of W/y_{f1} is smaller than approximately 6, the location of the maximum scour hole is expected to be outside of the floodplain. Thus, based on the findings in this study, clear water abutment scour conditions can be classified as: Condition A - long setback abutment scour; Condition B - bankline abutment scour; and Condition C - short setback abutment scour. The detailed descriptions and classification in terms of the ratio of setback distance (W) to the approach flow depth in the floodplain, W/y_{f1}, in this study are given below:

- Condition A ($W/y_{f1} > 6$, $L_a/B_f = 0.53$ and 0.77): In a long setback abutment, scour occurs in the floodplain only, well removed from the main channel;
- Condition B ($W/y_{f1} = 0$, $L_a/B_f = 1.0$): For a bankline abutment, maximum scour occurs in the main channel of a compound channel;
- Condition C ($W/y_{f1} < 6$, $L_a/B_f = 0.88$): In a short setback abutment, scour occurs on the floodplain in the initial stage, but maximum scour at equilibrium occurs in the main channel because the setback distance is short.

3.2. Velocity Flow Field around the Abutment

The unique shape of the scour hole around an abutment, which results from a longitudinal and diagonal displacement of the deepest scour point relative to the abutment face, can be explained by the velocity flow field around the abutment. Thus, to understand the complex flow physics and resulting sediment transport, velocities were measured at C.S. 1 and C.S. 2. Figure 5 shows the resultant velocity vectors in y direction (v) and z direction (w) in the floodplain and in the main channel for runs 1, 3 (F) and 13, 15 (OT). As shown in Figure 5, in the cross-sectional velocity plot at upstream face of the bridge (C.S. 1) and the downstream face of the bridge (C.S. 2), higher magnitude velocity vectors are observed around the abutment resulting from local lateral flow contraction where deeper scour can occur as the contracted flow curved around the abutment. For runs 13 and 15, the velocity is even higher than in runs 1 and 3 because of the higher lateral flow contraction in longer abutment. In addition to the existing lateral flow contraction, for the cases with higher discharge (SO and OT case), the submergence of the upstream face of the bridge produced vertical flow contraction (i.e., the downward component of the velocity vectors in Figure 5c,g). However, at C.S. 2 (downstream of the bridge section in Figure 5d,h), the upward components of velocity are observed. The downward velocity in upstream and upward velocity in downstream flow motion through the bridge induced by the bridge deck resulted in vertical contraction scour because the flow is accelerated, then decelerated. For Run 2 and Run 3, as shown in Figure 4, the lower bed elevation along the location of bridge deck in the floodplain is the result of vertical flow contraction by the unique velocities motions under the bridge. For the cases in other runs, similar observations as in runs 2 and 3 for the contours cannot be discovered because of the extent of abutment scour hole. However, the existing lateral flow contraction as well as vertical flow contraction definitely shows the higher scour depth through the bridge.

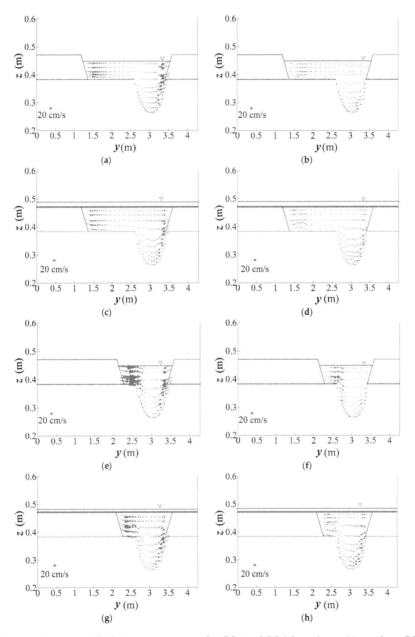

Figure 5. Cross-sectional velocity vectors measured at C.S. 1 and C.S. 2 for each run. (**a**) *v-w* plot at C.S. 1 (Run 1, F), (**b**) *v-w* plot at C.S. 2 (Run 1, F) ,(**c**) *v-w* plot at C.S. 1 (Run 3, OT), (**d**) *v-w* plot at C.S. 2 (Run 3, OT), (**e**) *v-w* plot at C.S. 1 (Run 13, F), (**f**) *v-w* plot at C.S. 2 (Run 13, F), (**g**) *v-w* plot at C.S. 1 (Run 15, OT), (**h**) *v-w* plot at C.S. 2 (Run 15, OT).

In the main channel, the *v-w* velocity plots also show unique flow motions. The narrow bridge opening induced by the abutments forced the water to re-enter through the bridge opening, causing lateral directional velocity. As shown in the Figure 5, for the setback abutment in the left floodplain, the flow moved towards the main channel, but the amount of shifting velocity was small compared to that of the bankline abutment in the right floodplain. Because the direction of lateral velocity vectors from the left side of abutment and the right side of abutment are opposite, the strong momentum transfer occurs when they meet at one point, and the interaction resulted in large counter clock-wise secondary current observed within the main channel. This secondary current initiates another scour hole close to the toe of the left-side slope within the main channel [23]. As shown in Figure 4, however, maximum scour depth around the bankline abutment occurred in the bottom of the main channel near the toe of the right bank. In fact, there appears to be interaction between the two scour holes from the left and right abutments during the initial scour development, but there was one remaining scour hole left in the equilibrium stage. Kara et al. [24] have applied a 3D numerical model to the problem of compound channel flows because the flow and turbulence distributions are so important to the prediction of scour when a bridge abutment is placed in a compound channel. Their results show the important contribution of secondary currents and turbulent stresses to the apparent shear stress at the main channel/floodplain interface when the momentum equation is depth-averaged. Both the secondary current and turbulent stress contributions to the apparent shear stress increase as the relative depth in the floodplain decreases.

Figure 6 shows the vertical velocity profiles for each run measured in the floodplain (Figure 6a) and in main channel (Figure 6b). It is interesting to note that the shape of velocity profiles is different for the case with submerged flow compared to in the free flow case. The vertical velocity measurement location (y_x) on the ordinate axis has been non-dimensionalized by total water depth ($\overline{Y}$) and the point velocity measurements (V_x) was normalized by depth-averaged velocity ($\overline{V}$) in Figure 6. Submergence of the bridge during extreme hydrologic events produced vertical flow contraction leading to more complex flow field through the bridge than in the free flow cases. As shown in Figure 6, the presence of bridge deck resulted in a "fuller" velocity profile than in the free flow cases and tends to shift the higher velocity closer to the bed. The degree of shifting of higher velocity closer to the bed is a key element to understanding the intensity of shear stress that causes erosion.

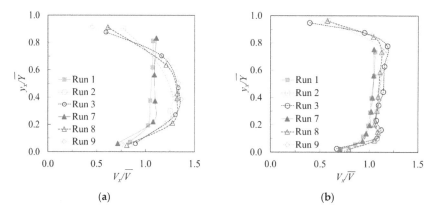

Figure 6. Vertical velocity profiles for (**a**) in the floodplain and (**b**) in the main channel

Thus, the degree of shifting of velocity at the bed is explored with non-dimensional parameters commonly used for the theoretical contraction scour. Hong et al. [2], with their experimental results, confirmed that maximum scour around an abutment can be considered as some amplification of the theoretical contraction scour, $(V_1/V_c)(q_2/q_1)$. Previously, this concept has only applied to the free flow cases. However, they observed that maximum scour depth even in different flow types can be calculated with the product of (V_1/V_c) and (q_2/q_1), as long as the flow contraction ratio(q_2/q_1) is predicted accurately. Thus, the effect of submerged orifice flow and overtopping flow provided by the vertical flow contraction as well as the lateral flow contraction can also be parameterized by the flow contraction ratio. In this study, the velocity gradients (slope) between the origin and a point measured around 20% of water depth was decided using the vertical velocity profiles shown in Figure 6 and are plotted for the floodplain in Figure 7a and for the main channel in Figure 7b, respectively, according to the dimensionless variables, $(V_1/V_c)(q_2/q_1)$, suggested by the theoretical contraction analysis. As shown in Figure 7, as the dimensionless variables, $(V_1/V_c)(q_2/q_1)$, in the x-axis increases, the value of slope decreases because the higher flow contraction increases the velocity close to the bed, leading to the milder slope shown in Figure 6. The measured slope from each profile seems to follow the same trend, even if they have different flow types because the value of q_2/q_1 can be a viable indicator of the combined influence of vertical flow contraction as well as lateral contraction, as explained in the previous paragraph.

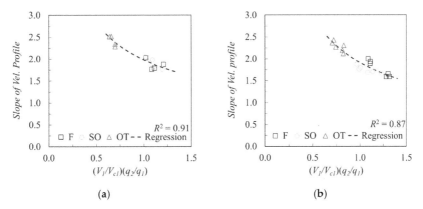

Figure 7. Variation of velocity gradient close to the bed as a function of $(V_1/V_{C1})(q_2/q_1)$ for (**a**) in the floodplain and (**b**) in the main channel

A least-squared regression analysis was conducted on the data given in Figure 7, and the best-fit equation is given by

$$Slope = 1.98 \times \left[\left(\frac{V_{f1}}{V_{fc1}}\right)\left(\frac{q_{f2}}{q_{f1}}\right) \right]^{-0.514}, \text{ for floodplain} \qquad (1)$$

$$Slope = 1.91 \times \left[\left(\frac{V_{m1}}{V_{mc1}}\right)\left(\frac{q_{m2}}{q_{m1}}\right) \right]^{-0.635}, \text{ for main channel} \qquad (2)$$

with coefficients of determination of 0.91 and 0.87, respectively. Equations (1) and (2) can be used to calculate bed shear stress because shear stress is decided by the velocity gradient (slope). Theoretically, scour is initiated when the bed shear stress induced by the flow is larger than the value of critical shear stress of the bed sediment size and the magnitude of scour depth is directly related to the value of shear stress. Thus, predicting shear stress is key driver to a better understanding of scour around an abutment. Furthermore, the provided equations can be used for bed shear stress closure of numerical modeling around bridge abutments for model developers.

4. Conclusions

A lot of research in past twenty years has focused on a particular type of bridge foundation scour, e.g. pier scour and/or abutment scour using simple experimental set-up. Thus, so far, engineers and researchers have been using SINGLE factors such as mean hydraulic variables, Reynolds stresses, turbulent kinetic energy (TKE), vorticity, and other measures of turbulent structure to understand/calculate the scour depth. However, recent extreme rainfall events associated with climate change can often result in submerged orifice flow with or without overtopping flow, in which the flow field through the bridge is more complex because of the simultaneous existence of local turbulence around the base of the structure as well as vertical flow contraction in addition to the lateral flow contraction. Thus, in this paper, for the investigation of the complex mechanism of scour under extreme hydrologic events, laboratory experiments were conducted using scale down river geometry and the detailed velocity flow field through the bridge were measured. The results show that, in addition to the local turbulence structure being wrapped around the base of abutment, higher cross-sectional velocity around an abutment due to the local flow acceleration cause the maximum scour close to the abutment. During large flooding and bridge submergence, in addition to the higher velocity and the local turbulence structure in the vicinity of the abutment, the unique down-ward and up-ward flow motion lead to the additional scour. Furthermore, the experimental results show that the shape of the velocity profiles is "fuller" compared to the free flow cases, and the unique shape of velocity profiles resulted in higher velocity gradients close to the bed. Based on the measurements, the way of estimating higher velocity gradients close to the bed, which is the key element to exploring shear stress, is suggested with respect to the discharge contraction ratio. For broader impacts, the research is expected to contribute practical design for hydraulic engineers by suggesting the classification of scour conditions with respect to the non-dimensional value of set-back distance because the opening width, location of pier and abutment are important design criteria for the safety of bridges vulnerable to scour.

Even if this study provides new insights into the velocity flow field through the scour-critical bridges subject to the submerged orifice flow and submerged flow with overtopping, the entire erosion development process should be simulated numerically to find the changes in turbulent structure, flow contraction and their interactions with the bed in both time and space. The qualitative understanding of the flow pattern and the resulting sediment transport shown in this study can serve to stimulate and guide quantitative experiments and numerical simulation. Furthermore, based on the research for scouring under ice cover, very rough ice can push the maximum velocity even further towards the bed [25,26]. These findings alarmed us and suggest that additional studies should be conducted with different types of bridge deck because the value of roughness under the model bridge deck varies depending on the size and shape of the girders.

Author Contributions: K.S.Y., S.O.L., and S.H.H. provided descriptions and motivation. K.S.Y. and S.H.H. designed experiments and measured the results; S.O.L., S.H.H., and K.S.Y. analyzed the data and interpretation of results; K.S.Y. and S.H.H. prepared the manuscript; and all authors show their efforts in reviewing and editing of the manuscript.

Funding: This work was partially supported by the Korea Agency for Infrastructure Technology Advancement (KAIA) grant funded by the Ministry of Land, Infrastructure and Transport (Grant 18AWMP-B121095-03). Seung Ho Hong was supported from West Virginia University.

Acknowledgments: Authors would like to show our sincere appreciation to Terry W. Sturm in Georgia Tech for his support. The detailed velocity measurement data for additional research using numerical simulation are available upon request from an interested party.

Conflicts of Interest: The authors declare no conflict of interest.

References

1. Ettema, R.; Nakato, T.; Muste, M. *Estimation of Scour Depth at Bridge Abutments*; Final Report, NCHRP 24-20; TRB: Washington, DC, USA, 2010.
2. Hong, S.H.; Sturm, T.W.; Stoesser, T. Clear water abutment scour in a compound channel for extreme hydrologic events. *J. Hydraul. Eng.* **2015**, *141*, 04015005. [CrossRef]
3. Kwan, T.F.; Melville, B.W. Local scour and flow measurements at bridge abutments. *J. Hydraul. Res.* **1994**, *32*, 661–673. [CrossRef]
4. Melville, B.W.; Coleman, S.E. *Bridge Scour*; Water Resources Publications: Highlands Ranch, CO, USA, 2000.
5. Chrisohoides, A.; Sotiropoulos, F.; Sturm, T.W. Coherent structures in flat-bed abutment flow: Computational fluid dynamics simulations and experiments. *J. Hydraul. Eng.* **2003**, *129*, 177–186. [CrossRef]

6. Kim, H.; Roh, M.; Nabi, M. Computational modeling of flow and scour around two cylinders in staggered array. *Water* **2017**, *9*, 654. [CrossRef]

7. Zang, L.; Wang, P.; Yang, W.; Zuo, W.; Gu, X.; Yang, X. Geometric characteristics of spur dike scour under clear-water scour conditions. *Water* **2018**, *10*, 680. [CrossRef]

8. Chen, S.; Tfwala, S.; Wu, T.; Chan, H.; Chou, H. A hooked-collar for bridge piers protection: Flow fields and scour. *Water* **2018**, *10*, 1254. [CrossRef]

9. Zang, L.; Wang, H.; Zang, X.; Wang, B.; Chen, J. The 3-D morphology evolution of spur dike scour under clear-water scour conditions. *Water* **2018**, *10*, 1583. [CrossRef]

10. Yang, Y.; Qi, M.; Li, J.; Ma, X. Evolution of hydrodynamics characteristics with scour hole developing around a pile group. *Water* **2018**, *10*, 1632. [CrossRef]

11. Hong, S.H.; Sturm, T.W. Physical Model Study of Bridge Abutment and Contraction Scour under Submerged Orifice Flow Conditions. In Proceedings of the 33rd IAHR Congress: Water Engineering for a Sustainable Environment, Vancouver, BC, Canada, 9–14 August 2009.

12. Hong, S.H.; Sturm, T.W. Physical Modeling of Abutment Scour for Overtopping, Submerged Orifice and Free Surface Flows. In *Scour and Erosion, Proceedings of the Fifth International Conference on Scour and Erosion, San Francisco, CA, USA, 7–10 November 2010*; American Society of Civil Engineers: Reston, VA, USA, 2010.

13. Saha, R.; Lee, S.; Hong, S.H. A comprehensive method of calculating maximum bridge scour depth. *Water* **2018**, *10*, 1572. [CrossRef]

14. Hong, S.H. Prediction of Clear Water Abutment Scour Depth in Compound Channel for Extreme Hydrologic Events. Ph.D. Thesis, School of Civil and Environmental Engineering, Georgia Institute of Technology, Atlanta, GA, USA, 2013.

15. Ge, L.; Lee, S.; Sotiropoulos, F.; Sturm, T.W. 3D unsteady RANS modeling of complex hydraulic engineering flows. Part II: Model validation and flow physics. *J. Hydraul. Eng.* **2005**, *131*, 809–820. [CrossRef]

16. Garcia, C.M.; Cantero, M.I.; Nino, Y.; Garcia, M.H. Closure to "Turbulence measurements with acoustic doppler velocimeters". *J. Hydraul. Eng.* **2007**, *131*, 1062–1073. [CrossRef]

17. Hong, S.H.; Abid, I. Physical model study of bridge contraction scour. *KSCE J. Civ. Eng.* **2016**, *20*, 2578–2585. [CrossRef]

18. Hong, S.H.; Lee, S.O. Insight of Bridge Scour during Extreme Hydrologic Events by Laboratory Model Studies. *KSCE J. Civ. Eng.* **2018**, *22*, 12871–12879. [CrossRef]

19. Lee, S.O.; Hong, S.H. Reproducing field measurements using scaled-down hydraulic model studies in a laboratory. *Adv. Civ. Eng.* **2018**, *2018*, 9091506. [CrossRef]

20. Goring, D.; Nikora, V. Despiking acoustic Doppler velocimeter data. *J. Hydraul. Eng.* **2002**, *128*, 117–126. [CrossRef]

21. Chang, F.; Davis, S. Maryland SHA Procedure for Estimating Scour at Bridge Abutments, Part 2—Clear Water Scour. In *Proceedings of Water Resources Engineering '98*; ASCE: Memphis, TN, USA, 1998; pp. 169–173.

22. Chang, F.; Davis, S. Maryland SHA Procedure for Estimating Scour at Bridge Waterways, Part 1—Live Bed Scour. In *Stream Stability and Scour at Highway Bridges*; Richardson, E., Lagasse, P., Eds.; American Society of Civil Engineers: Reston, VA, USA, 1999; pp. 4001–4011.

23. Hong, S.H.; Abid, I. Scour around an erodible abutment with riprap apron over time. *J. Hydraul. Eng.* **2019**, in press.

24. Kara, S.; Stoesser, T.; Sturm, T.W. Flow dynamics through a submerged bridge opening with overtopping. *J. Hydraul. Res.* **2015**, *53*, 186–195. [CrossRef]

25. Zabilansky, L.J.; Hains, D.B.; Remus, J.I. Increased Bed Erosion Due to Ice. In *Cold Regions Engineering 2006: Current Practices in Cold Regions Engineering*; American Society for Civil Engineers: Reston, VA, USA, 2006; pp. 1–12.

26. Hains, D.; Zabilansky, L. *Laboratory Test of Scour under Ice: Data and Preliminary Results*; ERDC/CRREL TR-04-9; U.S. Army Engineer Research and Development Center: Hanover, NH, USA, 2004.

Article

Effect of Flaring Gate Piers on Discharge Coefficient for Finite Crest-Length Weirs

Zhong Tian, Wei Wang *, Ruidi Bai and Nan Li

State Key Laboratory of Hydraulics and Mountain River Engineering, Sichuan University,
Chengdu 610065, China; tianzhong@scu.edu.cn (Z.T.); bairuidiscu@163.com (R.B.); nancyli1220@163.com (N.L.)
* Correspondence: profwangwei@sina.com

Received: 13 August 2018; Accepted: 26 September 2018; Published: 28 September 2018

Abstract: The use of flaring gate piers (FGPs) along with finite crest-length weirs changes the shape of plunging jets and increases the efficiency of energy dissipation in some projects; however, the FGPs may affect the discharge capacity. In this study, the flow pattern and discharge coefficient were experimentally investigated under different conditions by varying the weir lengths L_w, contraction ratio β, contraction angle θ, and water heads H. A comparative analysis of the weirs with and without FGPs was performed. For the finite crest-length weirs with FGPs, the water-surface profiles in the flow channel were backwater curves. Moreover, the plunging jets leaving the weir became narrower and then subsequently diffused largely in the transverse and longitudinal directions in air. The discharge coefficients of the weirs with FGPs were approximately equal for various weir lengths. Moreover, following the earlier studies on traditional finite crest-length weirs, a discharge-coefficient equation was developed for the weir with an FGP in this study. The results showed that in the weirs with FGPs, the discharge coefficients clearly increased with the increase in the contraction ratio and water head, but the changes in their values along with the contraction angle were neglected.

Keywords: finite crest length weir; FGP (flaring gate pier); discharge coefficient; subcritical flow

---

1. Introduction

Generally, based on the relative length of the weirs, H/L_w, where H is the head on the weir crest and L_w is the weir length in the direction of flow [1–5], weirs are divided into two groups: the sharp-crested weirs ($H/L_w > 2$) and finite crest-length weirs ($H/L_w < 2$). The finite crest-length weirs can be further classified into three categories: the short-crested ($0.4 \leq H/L_w < 2$), broad-crested ($0.1 \leq H/L_w < 0.4$), and long-crested weirs ($0 < H/L_w < 0.1$). The discharge characteristics of the flows over the finite crest-length weirs with different cross sections, such as rectangular, trapezoidal, and triangular, have been studied. According to a systematic experimental study on the finite crest-length weirs having a rectangular shape with $0 < H/L_w \leq 2$ and $0 < H/P < 1$ (where P is height of the weir), three discharge equations have been developed for the short-crested, broad-crested, and long-crested weirs [6]. Ramamurty et al. [7] studied the influence of upstream rounding on the discharge coefficient. Sarker and Rhodes [8] used computational fluid dynamics (CFD) modeling to model the free-surface profiles of the broad-crested weirs; the experimental results were in good agreement with those of the numerical models. In order to model the free surface flow of real reservoirs, Andersson et al. [9] performed a physical scale model test and three-dimensional simulation of the spilling from a reservoir, which gives a better understand of the flow pattern for the short-crested weir. Chen et al. [10] studied the discharge coefficient of a rectangular short-crested weir with varying slope coefficients and developed an equation for calculation of the discharge coefficient. Recently, Azimi et al. [11] conducted a series of laboratory experiments to study the correlations for nine different types of weirs, which extended the earlier studies on the finite crest-length weirs having a trapezoidal shape, considering sloping crests and either upstream ramps, downstream ramps, or both. In addition, the study was extended to triangular weirs (in which the crest length is zero) with either upstream ramps, downstream ramps, or both. Jan et al. [12] proposed a linear combination of traditional discharge equations of simple triangular and rectangular weirs to develop the discharge equations for the complex broad-crested weirs.

A new concept of energy dissipation by combining the flaring gate pier (FGP) with the weir was proposed in China [13]. The FGP is usually employed on a weir to achieve 3D flow by making the flow emerge from the FGP with lateral contraction and longitudinal dispersion; hence, a shorter length of the stilling basin and a larger energy dissipation ratio could be obtained [14]. FGPs have been extensively applied in roller-compacted concrete dams. During the construction, some sections of the dam serve as gaps for releasing floodwater, and the FGP is integrated in the system to reduce the erosion of downstream channels. Hence, the finite crest-length weirs are integrated with the FGP, and the overflows are largely subcritical when they are constructed to a certain height. The appraisal of the diversion schemes for flooding is related to the construction period, quality, cost, and safety of the whole project; hence, the diversion schemes need to be designed carefully in advance. It is important to study the discharge characteristics of the finite crest-length weirs with the FGP. In previous studies, the FGP was almost integrated on the ogee-overflow weirs (curved short-crested weirs) and the flow over the weir was largely supercritical [15,16]. Li et al. [17] experimentally studied the effects of the FGP on the discharge capacity and flow pattern of the ogee-overflow weirs. The results showed that the effect of the FGP on the discharge capacity is closely related to the flow pattern; four types of flow patterns occurred, depending on the weir head and the contraction ratio. Recently, Li et al. [18] showed that the effect of the variations in the design parameters of the FGP on the discharge capacity of the surface spillway is insignificant when the flow in the channel is supercritical.

Previous studies focused on the hydraulic characteristics of FGP on short-crested weirs. The purpose of this paper is to discuss the flow coefficient after setting an FGP on finite crest-length weirs (including short-crested weirs and broad-crested weirs). A comparative analysis of the weir with different parameters of the FGP was performed.

2. Experimental Setup and Methodology

The experiments were conducted in the State Key Laboratory of Hydraulics and Mountain River Engineering, Sichuan University. The experimental setup comprised a flow measurement section, a large feeding tank (6 m in length and 2 m in width, to ensure steady-state flow condition), a test section, and a flow return system. A right-angled triangle weir is installed upstream of the reservoir, and an electromagnetic flowmeter is installed in the return pipe downstream for flow calibration. Because of the defined relationship between the discharge Q and the head H, the right-angled triangle weir is used as a flow measuring device, for which the discharge calculation formula is

$$Q = 1.33H^{2.465} \tag{1}$$

The head of the right-angled triangle weir is measured by a water level stylus, which is installed 1.0 m upstream of the weir, and the accuracy is 0.1 mm. The water level of the reservoir is also measured by a water level stylus with accuracy of 0.1 mm. The water level gauge point was set at the side of the feeding tank (or reservoir) and is 2 m away from the inlet of the finite-crest length weir. Figure 1 shows the experimental setup.

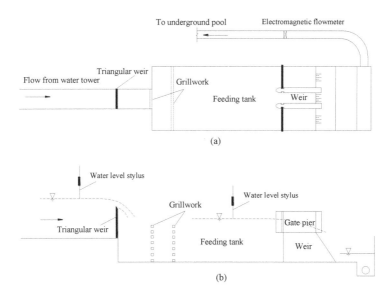

Figure 1. Experimental setup: (**a**) plane view; (**b**) side view.

The head of the right-angled triangle weir and water level of the reservoir were measured every 5 min for each discharge. Comparing discharge between the right-angled triangle weir and the electromagnetic flowmeter shows that the error is less than 1.4%.

The test model, made using Plexiglas, was a finite crest-length weir. The length (L_w), width (B), and height (P) of the type 3 weir were 0.601 m, 0.30 m, and 1.28 m, respectively, wherein the upstream face was upright and the downstream face had a slope of 1.2:1 (i). The channel width $B = 0.30$ m was reduced to $B' = 0.115$ m at the edge of the outlet for length $L_p = 0.271$ m when the FGP was integrated on the weir. The contraction angle was 15°, as shown in Figure 2. In this experiment, three types of finite crest-length weirs were employed, with weir lengths L_w of 0.601 m, 0.519 m, and 0.422 m and weir heights of 1.28 m, 1.378 m, and 1.507 m, respectively. In addition, nine types of FGPs were employed with contraction ratios $\beta = 0.38, 0.51, 0.64, 0.82$, and 1.00, and contraction angles $\theta = 7.0°, 9.6°,$ 12.7°, 15.0°, and 21.5°, respectively. The length of FGP (L_p in Figure 2) is variable, and the bulkhead slots stay at the same location. The ranges of H and H/L_w were approximately 0.112–0.547 m and 0.249–1.212 m, respectively.

The water-surface profiles were measured along the centerline of the weir. The coordinate system (X–Y) was established to describe the water-surface profiles efficiently, as shown in Figure 2c. The origin was at the center of the weir inlet. The location and average water depth of the measuring section were described by the X and Y values, respectively. X/L_w is the relative location of the measuring section.

To investigate the influence of the FGP on the discharge coefficient of the finite crest-length weir, 13 groups of experiments were performed, as listed in Table 1. The cases were divided into three sets: cases N1–N6 represent the effect of L_w, cases N5–N9 represent the effect of β, and cases N5 and N10–N13 represent the effect of θ. The hydraulic experiments were conducted under different conditions with 5–6 different weir heads (0.112–0.547 m) for each group, including discharge capacity, flow pattern, and water-surface profile in the flow channel.

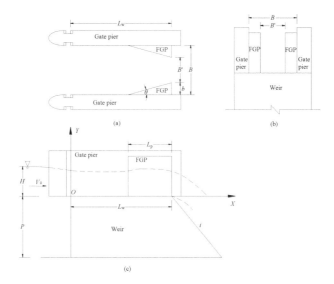

Figure 2. Definition sketch of the finite-crest length weir with FGP (Flaring Gate Pier): (**a**) plane view; (**b**) front view; (**c**) side view (L_w—the length of the weir, L_p—the length of the FGP, B—the width of the weir crest, b—the width of the weir outlet, P—the height of the weir, H—the head of the weir, V_0—approching velocity).

Table 1. Parameters of the finite-crest length weir and FGP.

Weir Type	Weir Height	Weir Length	Channel Width	Contraction Ratio	Contraction Angle	Case Name
	P (m)	L_w (m)	B (m)	$\beta = B'/B$	θ	
Type 1	1.493	0.422	0.30	0.51	15	N1
				1	0	N2
Type 2	1.379	0.519	0.30	0.51	15	N3
				1	0	N4
Type 3	1.28	0.601	0.30	0.51	15	N5
				1	0	N6
				0.38	15	N7
				0.64	15	N8
				0.82	15	N9
				0.51	7	N10
				0.51	9.6	N11
				0.51	12.7	N12
				0.51	21.5	N13

3. Results and Discussions

3.1. Flow Pattern Comparison with and without FGP

As shown in Figure 3a, when there was no FGP on the weir, the water depth gradually decreased along the channel and the plunging jet dropped into the stilling basin in a rigorous manner without the longitudinal spreading for a large discharge. In contrast, when the FGP was applied in the weir, the water on the upper side flowed slower than that on the bottom and the plunging jet became narrow just outside the weir; subsequently, the flow spread in the transverse and longitudinal directions in air for the contraction, as shown in Figure 3b,c.

Taking type 3 of the finite crest-length weir as an example, the Froude number (Fr) along the crest with FGP ranges from 0.3~0.5, thus it is a subcritical flow. Fr along the crest without FGP ranges from 0.5~1.2, and flow pattern changing occurs at the ending of the crest ($X/L_w > 0.8$). The water-surface profiles that described the change in depth along the channel are plotted in Figure 4. It was clear that the depth of water in the channel increased significantly after setting the FGP for the same discharge. When the depth of the water in the channel was low, the two water-surface profiles shifted toward the horizontal form. When the depth of the water in the channel was relatively high, the two profiles differed depending on whether the FGP was employed; i.e., the water-surface profile gradually decreased for the channel where the FGP was not installed, as shown in Figure 5a. The water surface upstream of the starting point of FGP is smooth; however, the backwater curve occurred at the beginning of the FGP when setting the FGP in the weir, as shown in Figure 5b, which could be regarded as a surface wave [2].

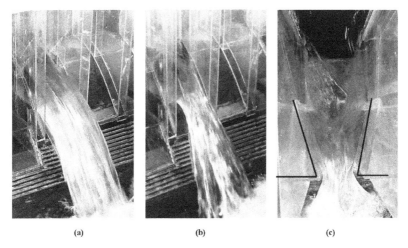

(a)   (b)   (c)

Figure 3. Photograph of the plunging jets in the present study (H = 0.3 m): (**a**) without FGP ($\beta = 1$); (**b**) with FGP ($\beta = 0.51$, $\theta = 15°$); (**c**) with FGP ($\beta = 0.51$, $\theta = 15°$).

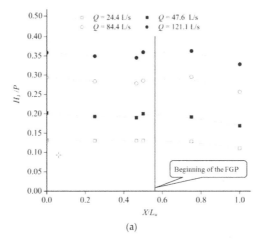

(a)

Figure 4. *Cont.*

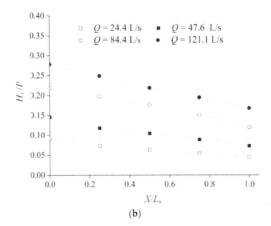

(b)

Figure 4. Comparison of water-surface profiles: (**a**) with FGP ($\beta = 0.51$, $\theta = 15°$); (**b**) without FGP ($\beta = 1$) (Q—the discharge of the weir, H_1—water depth of the weir crest).

(**a**) (**b**)

Figure 5. Photograph of water surface: (**a**) without FGP; (**b**) with FGP ($\beta = 0.51$, $\theta = 15°$) (Black arrow stands for the flow direction).

3.2. Discharge Capacity Comparison with and without FGP

Considering type 3 of the finite crest-length weir as an example, Figure 6 shows the head–discharge relationship of the weir for the cases with and without the FGP. The two discharge-stage curves, with slightly similar change rules, were both concave in shape. However, the curves for the weir with the FGP were always below that of the case without the FGP, which indicated that the discharge was reduced when the FGP was installed on the weir. The discharge coefficient C_d was the indicator of the discharge capacity of the weir, which was defined as follows [2]:

$$C_d = \frac{Q}{B\sqrt{2g}H^{\frac{3}{2}}} \qquad (2)$$

where Q is the discharge of the finite crest-length weir, which was measured using the right-angled-triangle weir in the model test; B is the width of the channel; g is gravitational acceleration; and H is the head on the weir. Because the height of the finite crest-length weir was considerable ($P/H > 1.33$), the influence of the approaching velocity on the test results could be neglected [19].

Figure 7 shows the relationship between the discharge coefficient (C_d) and the weir head (H). The results showed that the discharge coefficient gradually increased as the head increased, and the spacing of the two curves represented the reduction in the discharge coefficient after setting the FGP on the finite crest-length weir. This might be related to the flow pattern in the channel, because Li et al. [17]

showed that the discharge capacity reduces only if the flow was subcritical when the FGP was applied on the ogee-overflow structure.

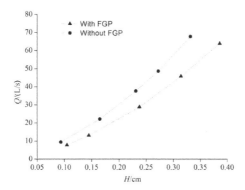

Figure 6. Head–discharge relationship of the weir for the cases with FGP (β = 0.51, θ = 15°) and without FGP.

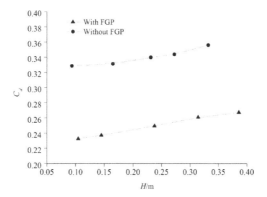

Figure 7. Relationship between the discharge coefficient (C_d) and the weir head (H) for the cases with FGP (β = 0.51, θ = 15°) and without FGP.

3.3. Influence of Weir Length on Discharge Capacity with and without FGP

Figure 8 shows the relationship between the dimensionless head on the weir (H/P) and discharge coefficient (C_d) for different weir lengths L_w, for the cases with and without the FGP. The results showed the effect of the weir length on the discharge coefficient. Moreover, a comparative analysis was performed for the cases with and without the FGP.

As shown in Figure 8a, for the case where the FGP is installed, the three C_d–H curves almost overlap. This implies that the effect of the variations in the weir length on the discharge coefficient of the weir is insignificant when the FGP is installed. In other words, the discharge coefficient of the weir with the FGP was largely unrelated to the weir length in a particular range. The inlet boundary condition could be considered to remain the same, and the boundary condition did not clearly change with the variation in the weir length. However, Rao and Muralidhar [6] and Hager and Schwalt [2] pointed out a particular feature of finite crest-length weirs, referred to as the undular flow on the crest when $H/L_w < 0.1$. Moreover, they showed that the value of C_d was constant, which was similar to the characteristics of the finite crest-length weir with the surface wave; a lower and relatively stable C_d was obtained for various weir lengths when setting the FGP.

It also could be seen that the distribution rule of the three C_d–H curves without the FGP was different from the aforementioned regularity. Two C_d–H curves for $L_w = 0.389$ m and $L_w = 0.317$ m almost overlap just above the one for $L_w = 0.451$ m, as shown in Figure 8b. It indicated that the discharge coefficient of the weir without the FGP increased as the weir length decreased in a particular range, but gradually tended to stabilize when the weir length continued to decrease.

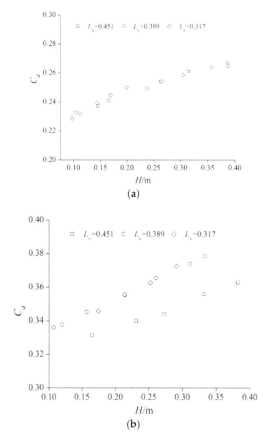

Figure 8. Discharge coefficient (C_d) versus weir head (H) together with different weir lengths (L_w): (**a**) with FGP ($\beta = 0.51$, $\theta = 15°$); (**b**) without FGP.

3.4. Discharge Coefficient of the Weir with FGP

Many studies have analyzed the relationship of the discharge coefficient for the finite crest-length weirs of different types, i.e., long-crested weirs, broad-crested weirs, short-crested weirs, and sharp-crested weirs [2,6,9–11]. They all indicated that the nondimensional parameter H/L_w was the dominant factor.

In the present experiment, the range of H/L_w was approximately 0.187–0.909. Keeping the other parameters constant and varying one parameter, the effects of β and θ were shown in Figure 9. Figure 9a shows the variations in C_d with β for each H/L_w at the same contraction angle θ. As indicated in the figure, the discharge coefficient significantly increased as the contraction ratio increases ($0.38 \leq \beta \leq 1$). Figure 9b illustrates the variations in C_d with θ for each H/L_w at the same contraction ratio β, which indicated that the discharge coefficient changes only slightly as the contraction angle increases $7° \leq \theta \leq 21.5°$.

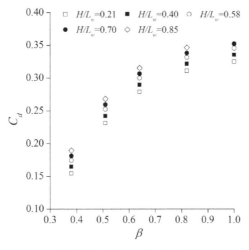

(a) Contraction angles $\theta = 7.0°$

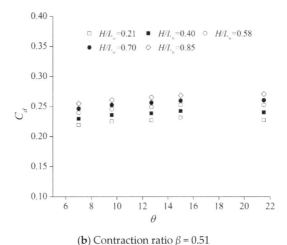

(b) Contraction ratio $\beta = 0.51$

Figure 9. Discharge coefficient C_d versus β (**a**) and θ (**b**).

Consequently, H/L_w and β were the dominant parameters of the discharge coefficient of the weir with FGP C_d. By the linear regression method, Equation (3) is obtained. The correlation coefficient R^2 is equal to 0.967, which shows that the accuracy of Equation (3) is higher. The relationship between discharge coefficient C_d and the two parameters described using Equation (3) is shown in Figure 10.

$$C_d = 0.056(H/L_w) - 0.107(\frac{1}{\beta}) + 0.427 \tag{3}$$

with $R^2 = 0.967$.

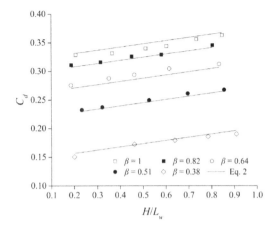

Figure 10. Compared of measured and Equation (3)-computed values.

4. Conclusions

In this study, 65 groups of experiments, including five different flow rates and 13 different geometries of the finite-crest length weir, were conducted to study the variation in the discharge coefficient and flow pattern of a finite crest-length weir with and without an FGP. Unlike the case of the weir without the FGP, for the weir with the FGP, the plunging jet was diffusive in the longitudinal directions in air and the water-surface profile in the flow channel was a backwater curve when the head was relatively high. The discharge capacity was significantly reduced after setting the FGP, which should be considered in engineering design. According to the compared experiments, it was found that the effect of the weir length on the discharge coefficient was different for cases of the weir with and without the FGP. In addition, based on the literature and experimental studies, this study proposed a discharge-coefficient equation for the weir with the FGP.

Author Contributions: Z.T. carried out the experimental investigation and writing; W.W. wrote the methodology and review; R.B. carried out the data curation; N.L. carried out data validation. All four authors reviewed and contributed to the final manuscript.

Funding: This work was supported by the National Key R&D Program of China (Grant No. 2016YFC0401603), the National Natural Science Foundation of China (Grant No. 51879178), and the Fundamental Research Funds for the Central Universities (Grant No. 20826041A4305).

Conflicts of Interest: The authors declare no conflicts of interest.

References

1. Horton, R.E. *Weir Experiments, Coefficients, and Formulas*; Proc. U.S. Geological Survey Water Supply; Government Printing Office: Washington, DC, USA, 1907.
2. Hager, W.H.; Schwalt, M. Broad-crested weir. *J. Irrig. Drain. Eng.* **1994**, *120*, 13–26. [CrossRef]
3. Fritz, H.M.; Hager, W.H. Hydraulics of embankment weirs. *J. Hydraul. Eng.* **1998**, *124*, 963–971. [CrossRef]
4. Bai, R.; Zhang, F.; Liu, S.; Wang, W. Air concentration and bubble characteristics sownstream of a chute aerator. *Int. J. Multiphase Flow* **2016**, *87*, 156–166. [CrossRef]
5. Bai, R.; Liu, S.; Tian, Z.; Wang, W.; Zhang, F. Experimental investigations on air–water flow properties of offset–aerator. *J. Hydraul. Eng.* **2018**, *144*, 04017059. [CrossRef]
6. Govinda Rao, N.S.; Muralidhar, D. Discharge characteristics of weirs of finite crest width. *Houille Blanche* **1963**, *18*, 537–545. [CrossRef]
7. Ramamurty, A.S.; Udoyara, S.T.; Rao, M.V.J. Characteristic of square-edged and round-nosed broad-crested weirs. *J. Irrig. Drain. Eng.* **1988**, *114*, 61–73. [CrossRef]

8. Sarker, M.A.; Rhodes, D.G. Calculation of free-surface profile over a rectangular broad-crested weir. *Flow Meas. Instrum.* **2004**, *15*, 215–219. [CrossRef]
9. Andersson, A.G.; Andreasson, P.; Staffan Lundström, T. CFD-modelling and validation of free surface flow during spilling of reservoir in down-scale model. *Eng. Appl. Comput. Fluid Mech.* **2013**, *7*, 159–167. [CrossRef]
10. Chen, Y.; Fu, Z.; Chen, Q.; Cui, Z. Discharge Coefficient of Rectangular Short-CrestedWeir with Varying Slope Coefficients. *Water* **2018**, *10*, 204. [CrossRef]
11. Azimi, A.H.; Rajaratnam, N.; Zhu, D.Z. Submerged Flows over Rectangular Weirs of Finite Crest Length. *J. Irrig. Drain. Eng.* **2014**, *140*, 06014001. [CrossRef]
12. Jan, C.D.; Chang, C.J.; Kuo, F.H. Experiments on discharge equations of compound broad-crested weirs. *J. Irrig. Drain. Eng.* **2009**, *135*, 511–515. [CrossRef]
13. Xie, S.Z.; Li, S.Q.; Li, G.F. Development of flaring gate piers in China. *Hongshui River* **1995**, *14*, 3–11. (In Chinese)
14. Guo, J.; Liu, Z.P. Field observations on the RCC stepped spillways with the flaring pier gate on the Dachaoshan Project. In Proceedings of the IAHR XXX International Congress, Theme D, Thessaloniki, Greece, August 2003; pp. 473–478.
15. Zhang, T.; Wu, C.; Liao, H.S. 3D numerical simulation on water and air two-phase flows of the steps and flaring gate pier. *J. Hydrodyn.* **2005**, *17*, 338–343.
16. Wang, B.; Wu, C.; Mo, Z.Y. Relationship of first step height, step slope and cavity in X-shaped flaring gate piers. *J. Hydrodyn.* **2007**, *19*, 349–355. [CrossRef]
17. Futian, L.; Peiqing, L.; Weilin, X.; Zhong, T. Experimental study on effect of flaring piers on weir discharge capacity in high arch dam. *J. Hydraul. Eng.* **2003**, *11*, 43–47. (In Chinese)
18. Li, N.W.; Liu, C.; Deng, J.; Zhang, X.Z. Theoretical and experimental studies of the flaring piers on the surface spillway in a high-arch dam. *J. Hydrodyn.* **2012**, *24*, 496–505. [CrossRef]
19. Ramamurthy, A.S.; Vo, N.D. Characteristics of circular-crested weir. *J. Hydraul. Eng.* **1963**, *119*, 1055–1062. [CrossRef]

Article

Capturing the Motion of the Free Surface of a Fluid Stored within a Floating Structure

Roman Gabl [1,2,*], Jeffrey Steynor [1], David I. M. Forehand [1], Thomas Davey [1] and Tom Bruce [1] and David M. Ingram [1]

[1] School of Engineering, Institute for Energy Systems, FloWave Ocean Energy Research Facility,
    The University of Edinburgh, Max Born Crescent, Edinburgh EH9 3BF, UK; Jeff.Steynor@ed.ac.uk (J.S.);
    D.Forehand@ed.ac.uk (D.I.M.F.); Tom.Davey@flowave.ed.ac.uk (T.D.); Tom.Bruce@ed.ac.uk (T.B.);
    David.Ingram@ed.ac.uk (D.M.I.)
[2] Unit of Hydraulic Engineering, University of Innsbruck, Technikerstraße 13, 6020 Innsbruck, Austria
*   Correspondence: Roman.Gabl@ed.ac.uk

Received: 06 December 2018; Accepted: 22 December 2018; Published: 29 December 2018

Abstract: Large floating structures, such as liquefied natural gas (LNG) ships, are subject to both internal and external fluid forces. The internal fluid forces may also be detrimental to a vessel's stability and cause excessive loading regimes when sloshing occurs. Whilst it is relatively easy to measure the motion of external free surface with conventional measurement techniques, the sloshing of the internal free surface is more difficult to capture. The location of the internal free surface is normally extrapolated from measuring the pressure acting on the internal walls of the vessel. In order to understand better the loading mechanisms of sloshing internal fluids, a method of capturing the transient inner free surface motion with negligible affect on the response of the fluid or structure is required. In this paper two methods will be demonstrated for this purpose. The first approach uses resistive wave gauges made of copper tape to quantify the water run-up height on the walls of the structure. The second approach extends the conventional use of optical motion tracking to report the position of randomly distributed free floating markers on the internal water surface. The methods simultaneously report the position of the internal free surface with good agreement under static conditions, with absolute variation in the measured water level of around 4 mm. This new combined approach provides a map of the free surface elevation under transient conditions. The experimental error is shown to be acceptable (low mm-range), proving that these experimental techniques are robust free surface tracking methods in a range of situations.

Keywords: free surface measurement; optical motion capturing; tank test; wave gauge; sloshing

1. Introduction

Designs for floating structures that hold internal fluids with a free surface must be capable of withstanding both the internal and external fluid pressures. The movement and stability of the structure is also affected by both fluid bodies. To investigate these effects, researchers must quantify sloshing within a free floating structure, which requires new methods of free surface tracking that are robust and repeatable. It is important that these measurement techniques are benchmarked and quantitatively understood if employed to quantify the free surface in experimental investigations that will be used to validate numerical methods for predicting such interactions, especially with larger motions.

Comparable research questions on the interaction between two separated water bodies can be found in oscillating water column (OWC) wave energy converters [1,2] as well as in ship design [3–5], especially in connection with the transport of liquefied natural gas (LNG) [6]. Sloshing experiments are an essential part of the design process, a problem typically explored using closed tanks mounted on motion platforms [7–9]. In most cases, such investigations focus on the impact forces exerted on the walls with respect to different filling levels. Hence, pressure transducers are the key measurement instrument [10–12].

For very large floating structures (VLFS), experimental data and numerical solutions are available, which investigate the deformation of VLFS depending on different mooring types [13–15]. Lee et al. [16] included multiple inner tanks filled with water in a VLFS and presented a numerical approach based on finite element method. This approach exhibits good agreement with experimental data but is limited to small motions. An example application of the full-scale structure would be as buoyant energy storage [17,18]. The size of such a device is comparable to ships and large platforms.

The motion of the free surface is a key indicator used in the comparison of experimental results and numerical simulations. The precision and accuracy of the free surface measurement is important in the validation of the numerical results. In general, a high spatial density of measurement points from two different and ideally independent systems is preferable to provide redundancy and reduce uncertainty [19]. It is also important that any additional mass on the model be minimised so that mass balance is not affected by the instruments. Finally, the movement of the floating tank should not be impeded by the experimental and measurement systems.

Classical approaches to detect and follow free surfaces are based on surface piercing wave gauges [20–25] or ultrasonic/acoustic probes, which measure the delay of the ultrasonic signal echo caused by the density change at the free surface [26,27]. Submerged acoustic instruments can measure the fluid velocity beneath the free surface at laboratory scale [28,29] as well as for field applications [30]. Such point measurement devices are very robust and precise. The observation of visible markers is an option for laboratory experiments [31,32] and can also be applied for field measurements [33].

Gomit et al. [34] present a method to detect the free surface depending on the 2D velocity field in a towing tank. In that case, a horizontal laser sheet is observed by up to three cameras and this requires optical access from the bottom of the tank. Similarly, vertical laser sheets used for particle image velocimetry (PIV) can be applied to measure the velocities as well as to detect free surfaces [35–40]. Belden and Techet [41] combined this method for air and water to detect the surface. These mentioned approaches are limited to a two-dimensional section, which results in a single slice through the surface and delivers results similar to a Light Detection And Ranging (LIDAR) measuring system [42,43]. Weitbrecht et al. [44] evaluate the accuracy of surface PIV, which use the free water surface as measurement plane without a laser. In this case the water depth is comparably shallow (approximately 4.6 cm) and the measurement of velocity vector field base on particles seeded continuously onto the water surface. Comparable big motion of the free surface in sloshing tanks leads to a high rate of mixing, which makes it harder to maintain the particle distribution. Akutina et al. [45] capture the movement of neutrally buoyant particles and address the problem of the measurement validation as well as reduce the influence of optical distortion.

Evers and Hager [46,47] present an approach to capture the curvature of the water surface caused by an impulse water wave in a tank. In their method, a light grid is projected on the free surface and the deformation is captured with cameras. The 2D images of the grid deformation are then transformed into a calculated 3D surface. This approach requires the addition of pigment to the water [48,49], which is unlikely to be a viable option when extending the approach to a large wave basin. Additionally, the floating structure would shadow the free surface and hence measurement near the structure walls would not be possible.

This paper presents a new approach to capture the free surface motion inside a floating tank with bespoke floating markers using a 3D motion capturing system. Comparable wave gauges have been previously used in wave run-up experiments on a slope [38,50,51] or on fixed cylinders [52] but in these cases the wave gauges were always stationary. Wave gauges made of copper tape are also investigated to determine if the water run-up within or around a floating body can be captured accurately even if the body orientation changes in relation to its initial configuration. Experiments with an open-top clear acrylic cylinder filled with water (Figure 1) under static and transient conditions have been conducted. The aim of this work is to evaluate these two approaches for a use in combination with a floating structure.

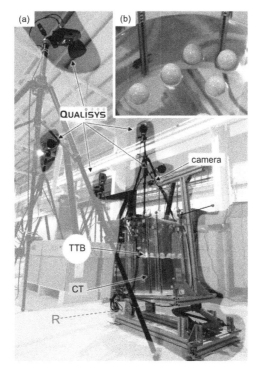

Figure 1. Experimental set-up—acrylic cylindrical tank mounted on an inclinable framework including table tennis balls (TTB) and wave gauges made of copper tape (CT)—rotation axis *R*—(**a**) overview of the inclined structure and (**b**) detail of the inner water surface.

2. Experimental Method

2.1. Overview

A cylindrical tank with an inner diameter *D* of 0.49 m is investigated (Figure 1). The transparent acrylic structure has a material thickness of 5 mm for the cylinder walls and 7 mm for the floor plate. If the cylinder is filled to its maximum height *H* of 0.5 m, it contains 94.28 L of water, ignoring any errors in circularity. For the presented experiments, the tank is fixed on a framework (rather than free floating), which allows an inclination around one fixed rotation axis *R* (Figures 1 and 2). The position of the structure is determined by a four camera Qualisys motion capturing system.

The water level in the tank is given by a measurement scale on the cylinder wall and captured by a camera, which is mounted at a constant position relative to the opening of the cylinder (Figure 1). This is used as a reference for the two measurement methods under investigation for the free surface motion: (a) floating infra-red reflective markers (Section 2.2) and (b) wave gauges made of conductive copper tape (Section 2.3). The first approach is based on the motion capturing system and is a new implementation, which enables this system to also capture a changing water surface. The investigation also aims to demonstrate that wave gauges made of light weight copper tape can be bonded to a moving body and deliver reliable measurements under different inclinations without the need for a recalibration.

2.2. Floating Infra-Red Reflective Markers

The movement of the containing cylindrical tank is captured with a commercial optical motion capture system from Qualisys. All six degrees of freedom (DoF) are determined based on the relative position of the markers, which are mounted on the rigid acrylic structure [53–56]. The Qualisys system is also capable of simultaneously tracking multiple independently moving reflective markers in the three translational degrees of freedom. Hence, the possibility of capturing the free water surface with additional floating markers randomly located inside the structure can be explored.

All known commercially available passive spherical reflective markers are either not suitable for wet conditions (i.e., they lose their reflectivity) or are not buoyant. Consequently, a bespoke solution is chosen to make each floating reflective marker from a standard table tennis ball. The table tennis ball (mass of 2.7 g, diameter 40 mm) serves as the basic body, which has enough buoyancy and visibility for the application. Reflective tape is applied to provide the reflectivity to infra-red light. This set-up is comparable to the taped underwater markers offered by Qualisys, but much lighter, water resistant and buoyant. A preliminary feasibility study showed that such a marker ball can be tracked by the Qualisys system whilst afloat on the surface and subject to wave motions in the University of Edinburgh's Curved Wave Tank [57] and in the FloWave basin, which has a diameter of 25 m and is also located at the University of Edinburgh [58–63].

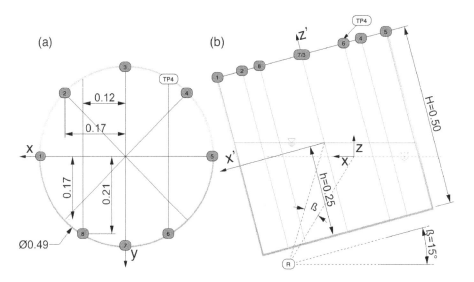

Figure 2. Basic geometry (**a**) plan view of the upright cylinder including the numbering of the wave gauges and (**b**) side view of the inclined geometry with an example pitch angle β—R is the rotation axis and $TP4$ the reference point introduced in Section 3.2—all dimensions in [m].

For the experiments, 12 marker balls are added to the water in the circular tank (Figure 1). They are left to freely float on the surface and are captured simultaneously with the rigid body markers. By distributing the balls well over the entire surface, nearly no clustering effects were observed. In the rare case that the balls do cluster, the system is capable of identifying correctly up to three markers, which may be in contact with each other. In post-processing, the markers have to be separately identified as the software cannot automatically recognise them. This is only possible for rigid bodies with a fixed distance between multiple markers. The identification of the marker balls has to be done only once for each measurement and further tracking is done automatically by the software. Alternatively, all unidentified trajectories can be exported and processed with additional software (for example Matlab). In this case, the initial fill level inside the cylinder can be chosen to autonomously identify the markers and separate from those, which are attached to the body.

As shown in Figure 1, the Qualisys system includes one Oqus 210c for video capturing and four Oqus 300+ cameras for marker tracking. The capture rate was set to 100 Hz for all tests.

Two sets of results from the motion capture system are presented here: (a) definition of the 6 DoF motion of the rigid body and (b) the coordinates of the free floating markers, which are recorded relative to the global coordinate system.

2.3. Copper Tape Resistive Wave Gauges

Wave gauges are a very reliable and accurate measurement system to detect and record the motion of water free surfaces. They are standard measurement instruments in tank testing [64] and can also be used under very complex conditions [38,50,59,60,65] as well as near submerged structures [21,22,66,67]. This measurement system correlates the resistance between two parallel rods with the submerged water depth. A direct electric current would lead to an anode/cathode arrangement leading to removal of ions in the water surrounding the gauges [20]. Hence the amplifier sends an AC signal with a typical excitation frequency of 10 kHz [68]. In general, the specification of such a wave probe depends on the application as well as the maximum expected wave amplitude. Typically heights vary between 0.3 to 1 m with a rod diameter of 3 to 4 mm [68], but thinner wire can be utilised if necessary. Andresen and Frigaard [69] applied wave gauges made of a wire with 1 mm near a cylinder under wave conditions and even thinner, namely 0.13 and 0.4 mm, are used by Liu et al. [70] in a wind-wave tank.

In the case of a tank test the typical accuracy of the conductive probes are in the range of ± 0.25–0.5 mm respectively 0.1% [64,68]. Similar ranges of accuracy, namely ± 0.3 mm, can be achieved by for ultrasonic sensors for experimental conditions as presented by Longo [28]. In the case of aerated and breaking wave this range has to be expanded. An accuracy of 1 mm is found by Gomit et al. [34] for the measurement of the crest with a resistive probe made by HR Wallingford. Petti and Longo [71] corrected the systematic error in the swash zone between the actual front on the beach in relation to the measurement by the run up wire with a separate video observation. A fixed offset of 5 mm was found as a good compromise to reduce the geometrical error without excessive signal error [71]. Whittaker et al. [51] quantify the accuracy for runup experiments with ± 2 mm in the vertical direction based on effects of surface tension and calibration.

At the outset, the possibility of using conductive paint was investigated as part of preliminary tests. Those proved that it is possible to fabricate wave gauges by drawing on a surface with conductive paint (MG Chemicals 842AR-P Conductive Pen Silver). The hand drawn wave gauge was connected to the standard Edinburgh Designs' wave gauge box. Despite that the application with this method is inaccurate, the technique is very flexible and can be used in the future for specific applications that limit the use of tapes and rods. When dry, the conductive paint is water-resistant but remains very delicate and subject to mechanical damage. The variable cross section of the conductive paint leads to a non linear variation in gauge resistance that is not easily quantified, and therefore is only recommended for discrete water level measurements.

Based on their robust and exact mounting in comparison to the conductive paint, wave gauges made of copper tape were chosen [51,52]. Previous applications of such a tape can found as a switch [72,73] to indicate discrete water levels. In another study, Heller and Spinneken [38] bonded stainless steel strips on the surface of the inclined walls. For the current project, the scale for the water level inside the tank was used as an alignment for the conductive tape on each side. Separations of 30 mm, 24 mm, 20 mm, 15 mm, 10 mm, 5 mm, 4 mm and 1 mm between the 6.25 mm wide copper tapes were investigated. Preliminary dip tests showed that very small spacings between the two tapes lead to a delayed response to water level changes due to residual water films reducing the gauge resistance. Greater separations reduced the potential for stray conductive paths resulting from the residual water films that had led to erroneous measurements. Smaller separations provided a better signal to noise ratio and a reduced error in detecting oblique water surfaces run up. The 5 mm separation was chosen as the distance between the two tapes, resulting in a centerline separation of w = 11.35 mm. Whittaker et al. [51] used a bigger spacing of 20 mm. As part of the current tests, the durability of wave gauges based on conductive tape could be shown for normal laboratory conditions.

One Edinburgh Designs WG8USB Wave Gauge Controller is used for the presented work. The sample rate is 128 Hz [68]. Figure 2 shows the location of each wave gauge. In total, all the available 8 gauges are used, which is the capacity of a single controller. The rotation axis R for the following presented experiments (Section 3.2) is parallel to the y-axis. The wave gauges are arranged in four pairs with equal distance to the y-axis. In addition to the quadrants, the two wave gauges $WG2$ and $WG4$ are located at octants and have the equal $|x|$ and y distance of $D/(2 \cdot \sqrt{2})$ = 0.173 m. With an $|x|$ coordinate of $D/4$ for the $WG6$ and $WG8$, the y-value is $D \cdot \sqrt{3}/4$ = 0.212 m. For the evaluation of the inclination tests an additional marker $TP4$ (tank point, Figure 2) is needed. The $[x, y, z]$-coordinates of this point are $[-D/4, D \cdot \sqrt{3}/4, H - h + dh]$ with an offset dh of approximately 10 mm and height of the cylinder H of 0.5 m. The origin of the local coordinate system is located in the centre of the tank and on the initial water level h of 0.25 m. The measurements of the Qualisys system are based on a different global coordinate system, which is defined as part of the calibration. The results are transformed into a local coordinate system, which has its origin at the tank point $TP4$ (Figure 2).

3. Results

3.1. Influence of Variable Inner Water Depth h

In the first phase of the investigation, different water levels are measured with both methods to quantify the differences between the two approaches. For this set of tests, the location of the cylinder is fixed in the stationary position and the volume of fluid inside the tank is changed. This investigation is similar to the required calibration of the WGs and to the application of wave run-up experiments [38,51,52]. Consequently, the measurements at the WG can be used to verify the new approach based on the floating markers.

A total of 50 different steady states are investigated. The observed value is the change in water depth Δh inside the tank. The zero/reference value is set to the initial water depth h = 0.25 m. Both investigated measurement systems are calibrated independently and all depths are verified with a scale inside the tank captured by a camera.

All measurements for this variation of the water level, as well as the inclination tests in Section 3.2, are repeated three times and include 20 s recording time for each. One data point in Figure 3 represents the mean value over the entire test duration for all of the repetitions, for all eight wave gauges. The same averaging techniques are applied to the z-coordinate results of the Qualisys system for all 12 markers.

In Figure 3a the measurements are directly compared (in mm). Ideally, both measurement systems would report the same results. The absolute differences in Figure 3b indicates that the measurement based on the balls has a tendency to underestimate the height. For most of the cases, the differences are in the range of 4 mm and only a smaller water depth results in an increase up to around 7 mm. It is assumed that the influence of potentially biased measurements through the transparent cylinder is increased with a lower water level. The repetition of the Δh equal 0 mm allows checking for such a potential time-dependent effect. Theses repetitions were conducted spread over the whole experimentation by starting with this water level, which is first increase up to maximum, than reduce to the minimum level and back to the starting value of Δh equal 0 mm. A detailed analysis of those runs indicates two very good agreements near to the calibration process and the following repetitions have an increased difference. A change of the air temperature over the experimental time may cause an additional uncertainty in the water temperature of the comparably small volume in the cylinder and the used set-up (Figure 1).

In the case of the wave gauges made of conductive tape, the minimum standard deviation (of all eight wave gauges and three repetitions each 20 s long) can be found around the zero value. For both increasing and decreasing water levels, the standard deviation value increases up to an approximate value of ±3 mm (Figure 3c). For the standard deviation of the measurements based on the Qualisys system (Figure 3d), such a clear trend cannot be found in relation to Δh_Q-values, hence there is no connection between the system calibration and its origin. The reduction of the standard deviation for the highest water depths in the tank can be explained by improved marker visibility. In general, the standard deviation is in the same range as the observed absolute differences between the two measurement systems.

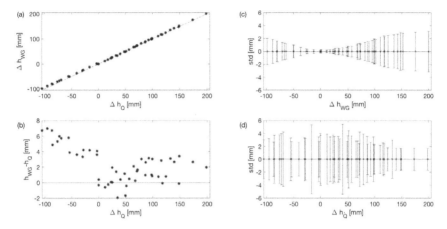

Figure 3. Comparison of the methods with a variable water depth h inside the cylinder—(**a**) comparison between water depth captured with the marker balls via Qualisys (Q) and wave gauges (WG); (**b**) absolute differences between the methods; (**c**,**d**) standard deviation (std) for each method.

Whittaker et al. [51] quantified the accuracy of this method with approximately ±2 mm in the vertical direction for a run-up experiment, which seem to be also applicable for the wave gauges made of copper tape. It seems to be reasonable to double this value for the floating balls based on the bigger standard deviation as well as the absolute differences between the two methods. On this basis, the next step is to investigate an inclined body in Section 3.2.

3.2. Influence of Variable Inclination Angle of the Tank

A fundamental difference between the two measurement systems is their point of reference. The locations of the infra-red markers are captured and reported based on a global reference frame which is independent of the position of the cylinder. The wave gauges are directly connected to the cylinder and thus measure the free surface in the local frame of reference of the cylinder. In contrast to the previous applications of onshore wave run-up probes [38,51,52], the orientation of the WGs changes according to the movement of the body on which they are bonded. Those tests allow to explore whether this effect on the measurement is significant or not.

It is necessary to transform the position of the floating markers to the local frame of reference of the cylinder or transform the wave gauge results into the global coordinate system. The required rotation matrix and the position vector is provided by the motion capturing system.

Figure 4a shows the mean value of the measurements of the two wave gauges laying on the y-axis ($WG3$ and $WG7$) vary with the rigid body's pitch angle as captured by Qualisys (β_Q). In the case of a rotation around the R-axis, which is parallel to the rotation axis y, those values should remain at 0 mm as well as the global mean value over all eight wave gauges. The other pairs of wave gauges (namely $WG1$ & $WG5$, $WG2$ & $WG4$, $WG6$ & $WG8$) are analysed in Figure 4b. The absolute distances in the x-direction for each pair is equal. Consequently, the mean value of each pair should also be

0 mm. Both analyses indicate that with an increasing β_Q-angle, a bigger difference to the theoretical position also occurs. This is possibly due to deformation of the structure resulting in a decrease in cross-sectional area, in turn increasing the height of the fluid in the structure. This will not be an issue for structures tested in the wave tank, as the surrounding water pressure will counteract the inner pressure.

The results of the three pairs of wave gauges with an x offset are used to back calculate the inclination of the structure β_{WG}, which should be identical to the pitch angle β_Q measured by the Qualisys system. Figure 4c presents the direct comparison of those values and Figure 4d the differences. A calculation of the inclination based on the wave gauge data would result in an under prediction of approximately $1°$.

For the analysis of the measurements based on the marker balls, the vertical transition caused by the rotation has to be included. Basic geometric dependences are shown in Figure 2 for the centre point on the water surface. A theoretical height increase $z_{Q,t}$ is calculated to compare the observation z_Q of the markers. The value $z_{Q,t}$ depends on the measured pitch angle β_Q of the cylinder and the exact position of the rotation axis R. It is assumed that the rotation axis R is parallel to the y-axis, and to specify the location of the axis R, a circle is fitted through the measured x- and z-coordinates of the tank point $TP4$.

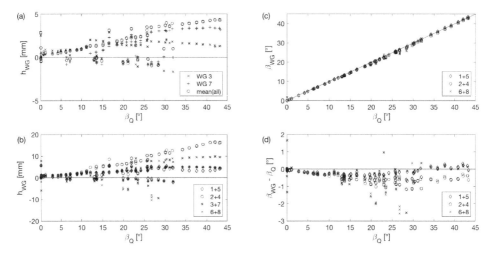

Figure 4. Influence of variable pitch angle β on the wave gauges (WG)—(**a**,**b**) analysis of the measured changes in the free surface; (**c**,**d**) comparison between the measured angle β_Q and the calculated β_{WG} based on the wave gauges pairs.

Figure 5a allows a comparison of the theoretical and measured values for the water surface and tank point $TP4$ in relation to the angle β_Q of the body. All values are transformed so that 0 mm indicates the initial water surface. The absolute differences between calculated and measured values are shown in Figure 5d. Both graphs indicate that, in a range of β_Q up to $30°$, the theory over-predicts the measurement and beyond that point it under-predicts.

For a further analysis, the same data-set is plotted in relation to the measured roll angle α_Q in Figure 5c and the yaw angle γ_Q Figure 5e. A comparable behaviour to the angle β_Q could be found for α_Q around $1°$, but not for γ_Q. This indicates that the complete system is not completely rigid and includes a small inclination around the x-axis and a part of the differences can be explained as a result. The standard deviation of the measurements of the marker balls (Figure 5b) show the same range as found by changing the water level h inside the tank (Figure 3d; Section 3.1).

The results of both stationary cases, namely the set of measurements with changing inner water level and variable inclination of the tank, show a reasonably small error of a few millimetres for each measurement system. The possible over-prediction of the water level in the case of inclined wave gauge in relation to the water surface should be considered for further comparisons with numerical simulations. Circularity measurements may verify if the increase in water level is due to deformation of the structure or a wave gauge error.

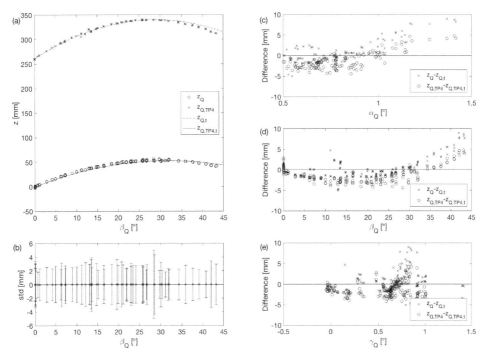

Figure 5. Influence of variable pitch angle β on the floating markers—(**a**) comparison of the theoretical with the measured z-coordinate of the reference point $TP4$ and the free surface; (**b**) standard deviation of the measured balls; (**c**–**e**) absolute difference of the z-coordinate in relation to all three rotation axes.

3.3. Transient Behaviour

The previous investigations assume steady-state conditions and focused on establishing a proof of concept for the adapted measurement system. However, the goal is to investigate the free floating structure motions and inner free water surface when influenced by periodic wave conditions in a wave tank. In this case the free water surface inside the cylinder will be non stationary. Different drop tests are conducted to show that both measurement concepts are capable of capturing transient water surfaces inside of the tank. To achieve this, the cylinder is fixed in an inclined position and the supporting mechanism is suddenly removed. Hence, the tank falls onto an end stop and the oscillation of the free surface inside the tank is stimulated.

The presented experiment is chosen because a small overtopping volume is provoked near the wave gauge $WG1$. Figure 6 shows the rotation from an initial pitch angle β_Q of 21.1° to 6.4°. For the other two rotations, the observed differences $\Delta\alpha$ and $\Delta\gamma$ to the starting condition (inclined cylinder) are very small. All systems are synchronised so that the approximated start of the movement is at $t = 0.25$ s. The end position of the structure has an inclination in pitch caused by the end stop, which is added to the framework. The first impact compresses this end stop and a nearly stable situation of the cylinder is reached after about one second. A small periodic movement of the structure is also observed, which is caused by the moving water body in the tank combined with the soft support of the end stop material.

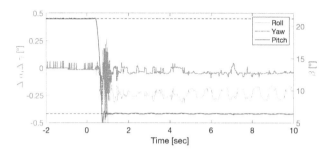

Figure 6. Measured angles of the rigid body caused by the sudden removal of the support structure.

Simultaneously two videos are captured, which give a very good overview of the behaviour of the water stored in the structure. The measured position of the markers can be directly connected with the video from the Oqus 210c. In contrast to this fixed position of observation, the second camera on the framework moves with the structure. This allows comparison of the visible scale inside the cylinder with the captured free surface changes [7,31]. Different random samples are analysed as part of the plausibility check and demonstrate a very good agreement (typically in the low range of a few millimetres) between both experimental methods. Therefore, the global coordinates of the marker balls had to be transformed into the local coordinate system of the cylinder.

Figure 7 presents the results of the wave gauges for the drop test. In each graphic, each pair of wave gauges with the same x offset are grouped, and the normalised water surface displacement η/h is plotted as the y-value. This non-dimensional elevation is also used for the second and third row in Figure 8 (two different side views) for the following analysis of the floating markers. As mentioned, the initial water height $h = 0.25$ m is equal to half of the total height H of the cylinder. The η/h value equal to 1 [-] in the first big peak of the wave gauge $WG1$ indicates the overtopping at this part of the cylinder. A wave period of approximately 0.7 s is observed for the introduced oscillation. The final conditions also include an inclination due the end stop material, which is added to the support structure, which alters the resting position of the structure, affecting the final value of all wave gauges except $WG3$ and $WG7$ because they lie on the y-axis.

The results show that the main oscillation direction is at first aligned with the movement of the cylinder in the x-z-plane and changes after a few seconds to the orthogonal direction in the y-z-plane. The $WG1$ and $WG5$ show a damped oscillation with a nearly constant decrease of the amplitude. The peak values for η/h at the $WG3$ and $WG7$ increase over the same time frame (Figure 7).

In addition to the wave gauge results, the elevation of each of the 12 free surface marker balls can be plotted. Figure 8 uses a different approach to show the three-dimensional motion of the markers. Four different time periods were chosen. For each period of record, the measured coordinates of all the 12 points were plotted in one graph. The top view (upper row) shows that the balls have the tendency to group together. Nevertheless, a good coverage is maintained. The side view of the first 5 s time period also captures the overflow with a η/h-value near to 1 [-], which fortunately did not result in the loss of a marker. The change of the main oscillation direction can also be observed in the measured data of the marker balls (going from 3rd to 4th column in Figure 8).

This simplified transient experiment showed the capability of both methods to capture the free surface inside the cylinder. The advantages and limitations of each approach are discussed in the following section.

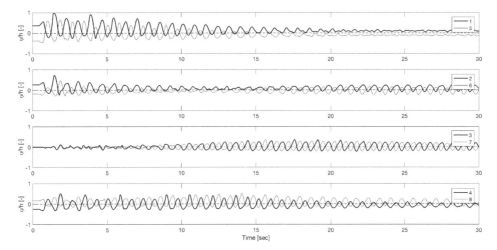

Figure 7. Results of the transient experiment—wave gauges (WG) in pairs as defined in Figure 2—normalised surface elevation η/h.

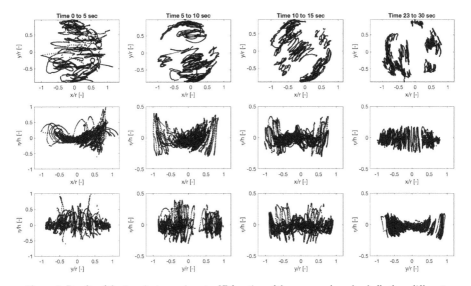

Figure 8. Results of the transient experiment—3D location of the measured marker balls; four different time periods (columns) and three different view directions (rows).

4. Discussion

Both measurement systems investigated here fulfill the basic requirements mentioned in Section 1 for the future use of capturing the motion of the free water surface inside a floating structure in a wave tank.

The addition of the bespoke marker balls into the tank is simple to implement, and makes use of the motion capture systems which are typical equipment in a laboratory with wave basins. Camera positions may have to be adapted and more carefully chosen, so that the inside of the floating cylinder is easily captured. A motion capture camera view from the top would be very advantageous for 3D capture of the markers, but this may not always be practical to realise.

The floating markers are capable of surviving very rough conditions in the cylinder and are very light. It is thus assumed that the influence of adding floating markers to the free surface is very small, but it cannot be excluded completely. Specific experiments with and without the markers should be compared for the individual applications in the future. For the current project, 12 markers were utilized, which covered approximately 8% of the available surface inside the cylinder. More floating markers could be applied, but as a result the interaction between the markers would also increase. The ideal number and size of markers should be optimised for each project.

Under certain conditions the capturing of one individual marker's position is interrupted because fewer than two cameras detect it. This can be caused by obscuring components (such as the mounted camera at the top) or extreme sloshing, including breaking waves. There is also the possibility that one marker is washed out of the structure. This is an abort criterion because the water volume inside the tank is reduced in such a case. Overtopping of the structure cannot be captured with this measurement system.

The novel characteristic of this system is that the markers cover the free surface inside the cylinder (Figure 8), but the exact location cannot be predicted beforehand and is not repeatable. This situation can be improved with a splitting of the available volume inside the tank, for example with a vertical mesh. This can limit the area which each marker can cover but also has a further effect on the fluid dynamics. Conventional wave gauges could be a better option instead of such highly restricted markers.

The wave gauges made of the copper tape are fixed to the cylinder structure and no additional support structure is needed. This prevents unnecessary mass additions and subsequent changes in the centre of mass. An electrical connection is still required, but the cable can be added to the mooring lines required to stabilise the floating body. This solution requires an insulating and robust surface to apply the copper tape. A minimum distance between two pairs has to be maintained so that no interference occurs (similar to normal wave gauges). The capillary effect between the two tapes on the wall may lead to erroneous measurement. Based on a preliminary investigation, a distance of 5 mm was chosen for this investigation (Section 2.3). Such a suitable choice of the distance between the two electrodes helps to minimise this effect. Furthermore, maximum local water heights in the floating structure are caused by run up on the walls which is quantifiable with this method. If data analysis of the wave gauges is carried out in the local coordinates of the cylinder, no additional post processing is needed to use the measured data. The major disadvantage of video recording the scale inside the tank is the post processing required to attain useful data.

It is well known that the measurement method of wave gauges is highly dependent on the electrical properties of the water. The water for the tests herein was taken from the 2.4 million litre basin at FloWave and not fresh from the main water supply, thus minimising any temperature adjustments or other changes. In comparison to the wave tank, the stored water volume inside the cylinder (Figure 1) is very small. The changes of the air temperature over a working day also influenced the temperature of the water in the tank. Consequently, some of the observed errors may be caused by this change in environmental conditions and the (water) temperature should be measured in comparable investigations in the future. The temperature changes of the water in the wave tank under normal conditions is nearly negligible and hence the conditions for a submerged structure are far better than in the present configuration of being supported on a framework.

5. Conclusions

This paper presents two different measurement methods for capturing the motion of the free water surface inside a floating structure by re-purpose existing systems: (a) floating markers, which are captured in addition to the movement of the rigid body and (b) wave gauges made of conductive copper tape fixed to the walls of the cylinder. The additional mass of both systems is very small compared to the structure. Both systems are tested under laboratory conditions and their durability is shown.

The conductive tape wave gauge proved very accurate and consistent in measuring the water height at the walls, where the maximum heights occur due to water run-up. It can be shown that the conductive tape wave gauges can also be applied to moving objects and that a change in the inclination in comparison to the calibration has a small effect on the accuracy. The standard deviation of the measurement system was found to be equal or smaller than 2 mm (variable water height, Section 3.1). Those measurements are directly connected to the movable structure, or local coordinate system. In contrast to this, the marker balls can freely float inside the water body and the measurement is in the global coordinate system. Hence the movement of the rigid body is captured in global coordinates, so that the free floating markers can be transformed into local coordinates for comparison with the copper tape wave gauge measurements. The standard deviation for this measurement system is in the range of 4 mm using markers with a diameter of 40 mm (Sections 2.2 and 3.1). A reduction of the marker size can result in a higher accuracy and should be considered in the future.

Different stationary and transient tests proved that both methods are capable of delivering measurements with a very high accuracy in low mm-range. The combination of both systems allows one to compare the two independent measurement methods and use the advantages of each approach. This has a significant benefit in providing useful data sets as a validation experiment for future numerical studies of free floating objects.

Author Contributions: R.G., J.S., D.I.M.F., T.D., T.B. and D.M.I. are responsible for the conceptualisation of the experimental investigation. R.G., J.S. and T.D. measured the data and analysed the data. R.G. and J.S. wrote the initial draft and D.I.M.F., T.D., T.B. and D.M.I. reviewed and edited the paper.

Funding: This research was funded by Austrian Science Fund (FWF) grant number J3918.

Acknowledgments: Further thanks go to the support team of Edinburgh Designs and Qualisys for their help.

Conflicts of Interest: The authors declare no conflict of interest.

Abbreviations

The following abbreviations are used in this manuscript:

LNG   liquefied natural gas
Q   Qualisys
VLFS   very large floating structures
WG   wave gauge

References

1. Chaplin, J.R.; Heller, V.; Farley, F.J.M.; Hearn, G.E.; Rainey, R.C.T. Laboratory testing the Anaconda. *Philos. Trans. R. Soc. A* **2012**, *370*, 403–424.

2. Farley, F.J.M.; Rainey, R.C.T.; Chaplin, J.R. Rubber tubes in the sea. *Philos. Trans. R. Soc. A* **2012**, *370*, 381–402. [CrossRef] [PubMed]

3. Hirdaris, S.E.; Bai, W.; Dessi, D.; Ergin, A.; Gu, X.; Hermundstad, O.A.; Huijsmans, R.; Iijima, K.; Nielsen, U.D.; Parunov, J.; et al. Loads for use in the design of ships and offshore structures. *Ocean Eng.* **2014**, *78*, 131–174. [CrossRef]

4. Jiao, J.; Ren, H.; Chen, C. Model testing for ship hydroelasticity: A review and future trends. *J. Shanghai Jiaotong Univ.* **2017**, *22*, 641–650. [CrossRef]

5. Sharma, R.; Kim, T.W.; Storch, R.L.; Hopman, H.; Erikstad, S.O. Challenges in computer applications for ship and floating structure design and analysis. *CAD Comput. Aided Des.* **2012**, *44*, 166–185. [CrossRef]

6. Bureau Veritas (BV). *Design Sloshing Loads for LNG Membrane Tanks—Guidance Note NI 554 DT R00 E*; Bureau Veritas (BV): Paris, France, 2010.

7. ITTC—Seakeeping Committee of the 28th ITTC. ITTC Quality System Manual Recommended Procedures and Guidelines—Procedure Sloshing Model Tests, Revision 00, Section: 7.5-02-07-02.7. Available online: https://ittc.info/media/7625/75-02-07-027.pdf (accessed on 27 December 2018).

8. Lyu, W.; Moctar, O.E.; Potthoff, R.; Neugebauer, J. Experimental and numerical investigation of sloshing using different free surface capturing methods. *Appl. Ocean Res.* **2017**, *68*, 307–324. [CrossRef]

9. Song, Y.K.; Chang, K.A.; Ryu, Y.; Kwon, S.H. Experimental study on flow kinematics and impact pressure in liquid sloshing. *Exp. Fluids* **2013**, *54*, 1592. [CrossRef]

10. Ariyarathne, K.; Chang, K.-A.; Mercier, R. Green water impact pressure on a three-dimensional model structure. *Exp. Fluids* **2012**, *53*, 1879–1894. [CrossRef]

11. Chuang, W.L.; Chang, K.A.; Mercier, R. Impact pressure and void fraction due to plunging breaking wave impact on a 2D TLP structure. *Exp. Fluids* **2017**, *58*, 68. [CrossRef]

12. Kim, S.Y.; Kim, K.H.; Kim, Y. Comparative study on pressure sensors for sloshing experiment. *Ocean Eng.* **2015**, *94*, 199–212. [CrossRef]

13. Kim, K.-T.; Lee, P.-S.; Park, K.C. A direct coupling method for 3D hydroelastic analysis of floating structures. *Int. J. Numer. Methods Eng.* **2013**, *96*, 842–866. [CrossRef]

14. Wei, W.; Fu, S.; Moan, T.; Lu, Z.; Deng, S. A discrete-modules-based frequency domain hydroelasticity method for floating structures in inhomogeneous sea conditions. *J. Fluids Struct.* **2017**, *74*, 321–339. [CrossRef]

15. Yoon, J.S.; Cho, S.P.; Jiwinangun, R.G.; Lee, P.S. Hydroelastic analysis of floating plates with multiple hinge connections in regular waves. *Mar. Struct.* **2014**, *36*, 65–87. [CrossRef]

16. Lee, K.H.; Cho, S.; Kim, K.T.; Kim, J.G.; Lee, P.S. Hydroelastic analysis of floating structures with liquid tanks and comparison with experimental tests. *Appl. Ocean Res.* **2015**, *52*, 167–187. [CrossRef]

17. Klar, R.; Steidl, B.; Sant, T.; Aufleger, M.; Farrugia, R.N. Buoyant Energy—Balancing wind power and other renewables in Europe's oceans. *J. Energy Storage* **2017**, *14*, 246–255. [CrossRef]

18. Klar, R.; Steidl, B.; Aufleger, M. A floating energy storage system based on fabric. *Ocean Eng.* **2018**, *165*, 328–335. [CrossRef]

19. Gabl, R.; Achleitner, S.; Neuner, J.; Aufleger, M. Accuracy analysis of a physical scale model using the example of an asymmetric orifice. *Flow Meas. Instrum.* **2014**, *36*, 36–46. [CrossRef]

20. Hughes, S.A. *Physical Models and Laboratory Techniques in Coastal Engineering*; Advanced Series on Ocean Engineering; World Scientific: Singapore, 1993. [CrossRef]

21. Lykke Andersen, T.; Frigaard, P.; Damsgaard, M.L.; De Vos, L. Wave run-up on slender piles in design conditions - Model tests and design rules for offshore wind. *Coast. Eng.* **2011**, *58*, 281–289. [CrossRef]

22. Mavrakos, S.A.; Chatjigeorgiou, I.K.; Lentziou, D.M. Wave Run-Up and Second-Order Wave Forces on a Truncated Circular Cylinder Due to Monochromatic Waves. In Proceedings of the 24th International Conference on Offshore Mechanics and Arctic Engineering, Halkidiki, Greece, 12–17 June 2005; pp. 231–238.

23. Teh, H.M.; Venugopal, V.; Bruce, T. Hydrodynamic characteristics of a free-surface semicircular breakwater exposed to irregular waves. *J. Waterway Port Coast. Ocean Eng.* **2012**, *138*, 149–163. [CrossRef]

24. Tripepi, G.; Aristodemo, F.; Veltri, P. On-Bottom Stability Analysis of Cylinders under Tsunami-Like Solitary Waves. *Water* **2018**, *10*, 487. [CrossRef]

25. Peruzzo, P.; De Serio, F.; Defina, A.; Mossa, M. Wave Height Attenuation and Flow Resistance Due to Emergent or Near-Emergent Vegetation. *Water* **2018**, *10*, 402. [CrossRef]

26. Chiapponi, L.; Longo, S.; Tonelli, M. Experimental study on oscillating grid turbulence and free surface fluctuation. *Exp. Fluids* **2012**, *53*, 1515–1531. [CrossRef]

27. Caron, P.A.; Cruchaga, M.A.; Larreteguy, A.E. Study of 3D sloshing in a vertical cylindrical tank. *Phys. Fluid* **2018**, *30*, 082112.

28. Longo, S. Experiments on turbulence beneath a free surface in a stationary field generated by a Crump weir: Free-Surface characteristics and the relevant scales. *Exp. Fluids* **2010**, *49*, 1325–1338.

29. Longo, S.; Liang, D.; Chiapponi, L.; Jimenez, L.A. Turbulent flow structure in experimental laboratory wind-generated gravity waves. *Coast. Eng.* **2012**, *64*, 1–15.

30. Sellar, B.; Harding, S.; Richmond, M. High-resolution velocimetry in energetic tidal currents using a convergent-beam acoustic Doppler profiler. *Meas. Sci. Technol.* **2015**, *26*, 085801. [CrossRef]

31. Bonakdar, L.; Oumeraci, H.; Etemad-Shahidi, A. Run-up on vertical piles due to regular waves: Small-scale model tests and prediction formulae. *Coast. Eng.* **2016**, *118*, 1–11. [CrossRef]

32. Tosun, U.; Aghazadeh, R.; Sert, C.; Özer, M.B. Tracking free surface and estimating sloshing force using image processing. *Exp. Thermal Fluid Sci.* **2017**, *88*, 423–433. [CrossRef]

33. Bechle, A.J.; Wu, C.H. Virtual wave gauges based upon stereo imaging for measuring surface wave characteristics. *Coast. Eng.* **2011**, *58*, 305–316. [CrossRef]

34. Gomit, G.; Chatellier, L.; Calluaud, D.; David, L. Free surface measurement by stereo-refraction. *Exp. Fluids* **2013**, *54*, 1540. [CrossRef]

35. Bregoli, F.; Bateman, A.; Medina, V. Tsunamis generated by fast granular landslides: 3D experiments and empirical predictors. *J. Hydraul. Res.* **2017**, *55*, 743–758. [CrossRef]

36. Chang, K.-A.; Liu, P.L.-F. Pseudo turbulence in PIV breaking-wave measurements. *Exp. Fluids* **2000**, *29*, 331–338. [CrossRef]

37. Fritz, H.M.; Hager, W.H.; Minor, H.E. Landslide generated impulse waves. 1. Instantaneous flow fields. *Exp. Fluids* **2003**, *35*, 505–519. [CrossRef]

38. Heller, V.; Bruggemann, M.; Spinneken, J.; Rogers, B.D. Composite modelling of subaerial landslide-tsunamis in different water body geometries and novel insight into slide and wave kinematics. *Coast. Eng.* **2016**, *109*, 20–41. [CrossRef]

39. Weigand, A. Simultaneous mapping of the velocity and deformation field at a free surface. *Exp. Fluids* **1996**, *20*, 358–364.

40. Park, J.; Im, S.; Sung, H.J.; Park, J.S. PIV measurements of flow around an arbitrarily moving free surface. *Exp. Fluids* **2015**, *56*, 56. [CrossRef]

41. Belden, J.; Techet, A.H. Simultaneous quantitative flow measurement using PIV on both sides of the air-water interface for breaking waves. *Exp. Fluids* **2011**, *50*, 149–161. [CrossRef]

42. Blenkinsopp, C.E.; Turner, I.L.; Allis, M.J.; Peirson, W.L.; Garden, L.E. Application of LiDAR technology for measurement of time-varying free-surface profiles in a laboratory wave flume. *Coast. Eng.* **2012**, *68*, 1–5. [CrossRef]

43. Montano, L.; Li, R.; Felder, S. Continuous measurements of time-varying free-surface profiles in aerated hydraulic jumps with a LIDAR. *Exp. Thermal Fluid Sci.* **2018**, *93*, 379–397. [CrossRef]

44. Weitbrecht, V.; Kühn, G.; Jirka, G.H. Large Scale PIV Measurements at the Surface of Shallow Water Flows. *Flow Meas. Instrum.* **2002**, *13*, 237–245. [CrossRef]

45. Akutina, Y.; Mydlarski, L.; Gaskin, S.; Eiff, O. Error analysis of 3D-PTV through unsteady interfaces. *Exp. Fluids* **2018**, *59*, 53.

46. Evers, F.M.; Hager, W.H. Videometric Water Surface Tracking: Towards Investigating Spatial Impulse Waves. In Proceedings of the 36th IAHR World Congress, Delft, The Netherlands, 28 June–3 July 2015; pp. 6618–6623.

47. Evers, F.M.; Hager, W.H. Spatial impulse waves: Wave height decay experiments at laboratory scale. *Landslides* **2016**, *13*, 1395–1403. [CrossRef]

48. Mignot, E.; Moyne, T.; Doppler, D.; Riviere, N. Clear-water scouring process in a flow in supercritical regime. *J. Hydraul. Eng.* **2016**, *142*, 04015063-1. [CrossRef]

49. Przadka, A.; Cabane, B.; Pagneux, V.; Maurel, A.; Petitjeans, P. Fourier transform profilometry for water waves: How to achieve clean water attenuation with diffusive reflection at the water surface? *Exp. Fluids* **2012**, *52*, 519–527. [CrossRef]

50. Heller, V.; Spinneken, J. On the effect of the water body geometry on landslide-tsunamis: Physical insight from laboratory tests and 2D to 3D wave parameter transformation. *Coast. Eng.* **2015**, *104*, 113–134.

51. Whittaker, C.N.; Fitzgerald, C.J.; Raby, A.C.; Taylor, P.H.; Orszaghova, J.; Borthwick, A.G.L. Optimisation of focused wave group runup on a plane beach. *Coast. Eng.* **2017**, *121*, 44–55. [CrossRef]

52. Niedzwecki, J.M.; Duggal, A.S. Wave runup and forces on cylinders in regular and random waves. *J. Waterway Port Coast. Ocean Eng.* **1992**, *118*, 615–634. [CrossRef]

53. Kang, Z.; Ni, W.; Zhang, L.; Ma, G. An experimental study on vortex induced motion of a tethered cylinder in uniform flow. *Ocean Eng.* **2017**, *142*, 259–267. [CrossRef]

54. O'Connell, K.; Thiebaut, F.; Kelly, G.; Cashman, A. Development of a free heaving OWC model with non-linear PTO interaction. *Renew. Energy* **2018**, *117*, 108–115. [CrossRef]

55. Sjökvist, L.; Wu, J.; Ransley, E.; Engström, J.; Eriksson, M.; Göteman, M. Numerical models for the motion and forces of point-absorbing wave energy converters in extreme waves. *Ocean Eng.* **2017**, *145*, 1–14. [CrossRef]

56. Usherwood, J.R. The aerodynamic forces and pressure distribution of a revolving pigeon wing. *Exp. Fluids* **2009**, *46*, 991–1003. [CrossRef] [PubMed]

57. Gyongy, I.; Richon, J.-B.; Bruce, T.; Bryden, I. Validation of a hydrodynamic model for a curved, multi-paddle wave tank. *Appl. Ocean Res.* **2014**, *44*, 39–52. [CrossRef]

58. Draycott, S.; Noble, D.; Davey, T.; Bruce, T.; Ingram, D.; Johanning, L.; Smith, H.; Day, A.; Kaklis, P. Re-creation of site-specific multi-directional waves with non-collinear current. *Ocean Eng.* **2018**, *152*, 391–403. [CrossRef]

59. Draycott, S.; Sutherland, D.; Steynor, J.; Sellar, B.; Venugopal, V. Re-creating waves in large currents for tidal energy applications. *Energies* **2017**, *10*, 1838. [CrossRef]

60. Draycott, S.; Davey, T.; Ingram, D.M. Simulating Extreme Directional Wave Conditions. *Energies* **2017**, *10*, 1731.

61. Ingram, D.; Wallace, R.; Robinson, A.; Bryden, I. The design and commissioning of the first, circular, combined current and wave test basin. In Proceedings of the Oceans 2014 MTS/IEEE Taipei, Taiwan, 18–21 May 2015.

62. Sutherland, D.R.J.; Noble, D.R.; Steynor, J.; Davey, T.; Bruce, T. Characterisation of current and turbulence in the FloWave Ocean Energy Research Facility. *Ocean Eng.* **2017**, *139*, 103–115. [CrossRef]

63. Draycott, S.; Sellar, B.; Davey, T.; Noble, D.R.; Venugopal, V.; Ingram, D. Capture and Simulation of the Ocean Environment for Offshore Renewable Energy. *Renew. Sustain. Energy Rev.* **2019**. under review.

64. MARINET (2012) Work Package 2: Standards and Best Practice—D2.1 Wave Instrumentation Database. Revision: 05. Available online: http://www.marinet2.eu/wp-content/uploads/2017/04/D2.01-Wave-Instrumentation-Database.pdf (accessed on 23 November 2018).

65. Heller, V.; Chen, F.; Brühl, M.; Gabl, R.; Chen, X.; Wolters, G.; Fuchs, H. Large-scale experiments into the tsunamigenic potential of different iceberg calving mechanisms. *Sci. Rep.* **2018**. [CrossRef]

66. Mai, T.; Greaves, D.; Raby, A.; Taylor, P.H. Physical modelling of wave scattering around fixed FPSO-shaped bodies. *Appl. Ocean Res.* **2016**, *61*, 115–129. [CrossRef]

67. Wienke, J.; Oumeraci, H. Breaking wave impact force on a vertical and inclined slender pile—Theoretical and large-scale model investigations. *Coast. Eng.* **2005**, *52*, 435–462. [CrossRef]

68. EDL-Edinburgh Designs Ltd. Available online: http://www.edesign.co.uk/product/wavegauges/ (accessed on 23 November 2018).

69. Andersen, T.L.; Frigaard, P. Horns Rev II, 2D-Model Tests: Wave Run-Up on Pile. In *DCE Contract Reports, No. 3*; Department of Civil Engineering, Aalborg University: Aalborg, Denmark, 2006.

70. Liu, H.-T.; Katsaros, K.B.; Weissman, M.A. Dynamic response of thin-wire wave gauges. *J. Geophys. Res.* **1982**, *87*, 5686–5698. [CrossRef]

71. Petti, M.; Longo, S. Turbulence experiments in the swash zone. *Coast. Eng.* **2001**, *43*, 1–24.

72. Bruce, T.; van der Meer, J.W.; Franco, L.; Pearson, J.M. Overtopping performance of different armour units for rubble mound breakwaters. *Coast. Eng.* **2009**, *56*, 166–179. [CrossRef]

73. Pullen, T.; Allsop, W.; Bruce, T.; Pearson, J. Field and laboratory measurements of mean overtopping discharges and spatial distributions at vertical seawalls. *Coast. Eng.* **2009**, *56*, 121–140. [CrossRef]

Article

Study on the Best Depth of Stilling Basin with Shallow-Water Cushion

Qiulin Li [1], Lianxia Li [1,*] and Huasheng Liao [2]

[1] State Key Laboratory of Hydraulics and Mountain River Development and Protection, College of Water Resource and Hydropower, Sichuan University, Chengdu 610065, Sichuan, China; leeqlin@stu.scu.edu.cn

[2] Department Civil and Environmental Engineering, Michigan State University, East Lansing, MI 48824, USA; liao@egr.msu.edu

* Correspondence: lianxiali@scu.edu.cn; Tel.: +86-133-0802-1514

Received: 7 November 2018; Accepted: 4 December 2018; Published: 7 December 2018

Abstract: The depth of the stilling basin with shallow-water cushion (SBSWC) is a key factor that affects the flow regime of hydraulic jump in the basin. However, the specific depth at which the water cushion is considered as 'shallow' has not been stated clearly by far, and only conceptual description is provided. Therefore, in order to define the best depth of SBSWC and its relationship between the Froude number at the inlet of the stilling basin, a large number of experiments were carried out to investigate SBSWC. First of all, 30 cases including five different Froude numbers and six depths were selected for which large eddy simulation (LES) was firstly verified by the experiments and then adopted to calculate the hydraulic characteristics in the stilling basin. Finally, three standards, based on the flow regime of hydraulic jump, the location of the main stream and the energy dissipation rate, were proposed to define the best depth of SBSWC. The three criteria are as follows: (1) a complete hydraulic jump occurs in the basin (2) the water cushion is about 1/10–1/3 deep of the stilling basin, and (3) the energy dissipation rate is more than 70% and the unit volume energy dissipation rate is as high as possible. It showed that the best depth ratio of SBSWC (depth to length ratio) was between 0.1 and 0.3 and it also indicated the best depth increased with the increase in Froude number. The results of the work are of significance to the design and optimizing of SBSWC.

Keywords: energy dissipation; hydraulic jump; Froude number; stilling basin with shallow-water cushion (SBSWC); large eddy simulation; best depth

---

1. Introduction

At present, the energy dissipation by hydraulic jump is used for downstream in many hydraulic projects [1], of which the stilling basin plays an important role in energy dissipation. However, the traditional stilling basin has disadvantages such as high underflow speed near the bottom [2–4], apparent damages by erosion and cavitation as well as insufficient energy dissipation rate [5]. Many researchers have made studies on such problems. The concept of a stilling basin with shallow-water cushion (SBSWC) was introduced [6]; namely, a shallow-water cushion is added to the ordinary stilling basin. In the structure of the new type stilling basin, the water cushion formed at the basin bottom can be used as the 'flexible bottom plate', which applies a flexible counterforce to the water stream in the steep slope section and 'absorbs' impact force of high speed flow on the floor of the stilling basin, so as to protect the bottom of the basin. After that, a series of studies on the impacts caused by inlet type, inflow angle, and low Froude number of the stilling basin with shallow-water cushion on its hydraulic characteristics were conducted [7]. For example, the concept of a stilling basin with double shallow-water cushions is an extension of SBSWC, which has the same advantages as SBSWC, but with better hydraulic characteristics [8], lower underflow speed, and better distribution of dynamic pressures of bottom plate of the stilling basin. The stilling basin with a drop sill is similar to SBSWC in shape [9–14]. However, the former one is focused on the connection type of a drop sill at the stilling basin inlet, while the latter one is focused on the energy dissipation mechanism of the stilling basin and the effect of the shallow-water cushion.

The depth of SBSWC is of critical importance to the energy dissipation effect of the stilling basin; however, no systematic study has been made on it for now. Besides, whether the water cushion is considered as deep or shallow has not been defined clearly since the concept of SBSWC was proposed. This paper attempt to propose an accurate and practical definition of the best depth of SBSWC in terms of three aspects; namely, flow regime, main stream location, and energy dissipation rate. If the water cushion depth is higher than the best depth, it is not necessary; if the water cushion depth whereas is lower than the best depth, the water cushion cannot play the role of the buffer action fully. Therefore, it is expected that in a stilling basin with the best depth, a complete hydraulic jump (critical hydraulic jump and submerged hydraulic jump are considered complete hydraulic jumps) occurs and the main stream is not located too close or far from the bottom of the basin, and the energy dissipation rate is as high as possible.

This study was not about the thickness of a shallow water cushion, but to study the depths of a stilling basin with shallow-water cushion required at different Froude numbers. In order to obtain the best stilling basin depth (represented by dimensionless depth-to-length ratio, which was convenient for application) corresponding to the specified Froude number; 30 cases including five different Froude numbers and six depths were selected firstly in the study. Then the large eddy simulation (LES [15]) was adopted to calculate hydraulic characteristics in the stilling basin, and the reliability and accuracy of which were verified by the results of physical model. Large eddy simulation has a proportional advantage, such as (1) prediction of turbulent flow from laminar flow, (2) prediction of high-speed turbulence. Therefore, according to the simulation results of the flow regime of hydraulic jump, the location of the main stream and the energy dissipation rate, the definition or the standard of the best depth of SBSWC was proposed; finally the relationship between Froude number and the best depth of the basin was revealed.

2. Materials and Methods

2.1. Experimental Setup

Experiments were conducted in a smoothened glass flume. It was composed of several sections including a 363 cm-length inflow discharge chute, baffle, 120 cm-length stilling basin with shallow-water cushion, and 300 cm-length tail water section (Longitudinal cross-sectional view as shown in Figure 1). The discharge chute section and all downstream parts were with rectangular cross section, with a width of 30 cm. The inlet control is a gate that is 10 cm wide, which controls the head of water to the inflow discharge chute. The slope of the inflow discharge chute used in this study is fixed to 17°. At downstream, the water level was controlled, and outflow condition was applied.

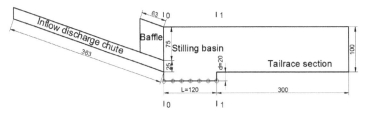

Figure 1. Schematic design of the model test (in cm).

2.2. Mathematical Model

In the present study, ANSYS 16.0 was utilized to investigate the flow characteristics over the stilling basin with shallow-water cushion with different Froude numbers. An iterative format selected an implicit time-marching scheme. The volume of fluid (VOF) [16,17] method was adopted to track and simulate the free surface and two-phase flow of air and water. The mathematical model of 3D large eddy simulation (LES) was used in the calculation, which is introduced below.

(1) Governing equations

In LES [18–22], the large-scale eddy was simulated directly by solving the momentum equation, and the small-scale eddy was expressed in the sub-grid scale model. The governing equation was,

$$\frac{\partial \rho}{\partial t} + \frac{\partial (\rho \overline{u_i})}{\partial x_i} = 0 \tag{1}$$

$$\frac{\partial (\rho \overline{u_i})}{\partial t} + \frac{\partial (\rho \overline{u_i u_j})}{\partial x_j} = -\frac{\partial \overline{p}}{\partial x_i} + \frac{\partial}{\partial x_j}\left(\mu \frac{\partial \overline{u_i}}{\partial x_j}\right) - \frac{\partial \tau_{ij}^{sgs}}{\partial x_j} + \rho g_i \tag{2}$$

where, the value with "–" represents the large-scale value obtained after filtering; ρ represents the density; U represents the velocity; t represents the time; p represents the pressure; g represents the acceleration of gravity; x represents the coordinate; i, j represents the coordinate orientation; and τ_{ij}^{sgs} represents the sub-grid stress, which indicates the impact caused by small-eddy movement on the large-eddy movement. Generally, the sub-grid stress is calculated by the eddy viscidity model,

$$\tau_{ij}^{sgs} - \frac{1}{3}\tau_{kk}^{sgs}\delta_{ij} = -2\mu_{sgs}\overline{S_{ij}} \tag{3}$$

where, μ_{sgs} represents the sub-grid turbulent viscosity coefficient, $\delta_{ij} = \begin{cases} 1, i = j \\ 0, i \neq j \end{cases}$; $\overline{S_{ij}}$ represents the strain rate tensor under the scale to be solved, which is defined as;

$$\overline{S_{ij}} = \frac{1}{2}\left(\frac{\partial \overline{u_i}}{\partial x_j} + \frac{\partial \overline{u_j}}{\partial x_i}\right) \tag{4}$$

The Smagorinsky–Lilly model is used to calculate the sub-grid turbulent viscosity coefficient: $\mu_{sgs} = \rho L_s^2 \sqrt{2\overline{S_{ij}} \times \overline{S_{ij}}}$, $L_s = \min\left(\kappa d, C_s V^{1/3}\right)$; where, represents the mixed length of the sub-grid scale; κ represents the Karman constant; d represents the distance to the nearest wall face; V represents the volume of computing control body; and C_s represents the Smagorinsky constant, which is 0.1 in this study [23].

(2) Discretization and algorithm

In this test, the finite volume method [24] was used to discrete of the governing equations, and second-order implicit scheme was used in the discretization of time item, and the PISO algorithm [25–27] was used to solve and control the coupling of speed and pressure in the equation. The VOF [16,17] method was adopted to track and simulate the free surface and two-phase flow of air and water.

(3) Computational domain and boundary conditions

(1) Computational domain

The computational domain and grid are shown in Figure 2 with the same dimensions as Figure 1. The longitudinal cross-section in Figure 2a is located at the center of Figure 2b. The grids in the stilling basin and tail water section were transited from coarse to fine from top to bottom, of which size is 2–3 cm, and the total grids of the model is 272,000.

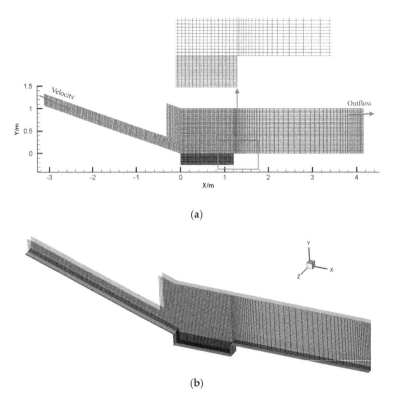

(a)

(b)

Figure 2. Schematic diagram for computational domain and grids of mathematical model. (**a**) Axial cross-sectional view; (**b**) The entire calculation domain.

(2)   Boundary conditions

The boundary condition of the inlet of the chute was set as velocity inlet condition which could be determined according to the flow and the area of inlet of the chute. The outflow condition [28] was applied at the outlet boundary. The viscous sub-layer of the near wall was treated by the wall function method [29]. Non-slip condition was applied on the fixed wall.

2.3. Verification of Mathematical Model

The calculation results of the large eddy simulation were compared with the test results of three cases based on the Froude number (Fr) at the inlet of stilling basin and the depth of SBSWC, which are (a) Fr = 6.92, d = 20 cm, (b) Fr = 7.64, d = 15 cm, and (c) Fr = 9.25, d = 25 cm, respectively.

The reliability and accuracy of LES for this study were verified by the computing and experimental results. Similar hydraulic jumps in the stilling basin were observed from the regime of both the test and computation (Figure 3) in all 3 cases, and the numerical model captured the flow profiles [30] accurately compared with test data (Figure 4) and the differences of the average pressure between test results and calculated results (Figure 5) were very small except for individual locations in case 3.

It can be seen from Figures 4 and 5 that, at most locations, the calculated results were fairly consistent with that of laboratory tests. The maximums of the relative error for flow profile and average pressure were 7.1% and 6.5%, respectively, indicating that the numerical simulations produced reliable and acceptable results.

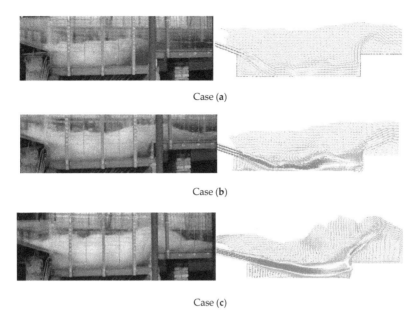

Case (**a**)

Case (**b**)

Case (**c**)

Figure 3. Schematic diagram for flow regimes under three cases.

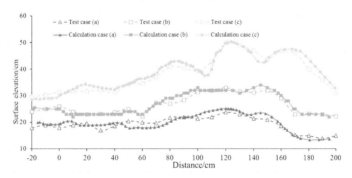

Figure 4. Comparison of flow profile.

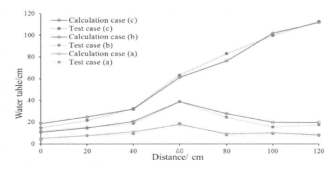

Figure 5. Comparison of average pressure.

3. Results and Discussion

A systematic study by means of LES was conducted to simulate the hydraulic characteristics such as flow regime, main stream location, and unit volume energy dissipation rate to determine the best depth of the shallow-water cushion on 30 cases including five typical Froude numbers and six different depths of shallow-water cushion shown in Table 1.

Table 1. The case of Froude number corresponding to the depth of the stilling basin.

Fr	d (cm)	d/L	Fr	d (cm)	d/L	Fr	d (cm)	d/L	Fr	d (cm)	d/L	Fr	d (cm)	d/L
	0	0.000		0	0.000		0	0.000		0	0.000		0	0.000
	5	0.042		5	0.042		5	0.042		5	0.042		5	0.042
7.64	10	0.083	6.92	10	0.083	8.05	10	0.083	9.25	10	0.083	11.24	10	0.083
	15	0.125		15	0.125		15	0.125		15	0.125		15	0.125
	20	0.167		20	0.167		20	0.167		20	0.167		20	0.167
	25	0.208		25	0.208		25	0.208		25	0.208		25	0.208

3.1. Flow Regime

The regimes of hydraulic jumps in the stilling basin at different conditions were presented in Figure 6. Since it was not possible to calculate every depth as a calculated working condition, the same six depths were selected for each Froude number, which were 0, 5, 10, 15, 20, and 25 cm respectively. Therefore, the calculated best depth in each corresponding Froude has an error of ±2.5 cm.

It is seen that the regimes and velocity distribution are different in the stilling basins for different depths with the same Froude number and with different Froude number but same depth. For the same Froude number, if the depth is not enough, no hydraulic jump takes place in the basin, and the regime of jump gets more complete as the depth increases; for example, when the depth of stilling basin is 0 cm or 5 cm, there is no hydraulic jump in the stilling basin; when the depth of stilling basin is 10 cm, 15 cm, 20 cm, or 25 cm, there is a hydraulic jump in the stilling basin. This pattern holds for all the groups (a)–(e) with different Froude numbers. For different Froude numbers, the depth required to form a complete hydraulic jump is different. For example, a complete hydraulic jump occurred in the depth of 15 cm, 20 cm, and 25 cm in the stilling basin when the Froude numbers are 7.64 and 6.92. In case that the Froude number is 8.05 and 9.25, the complete hydraulic jump occurred in the depth of 20 and 25 cm in the stilling basin. While in the case of Froude number 11.24, the complete hydraulic jump occurs in the depth of 25 cm in the stilling basin.

From the prospective of flow regime, the minimum depth of stilling basin with shallow water cushion (SBSWC) with different Froude numbers were depicted in Figure 7. It is indicated that the higher the Froude number, the deeper the basin required to form a complete hydraulic jump. In detail, in cases that the Froude number are 6.92, 7.64, 8.05, 9.25, and 11.24, the minimum depths of stilling basin with shallow water cushion are 15 cm, 15 cm, 20 cm, 20 cm and 25 cm.

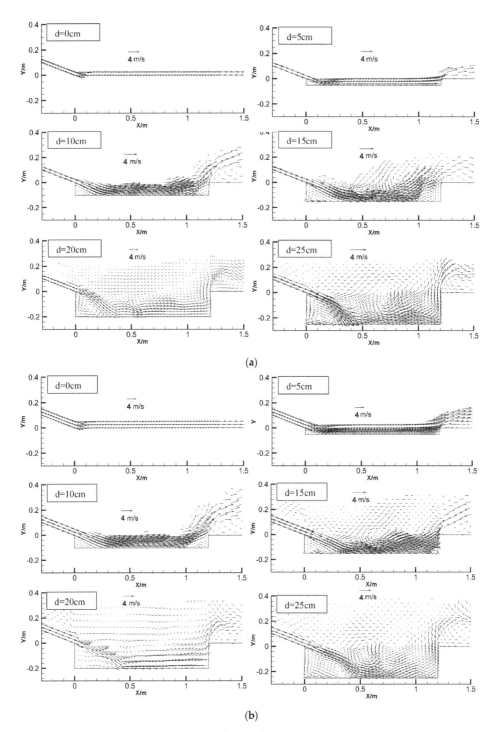

(a)

(b)

Figure 6. *Cont.*

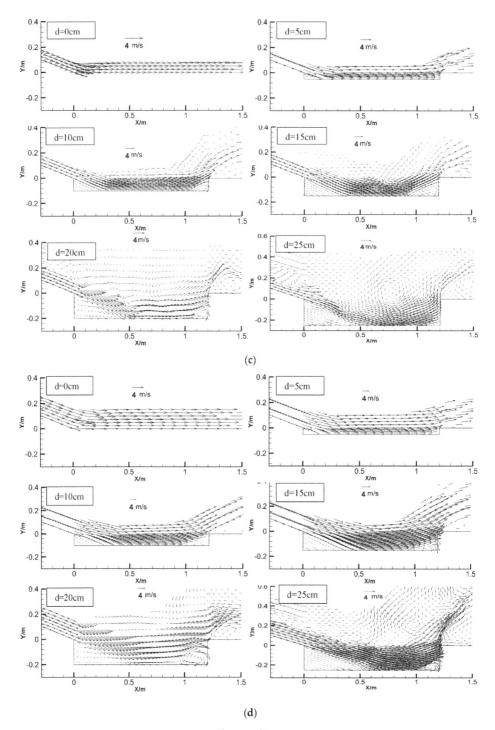

Figure 6. *Cont.*

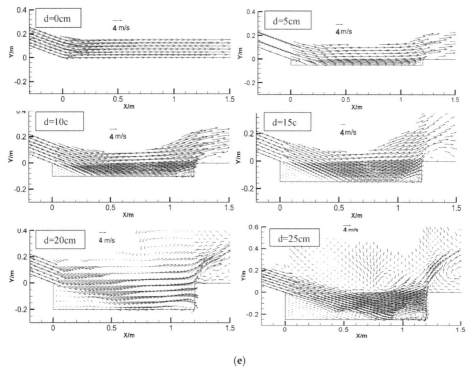

(e)

Figure 6. Flow regimes and velocity distribution at axial plane. (**a**) Fr = 6.92; (**b**) Fr = 7.64; (**c**) Fr = 8.05; (**d**) Fr = 9.25; (**e**) Fr = 11.24. Fr: Froude number.

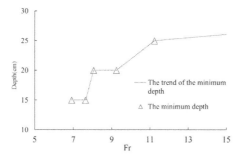

Figure 7. The minimum depth of the stilling basin for the complete hydraulic jump.

3.2. Main Stream Location

The main stream zone was defined as the area that the flow speed was larger than the average speed of cross section in this study. The water cushion was then defined as the water body under the lower boundary of the main stream and the dimensionless thickness of water cushion was defined as the thickness of water cushion referred to the depth of the basin.

The main stream location of the 30 cases were grouped and shown in Figure 8 in which the main stream was represented out by velocity iso-surface in some cases. It shows that the center of the main stream closest to the bottom of the basin is located at about 1/3 depth of the basin from the bottom in cases that hydraulic jumps occur which agrees with the results of previous study [7]. In cases that the Froude numbers are 6.92, 7.64, and 8.05, it can be clearly observed that the water cushions are formed between the main stream and the cushion bottom in the cases of the stilling basin depth of 15 cm, 20 cm and 25 cm. While in the case of Froude number 9.25 and 11.24 the water cushion locates between the main stream and the cushion bottom when the stilling basin depth is 20 cm or 25 cm.

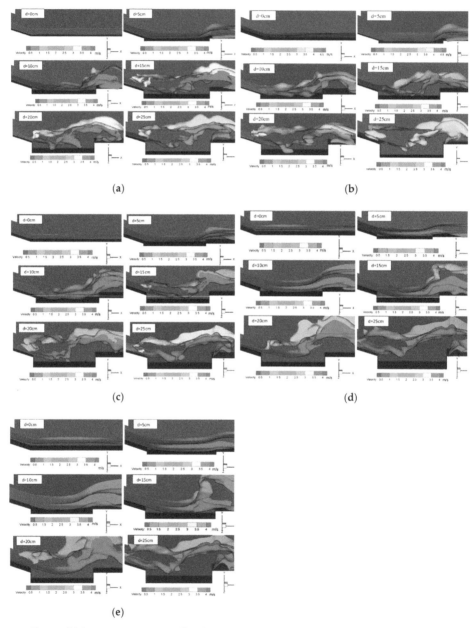

Figure 8. Main stream position in stilling basin. (**a**) Fr = 6.92; (**b**) Fr = 7.64; (**c**) Fr = 8.05; (**d**) Fr = 9.25; (**e**) Fr = 11.24.

Among the 30 cases, for each Froude number, the minimum dimensionless thickness (T) of water cushion (depth of water cushion over depth of stilling basin) varies with the depth of the stilling basin

(Table 2). It is seen that the thickness are different in the stilling basins with different depths with the same Froude number and with different Froude number but same depth. For the same Froude number, the deeper the stilling basin, the thicker the thickness of the water cushion. For the same depth, the bigger the Froude number, the thinner the thickness of the water cushion.

Table 2. Minimum of the dimensionless thickness of the water cushion.

Group	Fr	d = 0	d = 5 cm	d = 10 cm	d = 15 cm	d = 20 cm	d = 25 cm
1	6.92	0	0.08	0.09	0.13	0.15	0.16
2	7.64	0	0.08	0.08	0.13	0.14	0.14
3	8.05	0	0.07	0.09	0.12	0.13	0.13
4	9.25	0	0.06	0.07	0.09	0.11	0.12
5	11.24	0	0.06	0.04	0.08	0.10	0.11

3.3. Energy Dissipation Rate

The energy dissipation [31] rate of the stilling basin could be expressed as Equation (5),

$$e = \frac{E_0 - E_1}{E_0} \tag{5}$$

In which, E_0 and E_1 stand for the energy of the control section at the inlet and outlet of the stilling basin and taken the elevation at the tailrace section as the reference elevation in Figure 1.

Then the unit volume energy dissipation rate could be defined by Equation (6),

$$\omega = \frac{e}{\eta} \tag{6}$$

where, ω is the unit volume (dimensionless) energy dissipation rate; $\eta = \Omega/\Omega_0$ is dimensionless volume; $\Omega = LdB$ is the volume of the stilling basin, L, d and B are the length, the depth and the width of the stilling basin respectively, Ω_0 is the volume of the water between inlet and outlet of the stilling basin.

The energy dissipation rates of the 30 cases were calculated and listed in Table 3, in which, column 1 is the Froude number at inlet of the stilling basin; column 2 is the depth of the stilling basin; column 3 is the ratio of the depth to length of the stilling basin (the total length is 120 cm); columns 4 to 10 are the mean speed of different cross sections (the heads of 0, 20, 40, 60, 80, 100, and 120 indicate the distance of the cross sections away from the inlet of the stilling basin entrance with the unit of cm); column 11 is the energy dissipation rate; column 12 is the dimensionless volume of the stilling basin; and column 13 is the unit volume energy dissipation rate.

Using columns 2, 11, and 13 in Table 3, the energy dissipation rate and the unit volume energy dissipation rate related with the depth of the stilling basin were calculated and depicted in Figure 9 with the different Froude number as a parameter.

Generally speaking, the total energy dissipation (e) increases rapidly with the depth of the basin and changes little after reaching a certain value with the depth (Figure 9a); whereas, when the Froude number is smaller, the unit volume energy dissipation rate goes high with the increase of the depth first, and then drops down after reaching the maximum value, which means some of the water body does not involve the energy dissipation (Figure 9b); when the Froude number is larger, it goes higher with the increase of the depth. It also shows that the total energy dissipation reaches the maximum but the unit volume energy dissipation rate may not reach its maximum value, which indicates the depth of the basin is not good enough for the hydraulic jump; for example, the unit volume energy dissipation rate keep increasing with Froude number in the cases of Froude numbers equal to 9.25 and 11.24 which means the depth of 25 cm is sufficient for them getting to the maximum unit volume energy dissipation rates in the study.

Table 3. The unit volume energy dissipation rate of the stilling basin.

Fr (1)	d (cm) (2)	d/L (3)	V (m/s) at Different Sections							e (11)	η (12)	ω (13)
			0 (4)	20 (5)	40 (6)	60 (7)	80 (8)	100 (9)	120 (10)			
7.64	0.00	0.00	4.25	4.25	4.28	4.18	4.13	4.10	4.01	10.98%	0.00	/
	5.00	0.04	4.19	3.87	4.30	4.20	4.20	4.05	3.80	17.75%	0.76	0.30
	10.00	0.08	3.60	3.30	3.50	3.70	3.81	3.40	2.90	35.11%	0.72	0.55
	15.00	0.13	2.84	3.15	2.30	1.48	1.36	1.25	1.20	82.15%	0.56	1.46
	20.00	0.02	2.60	1.55	1.33	0.96	1.09	0.99	0.96	86.37%	0.43	2.10
	25.00	0.21	1.66	1.75	1.23	0.76	0.77	0.68	0.56	88.62%	0.45	2.05
6.92	0.00	0.00	4.37	4.38	4.37	4.29	4.22	4.20	4.14	10.25%	0.00	/
	5.00	0.04	4.32	4.17	4.11	4.08	3.90	3.90	3.83	21.40%	0.54	0.25
	10.00	0.08	4.07	3.89	3.85	3.88	3.80	3.50	3.03	44.58%	0.62	0.50
	15.00	0.13	4.03	3.85	3.13	2.77	2.15	1.53	1.36	88.61%	0.48	1.85
	20.00	0.02	3.30	1.73	1.55	1.13	0.91	0.73	0.98	91.18%	0.44	2.20
	25.00	0.21	2.30	1.83	1.44	1.16	0.88	0.72	0.66	91.77%	0.46	1.99
8.05	0.00	0.00	7.78	7.63	7.54	7.43	7.51	7.46	7.40	9.53%	0.00	/
	5.00	0.04	7.90	7.55	7.49	7.43	7.35	7.30	7.26	15.55%	0.42	0.37
	10.00	0.08	7.55	7.45	7.29	6.98	6.85	6.33	6.25	31.47%	0.42	0.60
	15.00	0.13	7.53	7.20	6.70	6.50	6.29	5.94	5.09	54.31%	0.46	1.19
	20.00	0.02	5.81	5.01	4.53	4.01	3.82	3.15	2.85	75.94%	0.36	2.01
	25.00	0.21	4.73	4.16	3.86	3.38	3.19	2.65	2.23	77.77%	0.37	2.10
9.25	0.00	0.00	10.39	10.14	10.01	10.02	9.97	9.91	9.88	9.58%	0.00	/
	5.00	0.04	10.36	9.89	9.93	9.76	9.71	9.64	9.63	13.60%	0.27	0.46
	10.00	0.08	10.20	9.77	9.53	9.52	9.44	9.29	8.61	28.75%	0.45	0.65
	15.00	0.13	10.42	9.56	9.44	9.36	9.20	8.09	7.43	49.16%	0.53	1.05
	20.00	0.02	9.44	9.49	9.26	8.79	7.85	6.25	5.30	68.48%	0.43	1.80
	25.00	0.21	7.89	6.18	6.03	5.82	5.33	4.85	3.88	75.82%	0.35	2.27
11.24	0.00	0.00	13.36	13.11	13.12	13	12.92	12.88	12.76	8.78%	0.00	/
	5.00	0.04	13.23	12.96	12.88	12.69	12.43	12.42	12.42	11.87%	0.26	0.50
	10.00	0.08	13.2	12.91	12.8	12.56	12.21	11.52	11.33	26.33%	0.43	0.70
	15.00	0.13	13.19	12.88	12.62	12.12	11.63	11.15	10.07	41.71%	0.31	0.92
	20.00	0.02	13.26	12.56	11.45	10.62	9.53	8.05	7.79	65.49%	0.32	1.60
	25.00	0.21	12.21	11.64	10.88	9.48	8.38	7.23	6.11	74.96%	0.32	2.34

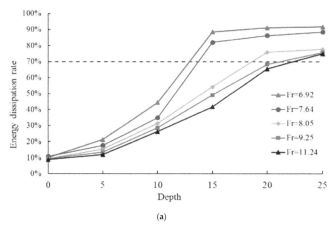

(a)

Figure 9. *Cont.*

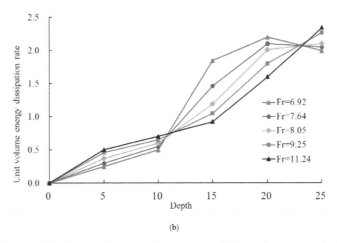

(b)

Figure 9. Energy dissipation rate. (**a**) Energy dissipation rate; (**b**) Unit volume energy dissipation rate.

3.4. The Best Depth of the Stilling Basin

For a given Froude number, it is expected that a complete hydraulic jump occurs in the stilling basin and the velocity near the bottom is relatively low and with a high energy dissipation rate in total and per volume. From the perspectives of flow regime, total energy dissipation rate and bottom velocity, the deeper a stilling basin is, more easily a complete hydraulic jump forms, higher the total energy dissipation rate and farther the main stream away from the bottom of the basin; however, from the perspective of the unit volume energy dissipation rate, the depth of the basin should not be too large, it otherwise is a waste as part of the water body in the basin is not involved in the energy dissipation. In addition, the thickness of the water cushion may be too large to regard the stilling basin as a 'shallow-water cushion' stilling basin and may be turned into a plunge pool which performs very different than the stilling basin with shallow water cushion in energy dissipation. Therefore, according to the results above of flow regime, flow profile, main stream location, energy dissipation, etc., the standards for defining the best depth of the stilling basin with shallow-water cushion can be proposed as (1) a complete hydraulic jump must occur in the basin (2) the thickness of water cushion is about 1/10–1/3 deep of the basin, and (3) the energy dissipation rate is more than 70% and the unit volume energy dissipation rate is as high as possible.

According to the definition for the best depth of SBSWC, three ranges of the depth were depicted in the 5 different Froude numbers respectively.in Figure 10. The blue line stands for the depth range for the standard (1) of flow regime according to the minimum depth for a complete hydraulic jump from Figure 6, and the green line stands for the depth range for the standard (2) of the relationship between the main stream and the thickness of the water cushion in Figure 8 and Table 2, and the red line stands for the depth range for the standard (3) of energy dissipation according to Table 3 and Figure 9. It is shown that all the 3 standards are necessary to obtain the range as neither the minimums nor the maximums of the range can be determined by the same standard alone for different Froude numbers. For example, for cases of Froude numbers equal to 6.92 and 9.25, the minimums are determined by the standards of flow regime and energy dissipation, respectively, while the maximums by energy dissipation and main stream location, respectively.

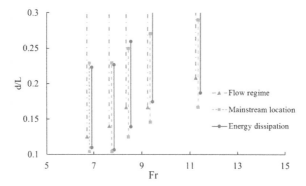

Figure 10. The range of the best depth of the stilling basin under three judgment conditions.

The overlap of all the three ranges for the same Froude number represented that the three standards were met in this integral (the integral between the maximum of the 3 minimums and the minimum of the maximums) and its mean is taken as the best depth of the stilling basin with shallow-water cushion shown in Figure 11 where the dimensionless depth (d/L, the depth over the length of the stilling basin) was adopted. It indicates that the ranges of the best depth are large for some cases while small for some other cases and that the best depth of the stilling basin increases with the increase of the Froude number.

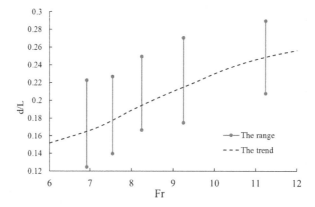

Figure 11. The range of the best depth of the stilling basin.

4. Conclusions

In this study, 30 cases consisting of five different Froude numbers and six depths were selected for which large eddy simulation (LES) was adopted to simulate the hydraulic characteristics in the stilling basin, including the flow regime of hydraulic jump, the location of the main stream and the unit volume energy dissipation rate. The following conclusions can be drawn:

1. After comparing large eddy simulation calculation results and model tests, the computation of the flow profile and hydraulic jumps were similar to the test, and the differences of the average pressure between test results and calculated results were very small except for individual locations. The numerical data and experimental results were in good agreement, indicating the reliability and accuracy of LES in the present work.

2. Based on the experimental results of flow regime of hydraulic jump, the location of the main stream and the energy dissipation rate, for the first time three standards of the best depth of stilling basin with shallow water cushion (SBSWC) are proposed in this paper; namely, (1) a complete hydraulic jump must occur in the basin (2) the thickness of water cushion is about 1/10–1/3 deep of the basin, and (3) the energy dissipation rate is more than 70% and the unit volume energy dissipation rate is as high as possible.

3. According to the definition for the best depth of SBSWC, three ranges of the depth were depicted in the 5 different Froude numbers respectively. Therefore, the overlapping sections are the range of the best depth of stilling basin with shallow water cushion for this study.

4. It showed that the best depth radio of stilling basin with shallow-water cushion (depth to length ratio) was between 0.1 and 0.3 and it also indicated that the best depth of the stilling basin with shallow water cushion increased with the increase of Froude number. The results in this paper are of significance for design and optimizing of SBSWC.

Author Contributions: Q.L. carried out the model simulations and result analysis as well as writing of the manuscript. L.L. provided the prototype field data and relevant engineering documents. H.L. took the leadership of whole project and participated in the discussion and decision-making process. All the authors participated and contributed to the final manuscript.

Funding: This research work was supported by the National Key Research Development Plan (No. 2016YFC0401705), National Natural Science Foundation of China (No. 51209154), (No. 51079091), and Sichuan Youth Innovation Team Project (No. 2016TD0020).

Conflicts of Interest: The authors declare no conflict of interest.

References

1. Chanson, H. Current knowledge in hydraulic jumps and related phenomena. A survey of experimental results. *Eur. J. Mech. B Fluids* **2009**, *28*, 191–210. [CrossRef]

2. Alikhani, A.; Behrozi-Rad, R.; Fathi-Moghadam, M. Hydraulic jump in stilling basin with vertical end sill. *Int. J. Phys. Sci.* **2010**, *5*, 25–29. Retrieved from http://www.academicjournals.org/IJPS.

3. Chanson, H. Hydraulic condition for undular-jump formations. *J. Hydraul. Res.* **2001**, *39*, 203–209. [CrossRef]

4. Zare, H.K.; Doering, J.C. Forced Hydraulic Jumps below Abrupt Expansions. *J. Hydraul. Eng.* **2011**, *137*, 825–835. [CrossRef]

5. Champagne, T.M.; Barlock, R.R.; Ghimire, S.R.; Barkdoll, B.D.; Gonzalez-Castro, J.A.; Deaton, L. Scour Reduction by Air Injection Downstream of Stilling Basins: Optimal Configuration Determination by Experimentation. *J. Irrig. Drain. Eng.* **2016**, *142*. [CrossRef]

6. Tianxiang, L.I. Investigation on Hydraulic Performances of Stilling Basin with Shallow-Water Cushion. Master's Thesis, State Key Laboratory of Hydraulics and Mountain River R & D in Sichuan University, Chengdu, China, 2006.

7. Li, L.X.; Liao, H.S.; Liu, D.; Jiang, S.Y. Experimental investigation of the optimization of stilling basin with shallow water cushion used for low Froude number energy dissipation. *J. Hydrodyn. B Ser.* **2015**, *27*, 522–529. [CrossRef]

8. Liu, D.; Li, L.X.; Huang, B.S.; Liao, H.S. Numerical Simulation and Experimental Investigation on Stilling Basin with Double Shallow-water Cushions. *J. Hydraul. Eng.* **2012**, 623–630. [CrossRef]

9. Hager, W.H.; Kawagoshi, N. Hydraulic jumps at rounded drop. *Proc. Inst. Civ. Eng. Part Res. Theory* **1990**, *89*, 443–470. [CrossRef]

10. Eroğlu, N.; Tokyay, N. Statistical approach to geometric properties of wave-type flow occurring at an abrupt drop. *J. Fac. Eng. Archit. Gazi Univ.* **2012**, *27*, 911–919.

11. Rice, C.E.; Kadavy, K.C. Riprap design upstream of straight drop spillways. *Trans. Asae* **1992**, *34*, 1715–1725. [CrossRef]

12. Ram, K.V.; Prasad, S.; Spatial, R. B-Jump at Sudden Channel Enlargements with Abrupt Drop. *J. Hydraul. Eng.* **1998**, *124*, 643–646. [CrossRef]

13. Sabzkoohi, A.M.; Kashefipour, S.M.; Bina, M. Investigation of effective parameters on stepped and straight drops energy dissipation using physical modeling. *J. Food Agric. Environ.* **2011**, *9*, 748–753. Retrieved from https://www.researchgate.net/publication/286803391.

14. Riazi, R.; Bejestan, M.S.; Kashkouli, H.; Khosrojerdi, A. Effect of roughness on characteristics of bed B-jump in stilling basin with abrupt drop. *Res. Crops* **2012**, *13*, 1137–1141. Retrieved from https://www.researchgate.net/publication/287059975.

15. Ilie, M. Fluid-structure interaction in turbulent flows; a CFD based aeroelastic algorithm using LES. *Appl. Math. Comput.* **2018**, *342*, 309–321. [CrossRef]

16. Wanik, A.; Schnell, U. Some remarks on the PISO and SIMPLE algorithms for steady turbulent flow problems. *Comput. Fluids* **1989**, *17*, 555–570. [CrossRef]

17. Wang, T.; Gu, C.G.; Yang, B. PISO algorithm for unsteady flow field. *J. Hydrodyn.* **2003**. [CrossRef]

18. Moeng, C.H. A Large-Eddy-Simulation Model for the Study of Planetary Boundary-Layer Turbulence. *J. Atmos. Sci.* **1984**, *41*, 2052–2062. [CrossRef]

19. Moin, P.; Kim, J. Numerical investigation of turbulent channel flow. *J. Fluid Mech.* **2006**, *118*, 1280–1284. [CrossRef]

20. Sarfaraz, M. Numerical Computation of Inception Point Location for Steeply Sloping Stepped Spillways. In Proceedings of the 9th International Congress on Civil Engineering, Isfahan University of Technology (IUT), Isfahan, Iran, 18–20 December 2012.

21. Smagorinsky, J.S.; Smagorinsky, J. General Circulation Experiments with the Primitive Equations. *Mon. Weather Rev.* **1963**, *91*, 99–164. [CrossRef]

22. Song, C.C.S. A weakly compressible flow model and rapid convergence methods. *J. Fluids Eng.* **1988**, *110*, 441–445. [CrossRef]

23. Canuto, V.M.; Cheng, Y. Determination of the Smagorinsky–Lilly constant CS. *Phys. Fluids* **1997**, *9*, 1368–1378. [CrossRef]

24. Issa, R.I. Solution of the Implicitly Discretised Fluid Flow Equations by Operator-Splitting. *J. Comput. Phys.* **1986**, *62*, 40–65. [CrossRef]

25. Hirt, C.W.; Nichols, B.D.; Romero, N.C. *SOLA: A Numerical Solution Algorithm for Transient Fluid Flows (LA-5852)*; Los Alamos Scientific Laboratory: Los Alamos, NM, USA, 1988. [CrossRef]

26. Nichols, B.D.; Hirt, C.W.; Hotchkiss, R.S. *SOLA-VOF: A Solution Algorithm for Transient Fluid Flow with Multiple Free Boundaries (LA-8355)*; Nasa Sti/recon Technical Report N.; Los Alamos Scientific Laboratory: Los Alamos, NM, USA, 1980.

27. Nichols, B.D.; Hirt, C.W.; Hotchkiss, R.S. A fractional volume of fluid method for free boundary dynamics. In *Seventh International Conference on Numerical Methods in Fluid Dynamics*; Springer: Berlin/Heidelberg, Germany, 1981.

28. Mossa, M.; Petrillo, A.; Chanson, H. Tail-water level effects on flow conditions at an abrupt drop. *J. Hydraul. Res.* **2003**, *41*, 39–51. [CrossRef]

29. Felder, S.; Chanson, H. Air-water flow patterns of hydraulic jumps on uniform beds macroroughness. *J. Hydraul. Eng.* **2017**, *144*. [CrossRef]

30. Pagliara, S.; Palermo, M. Hydraulic jumps on rough and smooth beds: Aggregate approach for horizontal and adverse-sloped beds. *J. Hydraul. Res.* **2015**, *53*, 243–252. [CrossRef]

31. Palermo, M.; Pagliara, S. Semi-theoretical approach for energy dissipation estimation at hydraulic jumps in rough sloped channels. *J. Hydraul. Res.* **2018**. [CrossRef]

Article

Design of A Streamwise-Lateral Ski-Jump Flow Discharge Spillway

Jun Deng, Wangru Wei *, Zhong Tian and Faxing Zhang

State Key Laboratory of Hydraulics and Mountain River Engineering, Sichuan University, Chengdu 610065, China; djhao2002@scu.edu.cn (J.D.); tianzhong@scu.edu.cn (Z.T.); zhfx@scu.edu.cn (F.Z.)
* Correspondence: weiwangru@scu.edu.cn

Received: 11 September 2018; Accepted: 1 November 2018; Published: 6 November 2018

Abstract: Spillway outlet design is a major issue in hydraulic engineering with high head and large discharge conditions. A new type of design for a streamwise-lateral spillway is proposed for ski-jump flow discharge and energy dissipation in hydraulic engineering. The water in the spillway outlet is constrained by three solid walls with an inclined floor, a horizontal floor on the bottom and a deflected side wall in the lateral direction. The water flow releases in a lateral direction into the plunge pool along the streamwise direction. It generates a free jet in the shape of "∩" in a limited area, causing the water to fully diffuse and stretch in the air simultaneously, and drop into the plunge pool to avoid excessive impact in the plunge pool. The formation mechanism for the flow pattern is analyzed, and the results show that the optimum inclination is an angle range of 30°~45° for a good performance of free ski-jump jet diffusion shape.

Keywords: ski-jump flow; spillway outlet; jet trajectory; energy dissipation; experimental model

1. Introduction

Energy dissipation is always a major issue in hydraulic engineering [1,2], especially for large discharge and complex geological conditions. Ski-jump energy dissipation has prevailed among the hydraulic engineering, considering the construction cost and dissipation performance [3–5]. The characteristics of this ski-jump are the free jets diffusing and aerating in the air [6,7] and then dropping into the water downstream, causing energy dissipation through shearing and disturbance. In the traditional design, water flow discharges in the spillway or steep chute and a flip bucket is set at the end of the spillway to make the water release into the flow [8,9]. The various geometries of flip buckets include a wide range of deflection angles, radius, takeoff angles, etc. [10–14].

In order to improve the energy dissipation rate and reduce the impact on the bottom of the plunge pool, the free jet is always made to widely diffuse in the air in both streamwise and lateral directions, and is dropped far away from the dam for safety [15,16]. Given the enormous flow discharge and velocity, the common ski-jump is not efficient, considering the energy dissipation performance and other damaging problems. In order to make the ski-jump flow diffuse in a certain direction, researchers and engineers developed a series of structure design to drive the water jet. The specific structures [17–19], such as a contract side wall or a bucket radius of flip bucket, can adjust the water jet to a certain direction. However, when the flow velocity exceeds 30–50 m/s, the hydrodynamic pressures on the structure is another important issue, which is extremely dangerous to outlet structure safety. The recent study on the flip bucket proposed a new design principle that is take full advantage of the natural pressure difference at the lateral direction to make more contribution, based on the motion characteristic of high-speed flow. The well performance of long-narrow ski-jump flow on the Jinping-I flood discharge tunnel showed much lower pressure on the structure of a leak-floor flip bucket than that of a lateral contraction of side-wall design. However, there have been limited applications of flip buckets for two reasons: (1) dissipation area and (2) free jet diffusion shape. For some projects, the dissipation area is limited, such as in valleys and in unsatisfactory souring conditions of river courses. The free jet diffusion shape influences the aeration and impact performance directly, and the effect is pivotal in the limited dissipation area.

In the present study, a new spillway design for a ski-jump streamwise-lateral discharge spillway is proposed for the two stated reasons. The spillway side wall close to the plunge pool is removed and an inclined floor is set along the streamwise direction. This makes the water flow release laterally as it moves downstream, and the free jet shape makes the most of the space as it diffuses in the air. The formation mechanism and characteristics of the water jet are presented, based on experiments and numerical simulation.

2. Physical Structure of the Streamwise-Lateral Discharge Spillway

The design of the streamwise-lateral discharge spillway is shown in Figure 1. The spillway is set on the right side of the plunge pool and the axis of the spillway is parallel to the plunge pool in the streamwise direction. The side wall of the spillway closed to the plunge pool (left wall in the present study) is removed (in contrast to the traditional spillway) and an inclined floor is set between the horizontal bottom floor and the right-side wall. An anti-arch design is connected to the straight part of the side wall, which is designed to deflect water into the plunge pool. The intersection part between the horizontal floor and the inclined floor is the diagonal line of the spillway. Here, X is the streamwise coordinate which is perpendicular to the cross-section Y-Z, where Y is the lateral coordinate and Z is the vertical coordinate, respectively.

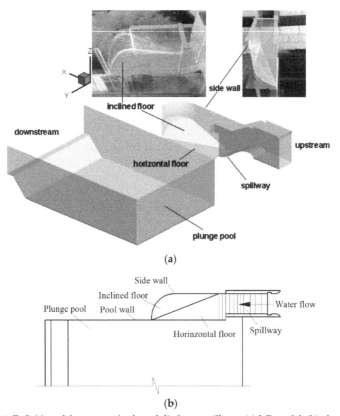

(a)

(b)

Figure 1. Definition of the streamwise-lateral discharge spillway: (**a**) 3-D model; (**b**) plan view.

The typical flow pattern of the physical model is shown in Figure 2. When the water flow moves downstream in the spillway, the pressure on the left side is reduced to the atmospheric pressure, while the pressure on the right side remains the hydrodynamic pressure. There is a pressure difference due to the specific design, and the water will flow transversely into the plunge pool and the streamwise-lateral discharge will be generated. When the water flow reaches the end of the spillway, the water is restrained by the inclined floor and the anti-arch side wall, and the free jet is then generated and stretches in the air above the plunge pool. The free jet diffuses pronouncedly in the transverse direction above the plunge pool in the very limited area, showing a "curling type" in the air. This will make a beneficial contribution to the energy dissipation through three aspects: (1) The water thickness is thin and the flow discharge per unit width dropping into the plunge pool is small; (2) the aeration of the free jet is fully developed as it diffuses in the air; and (3) the ratio of the shearing dissipation area in the plunge pool increases.

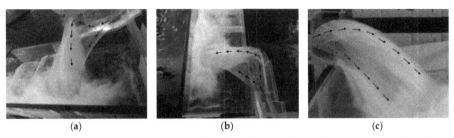

| (a) | (b) | (c) |

Figure 2. Typical pattern of the streamwise-lateral discharge spillway: (**a**) physical model from the downstream view; (**b**) physical model from the lateral view; (**c**) ski-jump flow.

To obtain the detained characteristics of configuration and diffusion of the streamwise-lateral ski-jump flow discharge spillway, it is practically impossible to see the jet configuration due to the failure of exiting measurements on the complex 3-D flow. The physical experiments can only provide some jet length and pressure distribution on the floor in order to describe the jet flow pattern. However, the numerical simulation can provide the entire jet field with detailed water profile and velocity distribution in the jet diffusion process. Thus, the physical model test and numerical simulation were used in this paper to systemically study the new design of streamwise-lateral ski-jump flow discharge spillway.

3. Investigation Methods and Definition of the Free Jet Flow Pattern

3.1. Physical Model

The study of the formation of the streamwise-lateral flow discharge spillway is based on energy dissipation in a certain hydropower station. The size of the plunge pool is 88 (length) $\times$ 100 (width) m^2, and the vertical distance from the horizontal floor to the bottom wall of the plunge pool is $H_0 = 43$ m. The length of the horizontal floor $L = 55$ m, with $W = 21$ m in width. The straight length of the side wall is $L_b = 36.4$ m, with the anti-arch radius $R = 19$ m and the deflection angle $\theta = 96°$ in the anti-arch structure. The angle α of the inclined floor ranges from $0°$ to $60°$, as shown in Figure 3. The flow discharge is 4016 m^3/s, with discharge per unit width $q_w = 251$ m^3/(s·m). The tail water depth is 18 m deep. The scale of a normal model is 1:60 based on the Froude criterion for experimental study. Considering that the roughness coefficient of the concrete in the prototype is 0.0140 and the roughness scale is $60^{1/6} = 1.979$, the theoretical value of the roughness coefficient is 0.0071. The physical model is made of transparent plexiglass with a roughness coefficient of 0.0079. The present physical model is approximately satisfactory for the similarity of boundary roughness between the model and prototype.

Considering the 1:60 scale of the physical model, the approach flow Froude number $Fr = V/(gh)^{0.5}$ is about 3.73 with a Reynolds number $Re = Vh/\nu$ of 1.83×10^5, where ν is the kinematic water viscosity,

g is the gravitational acceleration, and $V = 3.2$ m/s and $h = 0.075$ m are the mean velocity and depth of the approach flow on the horizontal floor, respectively. Thus, the present study focuses on the typical pattern of ski-jump flow in a streamwise-lateral discharge spillway, jet trajectory and distribution of mean pressure on the bottom floor of a basin, which satisfies the Froude criterion. The scale effect on air–water properties of supercritical flow cannot be neglected with regards to the surface tension and viscosity effects in high-speed flows [20].

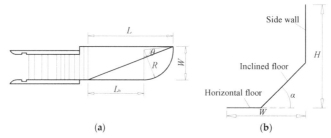

(a)   (b)

Figure 3. Sketch with relevant parameters: (**a**) plan view; (**b**) sectional view.

A thin plate wire with a graduated cylinder was set downstream for a water flow discharge measurement to the nearest 0.1 L/s, and the water level was read with a conventional point gauge to an accuracy of 0.1 mm. The jet trajectory of the ski-jump flow was measured at about 5 min, ensuring it did not change much with the time between observations. The pressure signal was measured using a series of pressure sensors (CY201, Test, Inc.: Chengdu, China) and an acquisition device (RS485-20, Test, Inc.: Chengdu, China). The accuracy of the pressure sensor was 0.1% of the 100 kPa full scale. The pressure time–domain process was sampled at 100 Hz for 10 s, which was adequate to evaluate the pressure fluctuation under the present hydraulic conditions. The pressure measurement system was calibrated by hydrostatic operation before each test was conducted to ensure measurement reliability. The time mean pressure P was obtained using the time–domain process p, and the maximum mean pressure p_m was determined as $p_m = P - P_0$.

3.2. Numerical Simulation

In the present work, the Fluent computer code (Fluent Inc.: Pittsburgh, PA, USA, 2005) was used to simulate the ski-jump flows of the streamwise-lateral discharge spillway. The Renormalization Group k-ε turbulent model was employed, which is an effective method to simulate the ski-jump flow movement in detail [21]. The governing equations are as follows:

Continuity equation:

$$\frac{\partial \rho}{\partial t} + \frac{\partial \rho u_i}{\partial x_i} = 0 \tag{1}$$

Momentum equation:

$$\frac{\partial \rho u_i}{\partial t} + \frac{\partial \left(\rho u_i u_j \right)}{\partial x_j} = -\frac{\partial p}{\partial x_i} + \frac{\partial}{\partial x_j}\left[(\mu + \mu_t) \times \left(\frac{\partial u_i}{\partial x_j} + \frac{\partial u_j}{\partial x_i} \right) \right] \tag{2}$$

k (turbulent kinetic energy) equation:

$$\frac{\partial \rho k}{\partial t} + \frac{\partial (\rho u_i k)}{\partial x_i} = \frac{\partial}{\partial x_i}\left[\left(\mu + \frac{\mu_t}{\sigma_k} \right) \times \frac{\partial k}{\partial x_i} \right] + G_k - \rho \varepsilon \tag{3}$$

ε (dissipation rate of turbulent kinetic energy) equation:

$$\frac{\partial \rho \varepsilon}{\partial t} + \frac{\partial (\rho u_i \varepsilon)}{\partial x_i} = \frac{\partial}{\partial x_i}\left[\left(\mu + \frac{\mu_t}{\sigma_\varepsilon} \right) \times \frac{\partial \varepsilon}{\partial x_i} \right] + C_{1\varepsilon}\rho \frac{\varepsilon}{k}G_k - C_{2\varepsilon}\rho \frac{\varepsilon^2}{k} \tag{4}$$

where t is time. ρ and μ are the average density of the volume fraction and molecular viscosity, respectively. p is the pressure and G_k represents the generation of turbulence kinetic energy due to the mean velocity gradients. μ_t is the turbulent viscosity which can be deduced for the turbulence intensity k and energy dissipation rate ε:

$$\mu_t = C_\mu \frac{k^2}{\varepsilon}, \; C_{1\varepsilon} = 1.42 - \frac{\eta\left(1 - \frac{\eta}{\eta_0}\right)}{1 + \beta\eta^3}, \; \eta = \frac{Sk}{\varepsilon}, \; S = \sqrt{2\overline{S_{ij}} \, \overline{S_{ij}}}, \overline{S_{ij}} = \frac{1}{2}\left(\frac{\partial u_i}{\partial x_j} + \frac{\partial u_j}{\partial x_i}\right) \tag{5}$$

where S is the modulus of the mean strain rate tensor. The detailed values are shown in Table 1.

The Volume of Fluid (VOF) method was used to simulate the free surface and air fraction of ski-jump flow. The governing equation for water fraction α_w is:

$$\frac{\partial \alpha_w}{\partial t} + u_i \frac{\partial \alpha_w}{\partial x_i} = 0 \tag{6}$$

$$\rho = \alpha_w \rho_w + (1 - \alpha_w)\rho_a \tag{7}$$

$$\mu = \alpha_w \mu_w + (1 - \alpha_w)\mu_a \tag{8}$$

where u_i is the velocity components and x_i is the coordinates (i = 1, 2, 3). j is the sum suffix. The air–water interface is tracked by solving the continuity equations. The sum of the water and air volume fraction was one in the controlling body. The ρ_w and ρ_a are the density of water and air, respectively. The μ_w and μ_a are the viscosity of water and air, respectively. When $\alpha_w = 0$, it means there is full of air; when $\alpha_w = 1$, it represents the air fraction of fluid is 0. When $0 < \alpha_w < 1$, it means the certain controlling volume. The total fraction in a certain controlling body is $\alpha_w + \alpha_a = 1$. Note that, although the VOF model can give an aeration result in the free surface area, the air concentration distribution and aeration development along the jet trajectory is different to the real situations due to the complex air–water interaction and bubble diffusion process. In the present study, the simulated results will be used for the analysis on formation of ski-jump flow released from the streamwise-lateral spillway, which is little affected by the aeration. The detailed hydraulic characteristics, including jet trajectory and impact pressure on the pool floor, will be analyzed based on the physical model results. The control volume method is introduced to discretize the partial differential equations, and the SIMPLER method, which has good convergence characteristics, is employed for numerical simulation [22,23].

In the present research, both structured and unstructured meshes were used. Unstructured mesh was employed in the streamwise-lateral spillway area due to the complex geometry condition. Structural mesh was used for other regions. Mesh size was refined until the computed results were independent of the size of the scale, based on the fact that the calculated water profile and the pressure on the spillway sidewall did not change more than 3%. The uncertainty analysis of different mesh densities showed that the number of meshes was reasonably set to 1,017,033. The approximate $0.5 \times 0.5 \text{ m}^2$ mesh size was selected to simulate the pattern of ski-jump flow in the air and its diffusion in the still basin. Based on the authors' previous studies [23–25] on the simulation of high-speed flows (flow velocity exceeded over 30 m/s) in hydraulic engineering, these mesh conditions can make sure the maximum uncertainties for the velocity and pressure characteristics in the whole calculated region were approximately smaller 10%, and the uncertainty in most locations was fairly small. The comparison between the experimental and simulated results with regards to the flow characteristics of the streamwise-lateral discharge spillway are shown in Table 2 for $\alpha = 45°$, where $Y = 0$ is the lateral cross-section at the straight side wall. The distributions of pressure on the side wall and water depth indicate that the simulation results are well in agreement with those of experimental measurements. The vector profile of the flow jet velocity in Figure 4 shows similar flow patterns for the experiments. According to the typical flow pattern of the streamwise-lateral discharge spillway, it is difficult to acknowledge the interior characteristics through experimental measurement, due to the irregular "curling" nappe shape and the complicated flow jet movement in the diffusion process. Because the computational water profiles and the pressure on the side wall are well in agreement with the test measurements, the flow patterns are similar between the simulation and the experiment and the complex high-speed fluid can be analyzed with the numerically simulated results. The characteristics of the jet profile and the pressure from the physical model were converted to the prototype according to the length scale. The simulated model was built up based on the real dimensions in the prototype, and the velocity distribution inside the ski-jump flow and stilling basin is the same as the prototype's dimensions.

Table 1. Constants of k-ε equations.

Parameters	η_0	β	C_μ	$C_{1\varepsilon}$	$C_{2\varepsilon}$	σ_k	σ_ε
Value	4.38	0.012	0.0845	1.42	1.68	0.7179	0.7179

Table 2. Comparison of tested and calculated data on the spillway flood discharge ($\alpha = 45°$, $q_w = 251$ m^2/s).

Pressure on the Sidewall of Spillway (m)						Water Depth along the Side Wall of Spillway (m)		
$Y = 15$ m			$Y = 18$ m					
H	Simulation	Test	H	Simulation	Test	X	Simulation	Test
8.9	23.7	23.4	6.7	25.4	24.3	4.1	11.1	10.3
11.3	17.0	17.6	9.2	19.6	19.1	14.1	10.5	10.3
12.4	13.6	12.6	11.7	13.4	13.6	23.8	11.3	11.2
14.8	8.5	9.7	15.3	6.3	5.6	33.1	13.0	13.0
17.1	4.8	4.5	17.8	2.4	2.7	41.5	15.4	14.6
19.5	1.9	1.6	21.5	1.4	0.9	49.5	19.7	19.3
23.0	0.5	0.6	22.8	1.0	0.9	55.1	25.5	25.4

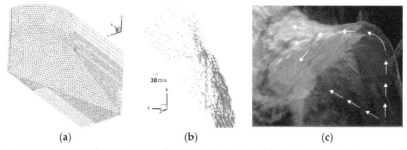

| (a) | (b) | (c) |

Figure 4. Comparison of numerical simulation with experimental flow: (**a**) a sample of computational mesh in the spillway area; (**b**) velocity vector profile; (**c**) experimental image.

4. Results and Discussion

4.1. Ski-Jump Flow Pattern

Figure 5 shows the simulated results about the water profile and velocity vector distributions of the ski-jump flow released from the streamwise-lateral spillway. At the initial area closed to upstream, the pressure near the left side reduces to atmosphere pressure without the left-side wall, while the water near the right side is constrained by the wall. Thus, the water on the left side moves laterally at the cross-section, and releases from the horizontal floor, flowing towards the plunge pool. Meanwhile, water flow on the right side moves along the streamwise direction and "climbs" along the inclined floor to the side wall of the spillway until it reaches the end of the spillway. Due to the coupling constraint of the inclined floor and the anti-arch wall in both streamwise and lateral directions, a thin and widely stretched nappe shape forms in the air, dropping into the plunge pool. In addition, the water moving laterally continues to move in the streamwise direction. Both of the movements contribute to the formation of three-dimensional ski-jump flow.

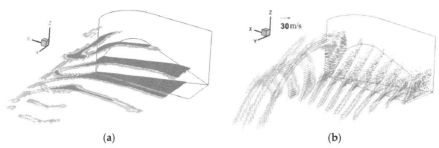

(a)   (b)

Figure 5. Jet diffusion of streamwise-lateral ski-jump flow (simulated results for $\alpha = 45°$): (**a**) water profile at different elevations; (**b**) velocity vector profile at different cross-sections.

Figure 6 shows the distributions of the detailed flow velocity vector in the air for various elevations. When the water flow leaves the spillway, the jet flow turns to the transverse direction pronouncedly under the effect of coupling constrained by the inclined floor and the anti-arch wall. In addition, the lateral discharge becomes smaller and the streamwise "curling" discharge increases with the decrease of elevation. The "∩" shape can clearly be seen. The water jet becomes fully stretched in the limited area before dropping into the plunge pool. The thickness of the flow nappe becomes thinner and thinner in the air. This greatly contributes to the aeration and air–water mixture of the flow jet in the diffusion and stretching process, which will accelerate flow velocity attenuation and improve shearing intensity in the plunge pool.

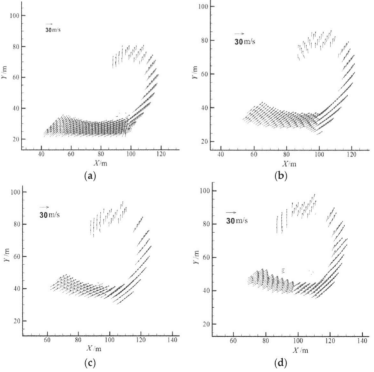

Figure 6. Velocity vector distributions of free jet flow: (**a**) $Z = -17$ m; (**b**) $Z = -22$ m; (**c**) $Z = -27$ m; (**d**) $Z = -32$ m.

4.2. Effect of Inclined Floor on Jet Length

To describe the free jet trajectory from the experimental results, the nearer jet length L_1 and the further jet length L_2 are introduced, as shown in Figure 7. The further jet length, L_2, is the longest distance in the lateral direction mainly affected by the coupling effect of the inclined floor and the anti-arch side wall of the spillway; and the nearer jet length, L_1, is the lateral jet length distance and is affected mainly by the pressure difference in the horizontal floor. The coefficient $\beta = (L_2 - L_1)/H_0$ describes the free jet diffusion in the air, deflecting from the spillway. The increase of β indicates that the diffusion performance improves and the stretch of water jet in the air space becomes fully developed. The angle α of the inclined floor influences L_1 and L_2 pronouncedly for the otherwise identical conditions. As shown in Figure 8a,b, with the increase of α to 30°, both L_1 and L_2 remain unchanged, which indicates that the effect of the inclined floor on the jet length is not distinct. As α further increases to 45°, L_1 increases while L_2 still remains significantly unchanged. The value of β decreases slightly which indicates that the water flow pattern is stable, as shown in Figure 8c. For $\alpha > 45°$, L_1 further increases while L_2 decreases pronouncedly. This results in a significant reduction in β, which indicates that the diffusion of the jet in the air is not sufficient in the lateral direction. Besides, the further jet length is around five times greater than the nearer jet length for $\alpha < 30°$, and when $\alpha = 60°$, the further jet length is around four times greater than the nearer jet length, indicating that the diffusion of the ski-jump flow nappe remained stable. Based on the effect of the inclined angle on the jet length, it is recommended that α is 30~45° for a good performance of free ski-jump jet diffusion.

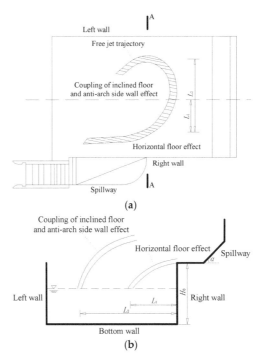

Figure 7. Sketch of free jet flow deflecting from the streamwise-lateral discharge spillway: (**a**) plane view; (**b**) lateral view (A–A).

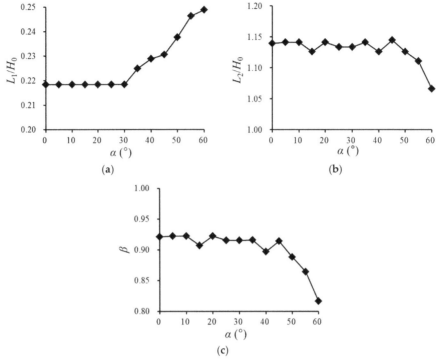

Figure 8. Effect of inclined angle α on free jet length (**a**): L_1; (**b**): L_2; (**c**): β.

4.3. Effect of Inclined Floor on Jet Impact Pressure

The pressure on the bottom floor of the plunge pool is measured experimentally in six areas, Area 1 to Area 6, based on the flow plane-shape, as shown in Figure 9. The maximum impact pressure p_m (defined as the difference between the measured and the mean static pressure in the plunge pool), scaled by the tail water depth P_0, is strongly affected by the inclined floor angle. When $\alpha = 0°{\sim}30°$, the increase of p_m/P_0 is not obvious. While for $\alpha > 30°$, the p_m/P_0 becomes larger with an increase of α. This is mainly due to the poor effect of the inclined floor constraint on the water flow streamline for a small α condition. Considering that the flow impinging on the anti-arch side wall results in a large amount of energy dissipation, the flow jet's impact pressure on the bottom floor of the plunge pool is relatively low. For a large α, the water flow is well coupling-constrained by both the inclined floor and the anti-arch side wall, and the streamline varies relatively smoothly when it becomes totally free from the streamwise-lateral discharge spillway. Larger α causes a lower energy dissipation in the spillway part. Consequently, the flow energy is mainly transported into the plunge pool, resulting in a high impact pressure on the bottom floor.

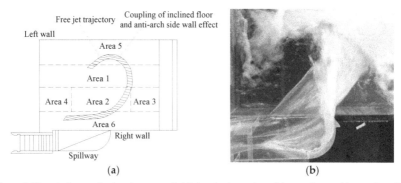

Figure 9. Flow jet impact in the plunge pool: (**a**) sketch plane view; (**b**) experimental image ($\alpha = 60°$).

The specific maximum impact pressure, p_{m}, in each area of the plunge pool describes the pressure and water discharge rate distributions through the stretched flow jet. For different inclined floor angles, the pressure distribution is obviously different, owing to the non-uniform flow jet stretching, as shown in Figure 10. For $\alpha < 30°$, the pressure profiles distribute smoothly in the whole plunge pool without obvious impact pressure on the bottom floor. The flow jets mainly drop into Area 1, Area 2 and Area 3, which are located in the center of the plunge pool. When α increases to $45°$, the flow dropping areas move to Area 1, Area 2 and Area 6, with an obvious increase of impact pressure. This indicates that more water discharge moves in a lateral direction, which is affected by the inclined floor. For $\alpha > 45°$, the pressure is extremely non-evenly distributed, with greater impact pressure in Area 6. When $\alpha = 50°\sim60°$, the local impact pressure is near 1.5~2.0 times greater than the mean static pressure in the plunge pool, and this will cause the bottom floor structure to be affected by negative unbalanced force conditions and may cause structural damage with flood discharge.

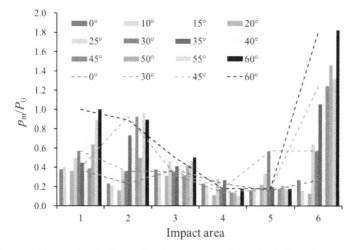

Figure 10. Pressure distributions in the plunge pool affected by the inclined floor angle α.

The ski-jump flow released from the streamwise-lateral spillway can be classified basically into three types. The classification is on the basis of the jet length and impact pressure corresponding to the effect of inclined floor and includes the following: (a) Weak effect: $\alpha < 30°$. The restraint of inclined floor on the ski-jump flow is weak and the jet is mainly affected by the anti-arch radius, which results in the unchanged of jet length (including both nearer and further jet length) and the distribution of impact pressure in the plunge pool. (b) Moderate effect: $30° < \alpha < 45°$. The restraint of the inclined floor becomes obvious and the jet diffusion is affected by both the inclined floor and anti-arch radius of side-wall. The nearer jet length increases with the distribution of impact pressure in the plunge pool is moderate. (c) Strong effect: $\alpha > 45°$. The restraint of inclined floor becomes significant and the stretch of water jet in the air becomes undeveloped. The distribution of impact pressure is extremely asymmetrical with high pressure in the region closed to the spillway.

Figure 11 shows the velocity vector distributions in the plunge pool when the jet flow drops into the pool ($\alpha = 45°$). It is clearly visible that two stable vortexes generate at the nearer and further flow dropping locations in the central part of the plunge pool. With the decrease of elevation, the vortexes' locations move slightly to the left side and the intensive velocity of the vortex attenuates (Figure 11a–c), dispersing near the pool bottom (Figure 11d). Moreover, the velocity attenuates quickly in both the X and Y directions. In terms of structural safety, there is no standing vortex penetrating the bottom floor of the plunge pool for the large flood discharge operation, although the flow pattern dropping into the pool is irregular with the velocity magnitude and direction gradient non-uniformly distributed. The average velocity of the tail-ridge area is about 3.41 m/s, indicating that the streamwise-lateral ski-jump flow is beneficial for energy dissipation due to the turbulent shearing and aeration without the risk of damage to the plunge pool structure.

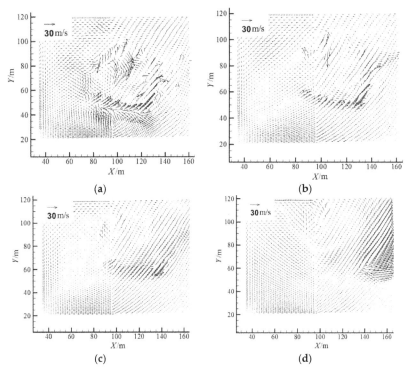

Figure 11. Velocity vector distributions in the plunge pool: (**a**) $Z = -38$ m; (**b**) $Z = -44$ m; (**c**) $Z = -50$ m; (**d**) $Z = -58$ m.

Figure 12 shows the velocity vector distributions at the cross-section ($X = 100$ m, the furthest lateral jet length for $\alpha = 45°$). The incident angles of nearer and further jet flow are almost the same, which are moderate to take full advantage of the water depth to dissipate the jet flow kinetic energy. The uniform distributions of velocity near the left side and the bottom wall of the plunge pool confirms that no obvious impact is caused by the ski-jump flow. The slight air–water surge is only at the dropping area and hardly influences the symmetric water flow pattern in the total plunge pool without a difference of water depth between the left and right side. This is because the thickness of the water jet dropping into the pool is thin and the aeration becomes fully developed as it diffuses into the air. Both of them will make efficient contributions to improving the energy dissipation and jet impact effect.

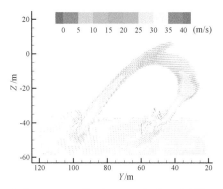

Figure 12. Lateral velocity vector profile of streamwise-lateral ski-jump flow (X = 100 m).

Based on the experimental tests and numerical simulation results, the flow characteristics of the streamwise-lateral spillway can be described as follows: (1) When the water flows on the spillway, some of the water discharge moves in a lateral direction without one side wall constraint, dropping into the plunge pool; (2) when the other water reaches the end of the spillway, a stretched ski-jump flow jet generates in the limited lateral space above the plunge pool, due to the coupling effect of the inclined floor and the anti-arch side wall; (3) when the water flow impinges into the plunge pool, submerged jet forms cause energy to dissipate through the drastic flow shearing and turbulent disturbance. The present study shows that for $\alpha = 30° \sim 45°$, the structural design of the streamwise-lateral discharge spillway performs well in controlling the ski-jump flow in terms of the high-speed and large flood discharge conditions, resulting in well jet diffusion in the air and the water discharge rate having uniform distribution.

5. Conclusions

A new type of design for streamwise-lateral ski-jump flow discharge in the spillway is introduced in this paper. In contrast to the traditional spillway, in this design, the side wall next to the plunge pool is removed, and an inclined floor is set between the horizontal bottom floor and the right-side wall with an anti-arch deflection design at the end. Lateral discharge generates along the streamwise flow discharge due to the natural pressure difference between the two sides. The free jet flow fully diffuses into the limited area, and is affected by the coupling constraint of the inclined floor and the anti-arch side wall.

The optimum free jet flow diffusion results occurred in the range of inclination angles of $30° \sim 45°$. The experimental and simulated results show the interior characteristics of the streamwise-lateral jet flow. This fully stretched free jet flow will make a beneficial contribution to the reduction in discharge per unit width dropping into the plunge pool, and an improvement in energy dissipation performance without an obvious impact on the plunge pool. Therefore, the application of the streamwise-lateral flow discharge spillway has a promising future in hydraulic engineering.

Author Contributions: This paper is a product of joint efforts of the authors who worked together on the experimental model tests. J.D. has a scientific background in applied hydraulics while W.W. and F.Z. conducted the experimental and numerical simulation investigations. J.D. and Z.T. generated the proposed effects of outlet structure on the hydraulic characteristics while J.D. and W.W. tested the methods and wrote this paper.

Funding: Resources to cover the Article Processing Charge were provided by the National Natural Science Foundation of China (Grant No. 51609162).

Conflicts of Interest: The authors declare no conflict of interest.

Water **2018**, *10*, 1585

References

1. Xu, W.L.; Luo, S.J.; Zheng, Q.W.; Luo, J. Experimental study on pressure and aeration characteristics in stepped chute flows. *Sci. China Technol. Sci.* **2015**, *58*, 720–726. [CrossRef]
2. Chanson, H. *Energy Dissipation in Hydraulic Structures*; IAHR Monograph; CRC Press: Boca Raton, FL, USA, 2015.
3. Liu, P.Q. *Energy Dissipater Theory of Modern Dam Construction*; Science Press: Beijing, China, 2010. (In Chinese)
4. Heller, V.; Hager, W.H.; Minor, H.E. Ski jump hydraulics. *J. Hydraul. Eng.* **2005**, *131*, 347–355. [CrossRef]
5. Hager, W.H.; Boes, R.M. Hydraulic structures: A positive outlook into the future. *J. Hydraul. Res.* **2014**, *52*, 299–310. [CrossRef]
6. Chanson, H. Aeration of a free jet above a spillway. *J. Hydraul. Res.* **1991**, *29*, 655–667. [CrossRef]
7. Lenau, C.W.; Cassidy, J.J. Flow through spillway flip bucket. *J. Hydraul. Div.* **1969**, *95*, 633–648.
8. Mason, P.J.; Arumugam, K. Free jet scour below dams and flip buckets. *J. Hydraul. Eng.* **1985**, *111*, 220–235. [CrossRef]
9. Juon, R.; Hager, W.H. Flip bucket without and with deflectors. *J. Hydraul. Eng.* **2000**, *126*, 837–845. [CrossRef]
10. Erpicum, S.; Archambeau, P.; Dewals, B.; Pirotton, M. Experimental investigation of the effect of flip bucket splitters on plunge pool geometry. *Wasserwirtschaft* **2010**, *4*, 73–80.
11. Mason, P.J. Practical guidelines for the design of flip buckets and plunge pools. *Water Power Dam Constr.* **1993**, *45*, 40–45.
12. Li, S.; Liang, Z. Gravity-affected potential flows past spillway flip buckets. *J. Hydraul. Eng.* **1988**, *114*, 409–427.
13. Rajan, B.H.; Shivashankara, R.K.N. Design of trajectory buckets. *J. Irrig. Power* **1980**, *37*, 63–76.
14. Steiner, R.; Heller, V.; Hager, W.H.; Minor, H.E. Deflector Ski Jump Hydraulics. *J. Hydraul. Eng.* **2008**, *134*, 562–571. [CrossRef]
15. Alias, N.A.; Mohamed, T.A.; Ghazali, A.H.; Noor, M.J.M.M. Impact of takeoff angle of bucket type energy dissipater on scour hole. *Am. J. Appl. Sci.* **2008**, *5*, 117–121. [CrossRef]
16. Wu, J.H.; Ma, F.; Yao, L. Hydraulic characteristics of slit-type energy dissipaters. *J. Hydrodyn.* **2012**, *24*, 883–887. [CrossRef]
17. Pfister, M.; Hager, W.H. Deflector-generated jets. *J. Hydraul. Res.* **2010**, *47*, 466–475. [CrossRef]
18. Li, N.W.; Liu, C.; Deng, J.; Zhang, X.Z. Theoretical and experimental studies of the flaring gate pier on the surface spillway in a high-arch dam. *J. Hydrodyn.* **2012**, *24*, 496–505. [CrossRef]
19. Zhang, T.; Chen, H.; Xu, W.L. Allotypic hybrid type flip bucket. II: Effect of contraction ratio on hydraulic characteristics and local scour. *J. Hydroelec. Eng.* **2013**, *32*, 140–146. (In Chinese)
20. Felder, S.; Chanson, H. Scale effects in microscopic air-water flow properties in high-velocity free-surface flows. *Exp. Therm. Fluid Sci.* **2017**, *83*, 19–36. [CrossRef]
21. Xu, W.L.; Liao, H.S.; Yang, Y.Q.; Wu, C.G. Computational and Experimental Investigation on the 3-D Flow Feature and Energy Dissipation Characteristics of Plunge Pools. *Chin. J. Theor. Appl. Mech.* **1998**, *30*, 35–42.
22. Deng, J.; Xu, W.L.; Zhang, J.M.; Qu, J.X.; Yang, Y.Q. A new type of plunge pool—Multi-horizontal submerged jets. *Sci. China Technol. Sci.* **2008**, *51*, 2128–2141. [CrossRef]
23. Deng, J.; Yang, Z.L.; Tian, Z.; Zhang, F.X.; Wei, W.R.; You, X.; Xu, W.L. A new type of leak-floor flip bucket. *Sci. China Technol. Sci.* **2016**, *59*, 565–572. [CrossRef]
24. Wei, W.R.; Deng, J.; Liu, B. Influence of aeration and initial water thickness on axial velocity attenuation of jet flows. *J. Zhejiang Univ. Sci. A* **2013**, *14*, 362–370. [CrossRef]
25. Yu, T.; Deng, J.; Xia, Y.; He, C.L.; Xu, W.L. Numerical simulation of turbulent trajectory flow with turnover nappe. *Water Resour. Hydropower Eng.* **2005**, *36*, 37–39. (In Chinese)

Article

Characteristics of Tidal Discharge and Phase Difference at a Tidal Channel Junction Investigated Using the Fluvial Acoustic Tomography System

Mochammad Meddy Danial [1,2], Kiyosi Kawanisi [1,*] and Mohamad Basel Al Sawaf [1]

[1] Department of Civil and Environmental Engineering, Graduate School of Engineering, Hiroshima University, 1-4-1 Kagamiyama, Higashi Hiroshima 739-8527, Japan; meddydanialstmt@gmail.com (M.M.D.); mbasel@hiroshima-u.ac.jp (M.B.A.S.)
[2] Faculty of Engineering, Tanjungpura University, Pontianak 78124, Indonesia
* Correspondence: kiyosi@hiroshima-u.ac.jp; Tel.: +81-82-424-7817

Received: 27 March 2019; Accepted: 20 April 2019; Published: 24 April 2019

Abstract: This study investigates the tidal discharge division and phase difference at branches connected to a channel junction. The tidal discharge at three branches (eastern, western, and northern branches) was continuously collected using the fluvial acoustic tomography system (FATS). The discharge asymmetry index was used to quantify the flow division between two seaward branches (eastern and western branches). The cross-wavelet method was applied to calculate the phase difference between the tidal discharge and water level. The discharge asymmetry index shows that the inequality of flow division is obviously prominent during the spring tide duration, where the eastern branch has the capability to deliver greater amounts of subtidal discharge, approximately 55–63%, compared with the western branch. However, the equality of flow division between the eastern and western channels can be observed clearly during the neap tide period. The wavelet analysis shows that the phase difference at the western branch is higher than at the eastern branch, because the geometry of the western branch is more convergent than that of the eastern branch. Accordingly, the amplitude of the tidal wave at the western branch is more magnified compared with that at the eastern branch. Moreover, the phase difference at the northern branch is greater than at the two seaward branches, implying that the phase difference is slightly increased after passing through the junction into the northern branch.

Keywords: tidal discharge; phase difference; tidal channel junction; flow division; fluvial acoustic tomography; wavelet analysis

---

1. Introduction

One of the most important features in a tidal channel network is the bifurcations, which are mostly located in the middle section of a delta. In the tidal channel network, there are many branches/junctions that are commonly bifurcated asymmetrically [1]. Because of the different geometrical shape of branches, the behavioral pattern and magnitude of discharge in the first branch are different from the second branch. In a tidal channel junction where there are three channels connected, the tides that propagate upstream in the channels affect each other, and the tidal energy can propagate in two directions. As a result, the magnitude of discharge and its phases in three tidal channels are not similar [2,3].

An investigation of hydrodynamics at the tidal channel junction related to the interaction between the tidal wave, upstream river, and the geometrical shape of branches has been carried out by a few researchers. For example, Buschman et al. [2], using numerical model analysis, pointed out that the inequality of subtidal flow division is affected by the geometrical shape of the channel such as depth, length, bed roughness, and river discharge. They also emphasized that the flow division of discharge at a junction cannot merely be estimated from the ratio of the wetted cross-sectional areas of the two branches, because the distribution of flow is also affected by spring and neap tide [2]. Sassi et al. [3], using numerical modeling, highlighted the effect of tide on river flow division, where the inequality increases with the bifurcation order. They also found that during the neap tides, the flow may enter the other branch, leading to an unequal discharge distribution. Zhang et al. [4] found that in general, the tides can modify the river discharge distribution over distributaries in the Yangtze estuary. In their numerical result, they also underline that the fortnightly tidal amplitude also contributes to the inequality of subtidal flow division. Moreover, his findings showed that the effect of tidal range on the inequality of flow division is significant. It is important to note that the previous researchers mentioned above focus on subtidal flow division.

However, previous studies mentioned above do not consider the phase difference between the tidal discharge and water level at a tidal channel junction. Indeed, investigating the behavior of phase difference in estuaries still receives little attention, specifically at a junction of estuaries. Horrevoets et al. [5] investigated the phase difference along the single channel of the estuary and suggested that river discharge can control the phase difference. They pointed out that the phase difference in the downstream area is mostly constant, whereas in the upstream part of estuary, the phase difference is not constant. Additionally, the phase difference can become negative in the upstream area. Savenije et al. [6] stated that the phase difference is a basic and significant parameter to describe the tidal wave propagation in an estuary. It is strongly affected by estuary shape and is also a function of the ratio between bank convergence and tidal wave length.

Nevertheless, based on the previous works mentioned above, the temporal variation of phase difference at a tidal channel junction is still unidentified. The change of phase difference in an estuarine system can identify the hydrodynamic processes such as amplification and damping of tidal wave [5,6]. Besides, when the phase difference is near quadrature (close to ~90°), the duration asymmetry of water level can induce asymmetries in tidal current magnitude in estuary channel [7]. More importantly, the phase difference can even influence subtidal transport of flow, sediment, and saltwater in the estuarine system [8,9].

Herein, the chief goal of this work is to shed light on the hydrodynamic aspects of a tidal channel junction monitored by means of an advanced hydro-acoustic system with high-frequency resolution, focusing on the following: (i) the temporal variation of flow division between two seaward branches that are connected to the tidal channel junction, and (ii) identifying the phase difference in tidal discharge in the channels that are connected to the junction.

To further explore the phase difference at a junction, two unidentified issues have been taken into account: (i) the temporal variation of the phase difference in each branch connected to the junction, and the dominant factors that influences the phase difference between the two seaward branches; (ii) the influence of spring-neap tide on the phase difference.

The structure of the conducted study is as follows. Section 2 introduces the field site, the overview of the fluvial acoustic tomography system (FATS), acquisition of FATS data, acoustic Doppler current profiler (ADCP) measurements, bathymetry survey, index-velocity method, and the wavelet analysis. Section 3 presents the results comprising the bathymetry, index velocity relation, tidal discharge result at the junction, discharge validation, streamflow division analysis between the eastern and western branches, and phase difference at the tidal channel junction. Section 4 presents the discussion. The last section outlines the main conclusions.

2. Materials and Methods

2.1. Field Site

The Ota River estuary is a small-scale estuary with a multi-channel network that connects to the Hiroshima Bay, located in Hiroshima City, Japan. Historically, the channel network of Ota River was formed on the Ota River delta [10]. The Ota River has important environmental qualities, and the downstream area is also known as a habitat of many creatures, particularly oysters.

As shown in Figure 1a, the Ota river has an apex junction with an asymmetric branch pattern and a north–south oriented branch. It consists of the Ota River floodway (the western-most branch), Tenma River, Kyu Ota River, and the Motoyasu and Enko Rivers (the eastern-most branch) as the five main branches.

The channel network of the Ota River estuary is characterized as a mixed-semidiurnal and mesotidal system with a tidal range varying from 2 to 4 m. The catchment area of the Ota River is approximately 1710 km^2. The freshwater discharge that is monitored at the Yaguchi gauging station mostly varies between 20 to 50 m^{3}s^{-1} during the low-flow condition, except during periods of heavy rainfall.

There are two sluice gates at the Ota River floodway positioned near the first bifurcation channel (see Figure 1a). The sluice gates of Gion consist of a movable weir with three gates that only open 10% during normal operation; the Oshiba gates consist of a movable weir with three gates that are completely open throughout the year.

The studied junction is in the middle part of the Ota River network channel, approximately 2.5 km downstream from the main bifurcation channel (apex junction) and approximately 5.8 km from the river mouth, as shown in Figure 1b. The tidal channel junction consists of three branches: the northern branch (upstream Kyu Ota River), the eastern branch (downstream Kyu Ota River), and the western branch (Tenma River). As an estuary system with multi-channel network, the saltwater intrusion in the Ota River can reach 11.5 km upstream from the river mouth. The bed materials of the channel network mostly consist of sand containing less silt and clay.

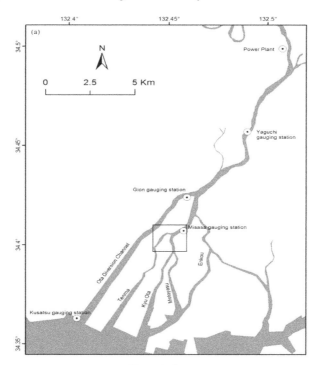

Figure 1. *Cont.*

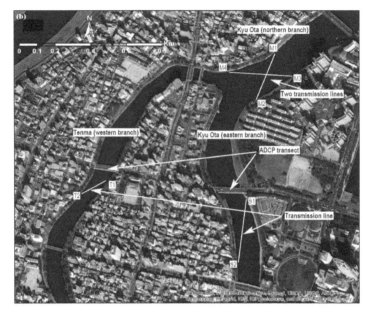

Figure 1. (**a**) Ota river channel network map. The black box denotes the study area. (**b**) Observation site and deployment of the fluvial acoustic tomography system (FATS) at the eastern, western, and northern branches. The yellow lines represent the transmission lines between the two transducers of FATS. S1–S2, T1–T2, and M1–M4 denote the locations of the transducers for FATS measurements. The red lines denote the acoustic Doppler current profiler (ADCP) transects for the comparison with FATS.

2.2. Measurement Method

2.2.1. The Overview of the Fluvial Acoustic Tomography System

The FATS developed at Hiroshima University was used to continuously measure the time series of tidal discharge at the tidal channel junction. The advantages of FATS are that it enables the measurement of discharge continuously in the long-term period that is suitable to be used even in such a tidal river with salt water intrusion, extremely shallow water depth area, and a larger width-to-depth ratio. The sound rays of FATS are capable of covering the whole cross-section of the stream and capturing the depth- and range-averaged velocity. As a result, the FATS can accurately estimate the cross-sectional averaged velocity [11]. To obtain the discharge (Q), the cross-sectional averaged velocity is multiplied by cross-sectional area (A_0), which varies in time and corresponds to water level fluctuation. For detailed information about the FATS measurement method and its reliability, readers can refer to previous studies [11–14].

2.2.2. Acquisition of FATS Data

Continuous measurements of discharge by the FATS were carried out from 8 to 20 June 2017 and 9 to 29 June 2017 for the western and eastern branches, respectively (see Table 1). Two pairs of FATS were deployed at the western and eastern branches located near a tidal channel junction, as shown in Figure 1b. The acoustic pulses (central frequency: 30 and 53 kHz) were transmitted concurrently from both transducers every 30 s. The transducers were installed at the height of 0.2 m above the channel bed using stands.

Table 1. Summary of the fluvial acoustic tomography system (FATS) experiment period.

Date of FATS Experiment	Deployment Site	Method of Discharge Estimation
9 to 29 June 2017	Eastern branch	Index-velocity method as described by Kawanisi et al. [11]
8 to 20 June 2017	Western branch	Index-velocity method as described by Kawanisi et al. [11]
6 to 29 November 2017	Northern branch	Two-crossing transmission lines as described by Kawanisi et al. [13] and Bahreinimotlagh et al. [14]

On the other hand, the monitoring of tidal discharge at the northern branch was conducted from 6 to 29 November 2017 to obtain additional data related to the phase difference investigation at the junction. In this case, we use two pairs of FATS with a two-crossing transmission lines configuration to estimate tidal discharge, as proposed by Kawanisi et al. [13] and Bahreinimotlagh et al. [14]. This method is applied to show the significant variation that occurs during field investigations.

The coordinates of the transducers and the distance between transducers for the western, eastern, and northern branches are presented in Table 2.

Table 2. Coordinates of transducers.

Code	River branch	Transducers	Latitude (°)	Longitude (°)	The Distance between Transducers (m)
Eastern branch	Kyu Ota river	S_1	34°24′00.60″	132°27′08.64″	246.363
		S_2	34°23′52.68″	132°27′07.32″	
Western branch	Tenma river	T_1	34°24′03.80″	132°26′43.63″	158.836
		T_2	34°24′01.09″	132°26′38.34″	
Northern branch	Kyu Ota river	M_1	34°24′20.36″	132°27′18.00″	289.744
		M_2	34°24′20.09″	132°27′06.66″	
		M_3	34°24′23.49″	132°27′14.52″	224.639
		M_4	34°24′16.68″	132°27′11.38″	

2.2.3. Index-Velocity Method

Application of the index velocity method for calculating continuous records of discharge has become increasingly common, particularly since the development of acoustic-based instruments such as acoustic velocity meters (AVMs) and horizontal acoustic Doppler current profilers (H-ADCPs) and acoustic tomography systems [11,15–17]. In this study, the index velocity method (IVM) is used to estimate the discharge based on the regression equation obtained from the relationship between the velocity parameter of ADCP measurement and velocity along the transmission line of FATS measurement, as described in the previous work of Kawanisi et al. [11]. Subsequently, the discharge can be performed as a product of the regression equation and cross-sectional area, which is a function of water depth.

2.2.4. ADCP Measurements and Cross-Sectional Area Determination

In this study, Teledyne RDI StreamPro ADCP was used to provide and establish reference discharge data for validating FATS measurements between two seaward branches. Each ADCP campaign was conducted approximately every 5–8 min to collect the data at the bridge near the transducers (see Figure 1b). However, for the validation of FATS measurement in the northern branch, the ADCP campaign was carried out along the transmission line between two transducers. StreamPro ADCP was set to operate in water mode 12 with a bin size of 12 cm, the number of bin sizes set to 30, with 6 pings per ensemble. In this experiment, ADCP discharges can be obtained using WinRiver II software, in which each of the bad bin ensemble parts did not exceed 2%, while bad ensembles did not exceed 3% in each transect.

To obtain the cross-sectional area (*Ao*) as a function of water level, water level data from three branches were recorded every 10 min using water level loggers (Hobo U20-001-01-Ti, Onset Co.) that were attached to the transducer stands. To obtain water level estimates, the measured pressure values should be normalized with respect to atmospheric pressure, as recorded by the barometer deployed on the riverbank. The accuracy of the water level sensors and the barometer was ±5 and ±3 mm H_2O, respectively. The post-processing of the water pressure and barometric data as parameter inputs to the software program Hoboware Pro [18].

2.2.5. Bathymetry Survey

The bathymetric survey of the three branches was conducted using an autonomous boat equipped with global positioning system (GPS) device (resolution 1/10000 s) and a single beam echo sounder at a frequency of 200 kHz (resolution: 0.01 m). Transect lines were identified according to the site accessibility conditions that were assessed in the field inspection. Thus, in the case of our monitoring program, we performed bathymetry transects for 53 sections, as depicted in Figure 2a. The directions of these lines are perpendicular to the stream direction, with the distance of each line being approximately 50 m (i.e., the distance between the left and right banks ≈50 m). Each transected line represents two points with the north–east coordinates based on the Japanese datum standard system. Output data of bathymetry transects obtained by the boat were stored as x, y, and z positions, where x and y represent the horizontal positions obtained from GPS recording, and z was the water depth measured by the echo sounder. The survey started at the upstream junction and finished at the downstream junction of both branches.

2.2.6. Wavelet Analysis

Wavelet transformation is an advanced analysis method in signal processing particularly used in investigating hydrodynamic processes in an estuary such as a tidal wave and its interaction with the river discharge [3,8,19]. In this study, three functions of wavelet were used to analyze the interaction between tidal discharge and water level, that is, continuous wavelet transformation (CWT), cross-wavelet transformation (XWT), and wavelet coherence (WTC).

CWT is used to detect variations in time series and their simultaneous representation in the time–frequency space, and also to determine the dominant period. Moreover, CWT analysis can display temporal and spatial evolutions of tidal frequency spectra [19]. XWT was utilized to analyze the phase difference between two-time series. The phase difference is basically the phase angle represented by the arrow, that is, the arrow pointing right means both signals travel in the same direction (in-phase), the arrow pointing left means both signals travel in the opposite direction (anti-phase), and the arrow pointing down means the first time series of signal leads the second one by 90°. WTC is used to find significant coherence between the two-time series and to show the confidence level. Details of wavelet are referred to in Grinsted et al. [20].

3. Results

3.1. Bathymetry at the Tidal Channel Junction

Figure 2a shows the bathymetry in three branches at a tidal channel junction where the bed level varies between −1.5 to −4.5 T.P. m, but mostly dominated by −3 T.P. m and −2 T.P. m for the eastern and western branches, respectively. From Figure 2a,b, it is revealed that the eastern branch is deeper and wider than the western branch. Moreover, sedimentation was found in the western branch around 800 m from transect 1. Similarly, the bathymetry in the northern branch shows that the width of the northern branch is wider compared with those of the eastern and western branches. Moreover, the bed level of the northern branch is relatively constant from −2.5 T.P. m to −3.0 T.P. m.

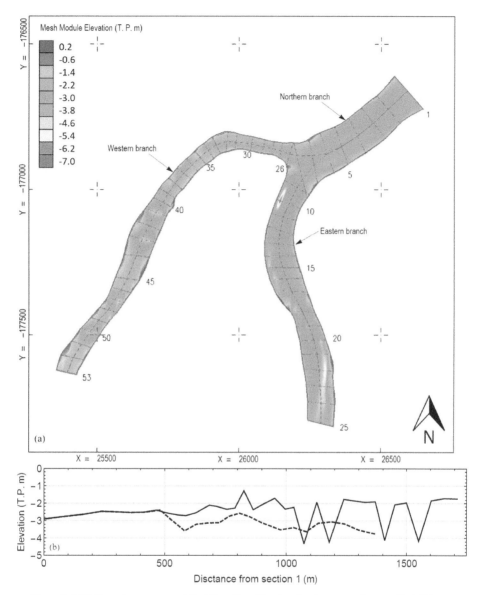

Figure 2. (**a**) Bathymetry map around the bifurcation. The elevation is measured as a vertical distance (T.P. m). The listed numbers in the range of 1–53 denote the transects of the cross-sectional line measured by the Coden boat RC–S3. (**b**) The longitudinal distributions of mean bed level. The black line and the black dashed lines represent the mean bed levels from the northern to the western and eastern branches, respectively.

3.2. Establishing an Index Velocity Rating and Validation of Discharge Measurement Obtained by FATS

We calculated tidal discharge by regressing the FATS index velocity with the velocity of ADCP, yielding discharge after multiplying it with the oblique cross-sectional area (A_0) along the transmission line, which varies in time as a function of water depth. Figure 3a,b show the linear regression relationships between the velocity of ADCP (Q_{ADCP}/A_0) and the FATS velocity along the transmission line (u_m) for the eastern and western branches, respectively. Both regressions show a high correlation with the R^2 ~0.99, though the data used for validation, particularly at the eastern branch, are limited. However, it is necessary to emphasize that having an additional number of ADCP transects in the eastern branch location is very difficult because of the high number of activities that take place along that branch (e.g., water taxi, fishing), which constrained our validation endeavors.

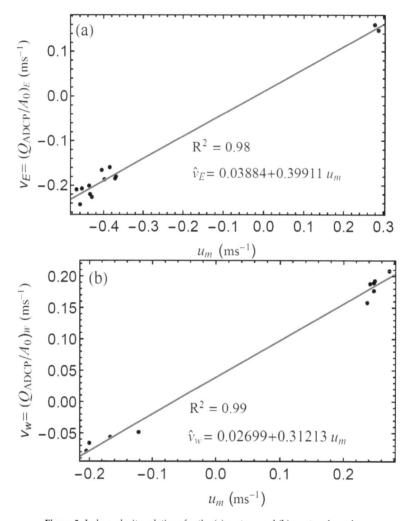

Figure 3. Index velocity relations for the (**a**) eastern and (**b**) western branches.

The index-velocity equations used for the calculation of the cross-sectional average velocity at eastern branch ($\hat{v}_E$) and the western branch ($\hat{v}_W$) are as follows:

$$\hat{v}_E = 0.03884 + 0.39911 \, u_m, \tag{1}$$

$$\hat{v}_W = 0.02699 + 0.31213 \, u_m. \tag{2}$$

Thus, the discharge of FATS for the western and eastern branches can be respectively computed as follows:

$$Q_E = \hat{v}_E \times A_0, \tag{3}$$

$$Q_W = \hat{v}_W \times A_0. \tag{4}$$

The FATS estimates were validated by the moving-boat ADCP measurements, as shown in Figure 4a–c for the eastern, western, and northern branches, respectively. The time series of tidal discharge for the eastern and western branches were obtained from Equations (3) and (4), respectively, whereas the discharge for northern branch was obtained from the two-cross paths of the FATS method adopted from Kawanisi et al. [13] and Bahreinimotlagh et al. [14].

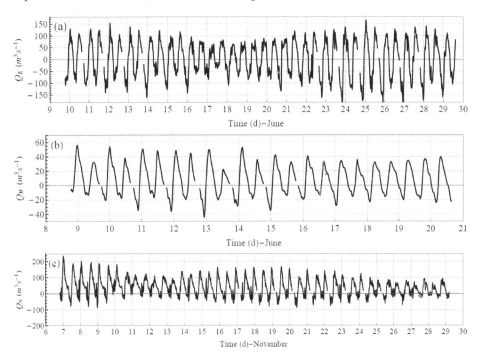

Figure 4. Temporal variations of (**a**) tidal discharge at the eastern branch, (**b**) tidal discharge at the western branch, and (**c**) tidal discharge at the northern branch. The red dots denote the discharge obtained using moving-boat ADCP measurements. Discontinuations of the discharge time series correspond to the missing period, as the result of the transducers were not covered by water during low tide.

It is important to note that particularly in the eastern branch, the validation using ADCP measurement is limited as a result of ship traffic. The relative difference of discharges estimated from ADCP measurements and FATS are mostly ranged from ~0.4% to ~10%, as shown in Table 3, Table 4, and Table 5. The results are consistent with those of previous works, where the relative differences between FATS and ADCP estimates range from 1% to 10% [12–14]. Moreover, we can evaluate how closely the FATS estimates match the ADCP results using the equation of root mean square error (RMSE):

$$\text{RMSE} = \sqrt{\frac{\sum_{i=1}^{n}[Q_{\text{FATS}}(t_i) - Q_{\text{ADCP}}(t_i)]^2}{n}}, \tag{5}$$

where $Q_{ADCP}(t_i)$ and $Q_{FATS}(t_i)$ correspond to the observed and calculated discharge at time t_i, respectively; and n is the number of data points. Thus, based on the data obtained from Table 3 to Table 5, the root means square error (RMSE) at the eastern branch is 5.0 m^{3}s^{-1} or only ~1.67% of the discharge range, whereas the RMSE of the western branch is 1.68 m^{3}s^{-1} or ~1.7% of the discharge range. Similarly, the root means square error (RMSE) at the northern branch is 5.3 m^{3}s^{-1} or ~1.65% of the discharge range. Thus, FATS measurement results are reliable estimates of the discharge in a tidal estuary.

Table 3. Comparison between FATS and acoustic Doppler current profiler (ADCP) measurements at the eastern branch.

Date	Local Time	Q_{FATS} (m^{3}s^{-1})	Q_{ADCP} (m^{3}s^{-1})	$\Delta Q = Q_{FATS} - Q_{ADCP}$ (m^{3}s^{-1})	$\Delta Q/Q_{ADCP}$ (%)
June 20, 2017	14:44:05	−57.56	−55.67	−1.89	3.40
June 20, 2017	14:55:13	−58.53	−55.67	−2.86	5.14
June 20, 2017	15:07:31	−59.09	−58.81	−0.28	0.48
June 20, 2017	15:16:31	−60.31	−58.41	−1.90	3.25
June 20, 2017	15:27:11	−65.81	−58.18	−7.63	13.11
June 27, 2017	13:29:25	108.38	99.55	8.83	8.87
June 27, 2017	13:41:38	109.60	104.65	4.95	4.73

Table 4. Comparison between FATS and ADCP measurements at the western branch.

Date	Local Time	Q_{FATS} (m^{3}s^{-1})	Q_{ADCP} (m^{3}s^{-1})	$\Delta Q = Q_{FATS} - Q_{ADCP}$ (m^{3}s^{-1})	$\Delta Q/Q_{ADCP}$ (%)
June 16, 2017	12:45:11	−11.75	−12.2	0.45	-3.69
June 16, 2017	12:59:19	−10.26	−9.13	−1.13	12.38
June 16, 2017	13:09:38	−8.05	−7.37	−0.68	9.23
June 16, 2017	13:14:28	9.80	8.91	0.89	9.99
June 16, 2017	14:14:28	25.19	24.26	0.93	3.83
June 16, 2017	14:53:59	31.85	31.47	0.38	1.21
June 16, 2017	15:13:13	31.17	27.17	4.00	14.72

Table 5. Comparison between FATS and ADCP measurements at the northern branch.

Date	Local Time	Q_{FATS} (m^{3}s^{-1})	Q_{ADCP} (m^{3}s^{-1})	$\Delta Q = Q_{FATS} - Q_{ADCP}$ (m^{3}s^{-1})	$\Delta Q/Q_{ADCP}$ (%)
November 15, 2017	12:46:48	52.98	49.68	3.3	6.64
November 15, 2017	13:22:43	43.67	42.95	0.72	1.68
November 20, 2017	12:15:03	88.61	103.34	−14.73	−14.25
November 20, 2017	13:14:28	95.68	104.48	−8.80	−8.42
November 20, 2017	14:06:36	78.95	78.64	0.31	0.39
November 27, 2017	11:37:19	−10.07	−10.75	0.68	−6.33
November 27, 2017	13:13:52	−17.80	−19.41	1.61	−8.29
November 27, 2017	14:34:03	−18.16	−20.15	1.99	−9.88
November 27, 2017	16:09:35	14.32	15.24	−0.92	−6.04
November 27, 2017	16:57:07	57.83	57.73	0.10	0.17
November 27, 2017	17:35:10	73.36	73.61	3.75	5.09
November 27, 2017	18:20:39	77.10	71.55	5.55	7.76
November 27, 2017	19:49:15	41.15	45.65	−4.50	−9.86
November 27, 2017	20:24:41	48.85	44.98	3.87	8.60

3.3. Time Series of Tidal Discharge in the Western, Eastern, and Northern Branches

Figure 5a shows the temporal variations of the water levels and discharges in the eastern and western branch. The time series of tidal discharge for the eastern and western branches were obtained from Equations (3) and (4), respectively. The tidal discharge shows the semidiurnal characteristic, that is, the high tide and low tide occur two times a day. The discharges in the branches show distinct characteristics, where the eastern and western branches range from −157 to 155 m³s⁻¹ and from −42 to 58 m³s⁻¹, respectively. From the comparison of the tidal discharges at the eastern and western branches, as shown in Figure 5a, it is revealed the tidal discharge of the eastern branch is approximately two to six times greater than that of the western branch during flood and ebb tides. This is caused by the difference in the geometry of both branches, that is, the water depth and the channel width of the eastern branch are much larger than that of the western branch, as can be seen in Figure 2. Moreover, the pattern of tidal discharge at the western branch is also much more asymmetric compared with that of the eastern branch, where the discharge during the ebb tide was higher than during the flood tide.

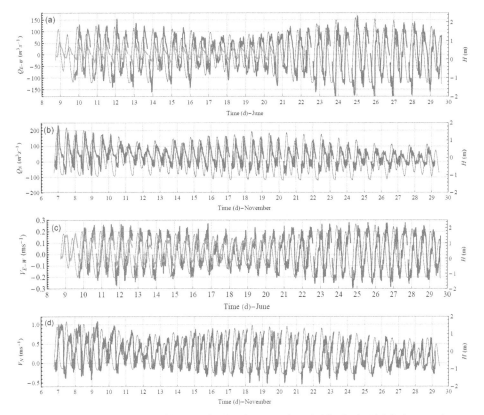

Figure 5. Temporal variations of (**a**) water level at the eastern branch (blue line), tidal discharge at the eastern branch (red line), and tidal discharge at the western branch (green line). (**b**) Water level at the northern branch (blue line) and tidal discharge at the northern branch (red line). (**c**) Water level at the eastern branch (blue line), tidal velocity at the eastern branch (red line), and tidal velocity at the western branch (green line). (**d**) Water level at the northern branch (blue line) and tidal velocity at the northern branch (red line). Positive discharge and velocity coincide with the seaward flow.

Similarly, Figure 5b shows the temporal variation of water level and tidal discharge at the northern branch located at the upstream junction, where the tidal discharge ranged from −90 to 230 m³s⁻¹. In addition, Figure 5c shows the temporal variation of the water level and tidal velocity for the eastern and western branch, whereas Figure 5d describes the tidal velocity and water level at the northern branch. The temporal variation of tidal velocity at the eastern, western, and northern branch ranges from −0.28 to 0.3 ms⁻¹, −0.12 to 0.19 ms⁻¹, and −0.5 to 1 ms⁻¹, respectively.

3.4. Subtidal Discharge Division between the Western and Eastern Branches

The purpose of quantifying subtidal discharge division is to explore the variation of flow division and the inequality of streamflow during the spring and neap tide. In this study, we use the subtidal discharge obtained by applying a Battle–Lemarie filter of wavelet function. The division of subtidal discharge between the eastern and western branches at a tidal channel junction can be quantified using the discharge asymmetry index (ψ), as proposed by Buschman et al. [2].

$$\psi = \frac{\langle Q_E \rangle - \langle Q_W \rangle}{\langle Q_E \rangle + \langle Q_W \rangle} \tag{6}$$

The tidal discharge asymmetry index (ψ) is zero for an equal discharge division, $\psi = 1$ when all river water flows through the eastern branch, and $\psi = -1$ when the western branch carries all the discharge. Therefore, a positive value is obtained when the discharge at the eastern branch is greater than that at the western branch, and vice versa.

Only data from 10 and 21 June, 2017 were used to calculate the discharge asymmetry index (ψ) using Equation (6), because of the difference in lengths of the data obtained from the FATS measurements between the eastern and western branches. Figure 6 shows the discharge asymmetry index and subtidal discharge variation in the eastern and western branch from 10 to 21 June, suggesting that the temporal variations in subtidal discharge of the eastern branch are slightly larger than those of the western branch, except within 18 and 19 June, the variation was almost the same during the neap tide. Figure 6, which captures a spring-neap fluctuation, shows the temporal variation in the discharge asymmetry index. The discharge asymmetry index fluctuates in the range of −0.02 to 0.26 from 10 to 21 June, indicating the inequality of flow division between the eastern and western branch, where the eastern branch can distribute larger discharge than that in the western branch during spring tide; the flow division between the western and eastern branch is nearly equal during the neap tide (18 to 19 June).

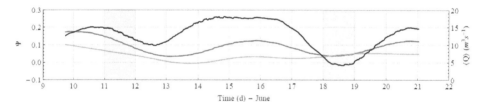

Figure 6. Discharge asymmetry index (black line). Temporal variations of subtidal discharges in the eastern (red line) and western branch (green line). The grey and blue highlighted boxes represent the spring and neap tide, respectively.

3.5. The Wavelet Analysis for the Interaction between the Water Level and Tidal Discharge in the Three Branches

In this section, three functions of wavelet analysis are used to investigate the time-series signal between tidal discharge and water level in three branches connected to a channel junction. In this study, XWT is used to determine the interaction between tidal discharge and water level, that is, propagating, standing wave, and mixed wave. Tidal discharge (Q) varies along the channel in estuaries and co-varies with the tidal wave. There are two methods to determine the type of wave in estuarine channels, that is, phase lag and phase difference [21,22]. The phase lag is the time lag between high water slack (HWS, i.e., when the discharge is zero) and high water (HW, i.e., when the water level is maximum) [8,21]. In contrast, the phase difference is the time lag between the peak tidal discharge and the HW [21,22]. Therefore, there is a relationship between the phase difference and the phase lag, that is, $\varepsilon + \varphi = \frac{\pi}{2}$, where ε is the phase lag and φ is the phase difference [8,23]. In this study, we use phase difference to describe the relationship between tidal discharge and water level [22,24]. If $\varphi = 90°$ is a standing wave, $\varphi = 0°$ is propagating, and $0° < \varphi < 90°$ is a mixed wave.

Figure 7 shows three figures consisting of CWT, XWT, and WTC analyses between the tidal discharge and water level at the eastern branches. From the figures, the dark red color background denotes a strong power spectrum, suggesting a dominant tidal signal. The vertical bar shows the energy of the tidal domain. From the CWT and WTC analyses, the semidiurnal signal is more dominant than the diurnal signal. Moreover, the semidiurnal domain is always present during the spring and neap tide, whereas the diurnal domain is only present during spring tide, but fades during the neap tide. Thus, the tide of the Ota River is characterized by the mixed-semidiurnal tide. Besides the semidiurnal and diurnal signal, we can also identify the quarter-diurnal in the eastern branch, which is discontinuous and weak. In contrast, the fortnightly signal cannot be detected. From XWT analysis, the phase difference between the tidal discharge and the water level at the eastern branch can be denoted by the black arrows pointing straight down at ~78°.

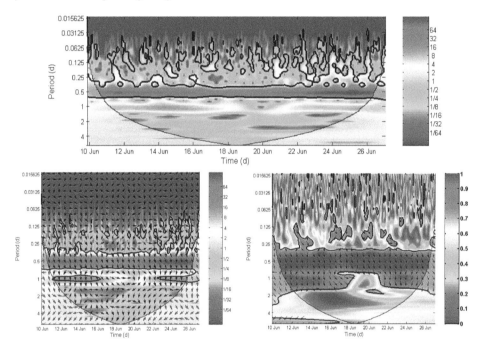

Figure 7. Wavelet analyses in eastern branch: (top) continuous wavelet transformation; (bottom left) cross-wavelet transform; (bottom right) wavelet coherence. The black contours represent the 0.95 confidence level against red noise, and the cone of influence (COI) where edge effects might distort the picture is shown as a lighter shade. The relative phase relationship is shown as arrows (with in-phase pointing right, anti-phase pointing left, and the first time series leading the second one by 90° pointing down). The wavelet power is $\log_2(A^2/v)$, where A is the wavelet amplitude and v is the variance of the original tidal signals. The y-axis in the CWT and XWT is on a $\log_2$ scale.

Similarly, Figure 8 shows three plots consisting of CWT, XWT and WTC analyses between the tidal discharge and water level at the western branches. The semidiurnal domain is always present, whereas the diurnal domain is only present during spring tide but diminishes during the neap tide. We can recognize the quarter-diurnal in the western branch which is discontinuous and weak, but more pronounced compared to the eastern branch. The fortnightly signal is also cannot be identified in the western branch.

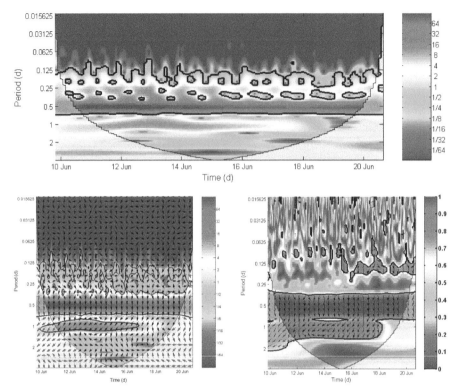

Figure 8. Wavelet analyses in western branch: (top) continuous wavelet transform. (bottom left) cross-wavelet transform. (bottom right) wavelet coherence. The black contours represent the 0.95 confidence level against red noise and the cone of influence (COI) where edge effects might distort the picture is shown as a lighter shade. The relative phase relationship is shown as arrows (with in-phase pointing right, anti-phase pointing left, and the first time series leads the second one by 90° pointing down). The wavelet power is $\log_2(A^2/v)$, where A is the wavelet amplitude, v is the variance of the original tidal signals. The y-axis in the continuous wavelet transformation (CWT) and cross-wavelet transformation (XWT) is on a $\log_2$ scale.

From XWT analysis, the phase difference between the tidal discharge and the water level at the western branch shows that the black arrows are pointing down at ~88.5°.

The phase difference in tidal discharge between the western and eastern branches was further examined to ensure that the tidal discharge phase of the eastern branch is greater than that of the western branch. Figure 9 confirms that the tidal discharge phase at the western branch is larger than the discharge phase at the eastern branch, with a phase difference of ~11°. Moreover, it also seems that the geometry of the branch has a greater effect on the behavior of the phase difference at the western branch compared with the eastern branch.

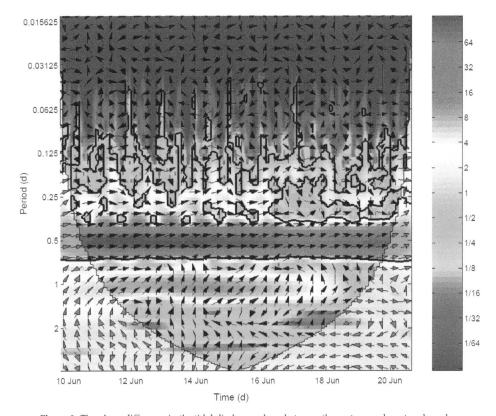

Figure 9. The phase difference in the tidal discharge phase between the eastern and western branches. The black contours represent the 95% confidence level for red noise, and the cone of influence (COI) where edge effects might distort the picture is shown as a lighter shade. The relative phase relationship is shown as arrows (with in-phase pointing right, anti-phase pointing left, and the first time series leading the second one by 90° pointing down). The wavelet power is $\log_2(A^2/v)$, where A is the wavelet amplitude and v is the variance of the original tidal signals. The y-axis is on a $\log_2$ scale.

As mentioned in Section 2.2.2, to gain a comprehensive understanding of phase difference behavior around the junction, an additional observation campaign at the northern branch using FATS is needed to collect the tidal discharge continuously. With this, the phase difference before and after the junction can be investigated completely (see Figure 1b).

The phase difference during semidiurnal is ~90°. This result confirms that the phase difference at the northern branch shows slightly increased behavior compared with the seaward branches (eastern and western branches). This result indicates that the behavior of phase difference is slightly increased after passing through the junction into the northern branch (a landward branch of the junction).

4. Discussion

4.1. Interpretation of Flow Division between Two Seaward Branches

From the bathymetry map presented in Figure 2, the depth and width of the eastern branch are greater than those of the western branch. However, the inequality of flow division at the tidal channel junction cannot be calculated using only a ratio of the wetted cross-sectional areas between the two branches [2], because the hydrodynamic process at the tidal channel junction is also strongly influenced by the spring and neap tide, as well as by the asymmetrical geometry shape between the two seaward branches.

The response of tidal dynamic to the local geometry between two seaward branches shows distinct characteristics, where the asymmetric discharge at the western branch is more pronounced compared with that at the eastern branch. The asymmetric tidal discharge observed at the western branch indicates the ebb discharge is larger than the flood discharge. On the contrary, the characteristic of tidal discharge at the eastern branch is nearly equal during the ebb and flood tide. As we can see in Figure 2, the western branch is shallower and narrower than the eastern branch. Therefore, the propagating tidal wave in shallow and narrow branches, and modulated with the river discharge, can generate the nonlinear effect as a result of interaction with the local topography [9]. This nonlinear effect leads to the generation and development of shallow-water constituents, which cause tidal distortion and asymmetry [25]. This asymmetric tidal discharge can induce a seaward sediment movement [26].

Regarding the flow division characteristic at the tidal channel junction of Ota River, our findings are depicted in Figure 6. The discharge asymmetry index ranges from −0.02 to 0.26 during the 10 to 21 June, suggesting that the eastern branch has the capability to deliver greater amounts of subtidal discharge, approximately 55–63% compared with the western branch. However, the flow division is nearly equal during the neap tide (i.e., 18 to 19 June). Figure 6 shows that the equality of asymmetry index between the eastern and western channels can be observed clearly during the neap tide period. Nevertheless, the inequality of flow division is obviously prominent during the spring tide duration. The fluctuation in the subtidal discharge at the eastern branch is greater than that at the western branch; the subtidal discharge at the eastern branch decreases during the neap tide, so that both subtidal discharges are nearly equal during the neap tide.

4.2. Interpretation of the Interaction between the Tidal Discharge and Water Level in the Three Branches

CWT analysis identifies the semidiurnal and diurnal signal in three branches connected to the junction. The quarter-diurnal signal can be detected, but is not dominant, because the signal energy is weak and discontinuous. This happened because the shallow water effect in the branch does not influence tidal distortion of D2, and thereby cannot produce enough energy to transform from D2 to D4 [19,25]. Moreover, the tidal channel junction is located in the middle of estuary, which is still near to the river mouth, implying that the tide needs to travel more in the landward direction to generate D4.

In contrast, the fortnightly tidal signal does not appear in the CWT analysis in the relationship between the tidal discharge and water level—possibly because of two reasons, that is, (i) the time-series data are too short for capturing fortnightly signal; or (ii) freshwater discharge is very low. For comparison with the work of previous researchers, for example, the work of Leonardi et al. [8], the fortnightly signal does not exist in their wavelet analysis over period of 1.5 months, with the freshwater discharge ranging from 110 to 8235 m^3s^{-1}. On the contrary, Sassi et al. [3] could recognize the fortnightly tide using longer time-series data over the period of six months, with the freshwater discharge ranging from 3000 to 5000 m^3s^{-1}. However, in our case, the time-series data are less than one month, with the freshwater discharge in the Ota estuary during normal condition ranging from 20 to 40 m^3s^{-1}. Thus, in the case of the Ota River, the absence of a fortnightly signal in the interaction between the water level and tidal discharge is the result of the short time-series data [19].

Two main questions arise about the characteristic of phase difference at the junction: (i) the first regarding the temporal variation of the phase difference between the two seaward branches (the eastern and western branches). As a follow-up question, what is the dominant factor that influences the phase difference between the two seaward branches? and (ii) is the phase difference influenced by the spring and neap tide?

To answer the first question, we must first determine the phase difference between the two branches. The phase difference between the tidal discharge and water level at the eastern and western branch for semidiurnal are ~78° and ~88.5°, indicating a mixed wave and mimic standing wave, respectively. It is important to note that the phase difference range of 78° to 88.5° implies that the tidal discharge leads the water level. The discharge phase difference between the eastern and western branches shows that the discharge phase at the western branch leads by ~15 min compared with the discharge phase at the eastern branch. In addition, the phase difference between the tidal velocity and water level for the eastern and western branches is ~71° and ~80°, respectively (figures not shown). Thus, the phase difference between the velocity and water level is consistent with the phase difference between the discharge and water level, where the phase difference in the western branch is larger than that in the eastern branch. We want to investigate further the reason that the phase difference at the western branch is higher than at the eastern branch. According to Savenije [21], the phase difference is strongly affected by the convergence shape effect, rather than river discharge and bottom friction. Accordingly, we quantify the convergence levels between the eastern and western branches using the exponential function form with three important parameters; namely, the mean water depth, river width, and cross-sectional area, as exemplified by Savenije et al. and Cai et al. [21,27].

Figure 10 shows the geometry of both the eastern and western branches. The geometry analysis presented in Figure 10 (right) obviously indicates that the western branch is more convergent compared with the eastern branch, as shown in Figure 10 (left), resulting in a larger phase at the western branch. From the results of phase difference characteristic and the geometry analysis between two seaward branches, it can be inferred that the amplitude of tidal wave at the western branch near the junction is more magnified compared with at the eastern branch [6,7].

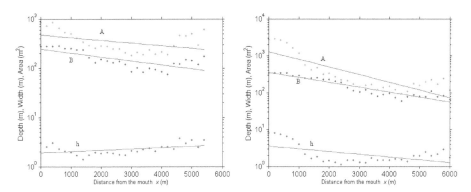

Figure 10. Geometry of the eastern branch (left) and western branch (right); showing the longitudinal variation of the cross-sectional area (A), the width (B), and the depth (h).

Figure 11 shows that the phase difference between the tidal discharge and water level at the northern branch (a landward branch of the junction) is ~90°, indicating a standing wave. Thus, it is important to note that the phase difference at the northern branch is slightly higher than that of the two seaward branches. This evidence indicates that the phase difference is slightly increased after passing through the junction into the northern branch. However, further investigation is needed to confirm this evidence.

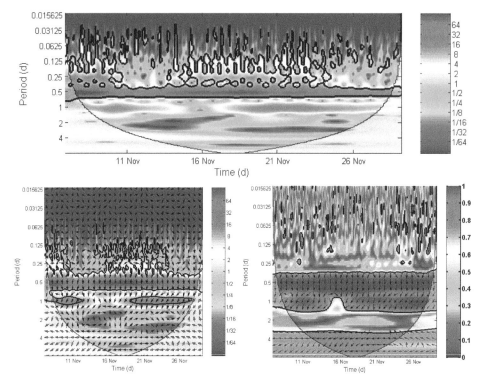

Figure 11. Wavelet analyses in the northern branch: (top) continuous wavelet transform; (bottom left) cross-wavelet transform; (bottom right) wavelet coherence. The black contours represent the 0.95 confidence level against red noise, and the cone of influence (COI) where edge effects might distort the picture is shown as a lighter shade. The relative phase relationship is shown as arrows (with in-phase pointing right, anti-phase pointing left, and the first time series leading the second one by 90° pointing down). The wavelet power is $\log_2(A^2/v)$, where A is the wavelet amplitude and v is the variance of the original tidal signals. The y-axis in the CWT and XWT is on a $\log_2$ scale.

The answer for the second question is that the effect of spring and neap tide on the phase difference in the semidiurnal domain is negligible. As shown in Figure 9, the phase difference between two seaward branches changes slightly during the spring and neap tide. The phase difference between the tidal discharge and water level is relatively constant, where during the spring tide, the phase difference is ~11°, whereas during the neap tide, the phase difference is ~9°. Therefore, the change of the phase difference between spring and neap tide is not significant. This phenomenon is probably because the phase difference is a function of a ratio between bank convergence and tidal wave length [21]. Moreover, the phase difference can be affected by two factors, that is, the river discharge [8,21] and topography of the estuarine channel, also known as estuary shape [21].

The upstream discharge from Yaguchi gauging station during the experiment is relatively constant, ranging from 20 to 40 m^3s^{-1} during June 2017, though the freshwater discharge from Yaguchi gauging station increased slightly from 20 to 90 m^3s^{-1} in November 2017. Thus, during this condition, the tidal processes of the Ota River can be classified as a tidally dominated estuary [8]. Additionally, this finding is still more or less consistent with the work of Leonardi et al. [8], who pointed out that the phase difference always tends to be close to 90° as long as the characteristic of flow is bidirectional.

5. Conclusions

Temporal variations of streamflow were successfully measured by FATS instruments at three branches connected to a tidal channel junction during a low-flow condition. To investigate the flow division of subtidal discharge between the two seaward branches (eastern and western branches), the discharge asymmetry index was used. Moreover, to characterize the tidal discharge behavior and its relationship with the water level and modulated with the freshwater discharge, the phase difference is investigated.

The response of tidal dynamic to the local geometry between two seaward branches shows that the asymmetric discharge at the western branch is more pronounced compared with the eastern branch. The asymmetric tidal discharge at the western branch indicates the ebb discharge is larger than the flood discharge. This happens because the interaction between tidal propagation and the local geometry of western branch that modulated by river discharge could generate the nonlinear effect. This nonlinear effect leads to generation and development of shallow-water constituents, which cause tidal distortion and asymmetry. This asymmetric tidal discharge can induce a seaward sediment movement. In contrast, the characteristic of tidal discharge at the eastern branch is nearly equal during the ebb and flood tide.

Regarding the flow division characteristic at the tidal channel junction of Ota River, the discharge asymmetry index varies from −0.02 to 0.26 during the studied period, suggesting that the eastern branch has the capability to deliver greater amounts of subtidal discharge, approximately 55–63% compared with the western branch, except during the neap tide where the flow division is nearly equal between the eastern and western branch. The temporal change in the flow division is induced by the fluctuation in the subtidal discharge at the eastern branch. The subtidal discharge decreases during the neap tide, so that both subtidal discharges are nearly equal.

The wavelet analyses of the temporal variation of the phase differences between the two seaward branches show slightly different behavior. A mimic standing wave characteristic ($\Delta\varphi = \sim 88.5°$) occurs at the western branch, whereas a mixed wave characteristic ($\Delta\varphi = \sim 78°$) occurs at the eastern branch. It is shown that the phase difference between tidal discharge and the water level is relatively constant during the neap and spring tide. The discharge phase at the western branch leads the discharge phase at the eastern branch by ~15 min. Additionally, the geometry analysis between the two seaward branches reveals that the western branch is more convergent compared with the eastern branch, and thus causes a larger phase difference compared with the eastern branch. As a result, the amplitude of tidal wave at the western branch near the junction is more magnified compared with eastern branch.

Using the same analysis, we can say the phase difference between the tidal discharge and water level at the northern branch (a landward branch of the junction) indicates a standing wave characteristic ($\Delta\varphi = \sim 90°$), which is slightly higher than the phase difference in the two seaward branches. This evidence implies that the phase difference is slightly increased after passing through the junction into the northern branch. Further study is needed to explore the characteristics of the phase difference in the larger area of the multi-channel estuary.

Although this study deals with the Ota River as a case-specific example, the presented results and discussion are not only aimed at demonstrating the dynamic of flow division and relationship between tidal discharge and water level at a junction, but also take advantage of utilizing the innovative hydro-acoustic system (FATS) for measuring discharge in high temporal resolution and exploring the hydrodynamic process at a tidal channel junction. Moreover, our findings should increase the knowledge for common hydrodynamic processes in multi-channel estuaries that have nearly similar environments over the world.

Author Contributions: Conceived the study, M.M.D.; Data curation, M.M.D. and M.B.S.; Validation, M.M.D.; Supervision, K.K.; Writing—original draft, M.M.D.; Writing—review & editing, M.B.S. and K.K.

Funding: This work was funded by JSPS KAKENHI, grant number JP17H03313.

Acknowledgments: The authors are grateful to Noriaki Gohda of the Hiroshima University/Aqua Environmental Monitoring Limited Liability Partnership (AEM-LLP) for his technical support. Our thanks are also addressed to the members of the Department of Civil and Environmental Engineering, Hiroshima University, for their assistance with the collection of field data in the estuary.

Conflicts of Interest: The authors declare no conflict of interest. The funders had no role in the design of the study; in the collection, analyses, or interpretation of data; in the writing of the manuscript; and in the decision to publish the results.

References

1. Edmonds, D.A.; Slingerland, R.L. Stability of delta distributary networks and their bifurcations. *Water Resour. Res.* **2008**, *44*, 1–13. [CrossRef]
2. Buschman, F.A.; Hoitink, A.J.F.; Van Der Vegt, M.; Hoekstra, P. Subtidal flow division at a shallow tidal junction. *Water Resour. Res.* **2010**, *46*, 1–12. [CrossRef]
3. Sassi, M.G.; Hoitink, A.J.F.; de Brye, B.; Vermeulen, B.; Deleersnijder, E. Tidal impact on the division of river discharge over distributary channels in the Mahakam Delta. *Ocean Dyn.* **2011**, *61*, 2211–2228. [CrossRef]
4. Zhang, W.; Feng, H.; Hoitink, A.J.F.; Zhu, Y.; Gong, F. Estuarine, Coastal and Shelf Science Tidal impacts on the subtidal flow division at the main bifurcation in the Yangtze River Delta. *Estuar. Coast. Shelf Sci.* **2017**, *196*, 301–314. [CrossRef]
5. Horrevoets, A.C.; Savenije, H.H.G.; Schuurman, J.N.; Graas, S. The influence of river discharge on tidal damping in alluvial estuaries. *J. Hydrol.* **2004**, *294*, 213–228. [CrossRef]
6. Savenije, H.H.G.; Veling, E.J.M. Relation between tidal damping and wave celerity in estuaries. *J. Geophys. Res.* **2005**, *110*, 1–10. [CrossRef]
7. Savenije, H.H.G.; Toffolon, M.; Haas, J.; Veling, E.J.M. Analytical description of tidal dynamics in convergent estuaries. *J. Geophys. Res.* **2008**, 1–27. [CrossRef]
8. Leonardi, N.; Kolker, A.S.; Fagherazzi, S. Advances in Water Resources Interplay between river discharge and tides in a delta distributary. *Adv. Water Resour.* **2015**, *80*, 69–78. [CrossRef]
9. Nidzieko, N.J. Tidal asymmetry in estuaries with mixed semidiurnal/diurnal tides. *J. Geophys. Res. Ocean.* **2010**, *115*, 1–13. [CrossRef]
10. Gotoh, T.; Fukuoka, S.; Tanaka, R. *Evaluation of Flood Discharge Hydrographs and Bed Variations in a Channel Network on the Ota River Delta*, 357th ed.; Chavoshian, A., Takeuchi, K., Eds.; IAHS: Wallingford, UK, 2013; ISBN 9781907161353.
11. Kawanisi, K.; Al Sawaf, M.B.; Danial, M.M. Automated real-time stream flow acquisition in a mountainous river using acoustic tomography. *J. Hydrol. Eng.* **2018**, *23*, 1–7. [CrossRef]
12. Razaz, M.; Kawanisi, K.; Nistor, I.; Sharifi, S. An acoustic travel time method for continuous velocity monitoring in shallow tidal streams. *Water Resour. Res.* **2013**, *49*, 4885–4899. [CrossRef]
13. Kawanisi, K.; Razaz, M.; Yano, J.; Ishikawa, K. Continuous monitoring of a dam flush in a shallow river using two crossing ultrasonic transmission lines. *Meas. Sci. Technol.* **2013**, *24*, 055303. [CrossRef]
14. Bahreinimotlagh, M.; Kawanisi, K.; Danial, M.M.; Al Sawaf, M.B.; Kagami, J. Application of shallow-water acoustic tomography to measure flow direction and river discharge. *Flow Meas. Instrum.* **2016**, *51*, 30–39. [CrossRef]
15. Sassi, M.G.; Hoitink, A.J.F.; Vermeulen, B.; Hidayat, H. Discharge estimation from H-ADCP measurements in a tidal river subject to sidewall effects and a mobile bed. *Water Resour. Res.* **2011**, *47*, 1–14. [CrossRef]
16. Levesque, V.A.; Oberg, K.A. *Computing Discharge Using the Index Velocity Method*; Techniques and Methods 3-A23; U.S. Geological Survey: Reston, VA, USA, 2012; Volume 148.
17. Hidayat, H.; Hoitink, A.J.F.; Sassi, M.G.; Torfs, P.J.J.F. Prediction of Discharge in a Tidal River Using Artificial Neural Networks. *J. Hydrol. Eng.* **2014**, *19*, 1–8. [CrossRef]
18. Kempema, E.; Stiver, J.; Ettema, R. *Effects of Warm CBM Product Water Discharge on Winter Fluvial and Ice Processes in the Powder River Basin*; 2010; Available online: http://www.uwyo.edu/owp/_files/project27finalreport.pdf (accessed on 23 April 2017).
19. Guo, L.; Van Der Wegen, M.; Jay, D.A.; Matte, P.; Wang, Z.B.; Roelvink, D.; He, Q. River-tide dynamics: Exploration of nonstationary and nonlinear tidal behavior in the Yangtze River estuary. *J. Geophys. Res. C Ocean.* **2015**, *120*, 3499–3521. [CrossRef]

20. Grinsted, A.; Moore, J.C.; Jevrejeva, S. Application of the cross wavelet transform and wavelet coherence to geophysical time series. *Nonlinear Process. Geophys.* **2004**, *11*, 561–566. [CrossRef]

21. Savenije, H.H.G. *Salinity and Tides in Fluvial Estuaries*, 1st ed.; Elsevier Science: Amsterdam, The Netherlands, 2005; ISBN 9780444521071.

22. Friedrichs, C.T.; Aubrey, D.G. Tidal propagation in strongly convergent channels. *J. Geophys. Res.* **1994**, *99*, 3321–3336. [CrossRef]

23. Cai, H.; Savenije, H.H.G.; Toffolon, M. Linking the river to the estuary: Influence of river discharge on tidal damping. *Hydrol. Earth Syst. Sci.* **2014**, *18*, 287–304. [CrossRef]

24. Jay, D.A. Green's law revisited: Tidal long-wave propagation in channels with strong topography. *J. Geophys. Res.* **1991**, *96*, 20585. [CrossRef]

25. Lu, S.; Tong, C.; Lee, D.; Zheng, J.; Shen, J.; Zhang, W.; Yan, Y. Propagation of tidal waves up in Yangtze Estuary during the dry season. *J. Geophys. Res. Ocean.* **2015**, *120*, 6445–6473. [CrossRef]

26. Guo, L.; Van der Wegen, M.; Roelvink, J.A.; He, Q. The role of river flow and tidal asymmetry on 1-D estuarine morphodynamics. *J. Geophys. Res. Earth Surf.* **2014**, *119*, 2315–2334. [CrossRef]

27. Cai, H.; Toffolon, M.; Savenije, H.H.G.; Yang, Q.; Garel, E. Frictional interactions between tidal constituents in tide-dominated estuaries. *Ocean Sci.* **2018**, *14*, 769–782. [CrossRef]

Article

Numerical Analysis on Hydraulic Characteristics of U-shaped Channel of Various Trapezoidal Cross-Sections

Ruichang Hu and Jianmin Zhang *

State Key Laboratory of Hydraulics and Mountain River Engineering, Sichuan University,
Chengdu 610065, China; hrc@stu.scu.edu.cn
* Correspondence: zhangjianmin@scu.edu.cn

Received: 9 November 2018; Accepted: 1 December 2018; Published: 5 December 2018

Abstract: Curved channel with trapezoidal cross-section is approximate to the common form in nature fluvial networks and its hydraulic characteristics are considerably complex and variable. Combined with volume of fluid (VOF) method, renormalization group (RNG) k-ε turbulence model was employed to numerically investigate the flow properties in the U-shaped channel with various trapezoidal cross-sections. Analyses were performed from the aspects of the water surface transverse slope in bend apex (WTS-BA), longitudinal velocity, secondary flow, shear stress and turbulent kinetic energy (TKE) under several scenarios, specifically, four types of radius-to-width ratio and seven types of slope coefficient with a constant aspect ratio. The calculated results suggested that the maximums of shear stress and TKE in the bend were observed in the convex bank and the maximal intensities of secondary flow were observed within the range of 60 to 75 degrees for various varieties. As the radius-to-width ratio increased, the maximums of shear stress, TKE and WTS-BA decreased; but increased with increasing slope coefficients. The intensity of secondary flow decreased as slope coefficients increased and the angle of maximum intensity of secondary flow moved to the upstream for the increasing radius-to-width ratios. In addition, a new equation concerning the vertical distribution of longitudinal velocity in trapezoidal cross-sectioned channel was presented.

Keywords: trapezoidal cross-section; U-shaped channel; radius-to-width ratio; slope coefficient; hydraulic characteristics

---

1. Introduction

The water surface transverse slope (WTS), violent turbulence, forceful secondary flow and shear stress aggravate the erosion and siltation of the channel [1]. Consequently, detailed study regarding the hydraulic characteristics of the channel is an essential aspect for solving problems related to river restoration and navigability.

The secondary flow was observed by the movement of impurities in straight and curved open channel [2] and then the investigations referring to the secondary flow and shear stress in the channel were extensively conducted [3–6]. The structure of secondary flow was changed for the varying shape of cross-section in straight channel, which was investigated using three slope coefficients of 0.58, 1 and 1.73 by Tominaga et al. [7]. And the spanwise distribution of boundary shear stress was extraordinarily affected by the secondary flow and the shape of the cross-section [8] in the straight trapezoidal channel. The value of shear stress increased in the bottom of the wall and a dramatical variation from the apex to the middle of the sidewall was observed with increasing slope coefficients [8,9] in straight trapezoidal channel.

The curved trapezoidal channel is approximate to the common form in nature. The phenomenon of multi-vortex was also observed in curved trapezoidal channel [2], however, the influence of slope coefficient had not been explained completely. Farhadi et al. [10] investigated the transformation law of multi-vortex in the cross-section for the varying Froude number with the slope coefficient of 2.375 and Termini and Piranino [11] suggested that counter rotating circulation was conspicuous only in the lower aspect ratio, while the effect of various slope coefficients and radius-to-width ratio on secondary flow had not been studied. Shukla et al. [12] predicted the free surface flow velocities and shear stress reasonably well, using the Reynolds-averaged Navier-Stokes and continuity equations and the calibration of empirical coefficients and constants in standard k-ε turbulence model was recommended to improve accuracy. The modified k-ε model adopted in a 180 degree curved trapezoidal channel by Mosalman et al. [13] demonstrated good agreement with the experimental data. Also, k-ε turbulence model has been used in other different numerical approaches, such as the grid-less models [14,15]. The investigations on the water-air flow properties were conducted by Peng et al. [16–20]. The secondary flow and vortex structures can also be investigated via the direct numerical simulations (DNS) [21–23].

The slope coefficient and radius-to-width ratio are the essential parameter for the investigation on hydraulic characteristics of the channel [8]. The research on small slope coefficient, which is common in city, is not perfect and the objective of the present work is to bridge the gap in previous literature. Simultaneously, the effect of flow properties on the radius-to-width ratios was performed. The present study verified the feasibility of RNG k-ε turbulence model by the existing experiment [24]. Next, the analysis of the effect of varying radius-to-width ratios and slope coefficients on the WTS-BA, longitudinal velocity, secondary flow, shear stress and TKE was presented.

2. Numerical Model

A large number of U-shaped channel of trapezoidal cross-sections were modeled using Gambit software [25]. Fluent [26] was used to perform the numerical simulation and the renormalization group (RNG) k-ε turbulence model originated by Yakhot and Orszag [27] was employed. The U-shaped channel with symmetrical arrangement of trapezoidal cross-section is consisted of a half circle bend of 180° with different centerline radiuses and two straight cross-sections up and down the bend of 4m in length.

The water depth of the inlet and outlet of the numerical model were 0.203 m and 0.19 m, respectively. 3D view of the numerical model and vertical view were shown in Figure 1a,b, respectively. The slope coefficient of the sidewall in the curved channel remains uniform and the shape of the cross-section or centerline radius were varied for the study. The cross-section with different slope coefficients were presented in Figure 1c. Simultaneously, the area of the cross-section which was 0.08 m^2 and the flow rate, 0.038 m^3/s, remained constant. And constant equivalent width (B_0) is 0.4 m, which was defined in Equation (1).

$$B_0 = Q/Hv \tag{1}$$

where H is water depth, v is the velocity and Q is the flow rate. In this study, H and v of the inlet of the U-shaped channel were adopted to calculate the equivalent width.

The slope coefficient (*m*) is the ratio of the projection of the wall on the horizontal plane to the vertical plane, as shown in Figure 1d. In order to distinguish the slope coefficient (*m*) from the unit meter (m), the slope coefficient (*m*) was shown in bold. The radius-to-width ratio (λ) can be defined as follows:

$$\lambda = R/B_0 \tag{2}$$

where R is centerline radius of the bed in bend, which is defined in Equation (3); B_0 is equivalent width.

$$R = 0.5(R_1 + R_2) \tag{3}$$

where R_1 and R_2 are the inner radius and outer radius of the bed in bend, respectively.

The relationships between bed width, width of the surface and equivalent width are as follows:

$$B_0 = B + mH \tag{4}$$

$$B_{up} = B + 2mH \tag{5}$$

where B is the bed width, B_{up} is the width of the surface and m is slope coefficient.

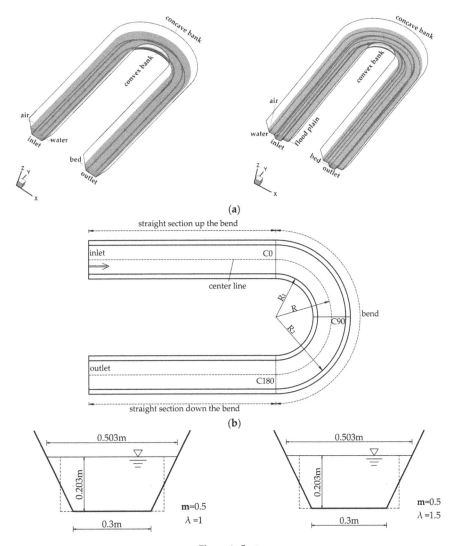

Figure 1. *Cont.*

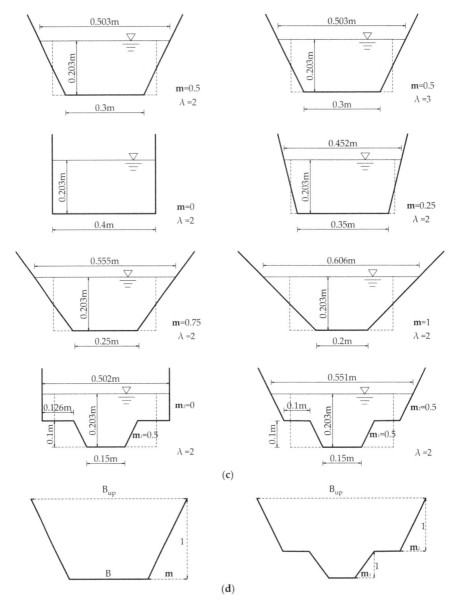

Figure 1. U-shaped channel and investigated cross-sections. (**a**) 3D view of the U-shaped channel; (**b**) vertical view of the channel; (**c**) geometry of the cross-section; (**d**) Definition of the slope coefficient.

The aspect ratio, the ratio of the equivalent width (B_0) to the inlet water depth (H) of U-shaped channel, remained 1.97 in this study. The numerical simulations were performed for four radius-to-width ratios (λ) of 1.0, 1.5, 2.0 and 3.0 and seven slope coefficients (*m*) of 0, 0.25, 0.5, 0.75, 1.0, 1.10 and 1.24. The test runs program were summarized in Table 1. The m_1 and m_2 of case9 were 0.5 and 0, respectively and the m_1 and m_2 of case10 were 0.5 and 0.5, respectively. For convenience, the slope coefficients (*m*) of case9 and case10 were calculated by the definition (shown in Figure 1d) and the values were 1.10 and 1.24, respectively.

Table 1. Summary of the test runs for the simulation, where m is slope coefficient; λ is radius-to-width ratio; θ is the central angle of the bend; B and B_0 are the width of the bed and equivalent width; R, R_1 and R_2 are centerline radius, inner radius and outer radius of the bend, respectively; Q is the flow rate; Fr is Froude number; Re is Reynolds number; B_0/H is the aspect ratio; and cells are the volume elements.

Series	m	λ	θ	B	R	R_1	R_2	Q	B_0	Fr	Re	B_0/H	Cells
			(°)	(m)	(m)	(m)	(m)	(L s^{-1})	(m)				
case1	0.50	1.0	180	0.30	0.40	0.25	0.55	38	0.40	0.33	4.98×10^4	1.73	180,000
case2	0.50	1.5	180	0.30	0.60	0.45	0.75	38	0.40	0.33	4.98×10^4	1.73	180,000
case3	0.50	2.0	180	0.30	0.80	0.65	0.95	38	0.40	0.33	4.98×10^4	1.73	180,000
case4	0.50	3.0	180	0.30	1.20	1.05	1.35	38	0.40	0.33	4.98×10^4	1.73	180,000
case5	0.00	2.0	180	0.30	0.80	0.60	1.00	38	0.40	0.33	4.66×10^4	1.73	180,000
case6	0.25	2.0	180	0.35	0.80	0.625	0.975	38	0.40	0.33	4.88×10^4	1.73	180,000
case7	0.75	2.0	180	0.25	0.80	0.675	0.925	38	0.40	0.33	4.95×10^4	1.73	180,000
case8	1.00	2.0	180	0.20	0.80	0.70	0.90	38	0.40	0.33	4.85×10^4	1.73	180,000
case9	1.10	2.0	180	0.15	0.80	0.725	0.875	38	0.40	0.33	4.03×10^4	1.73	262,500
case10	1.24	2.0	180	0.15	0.80	0.725	0.875	38	0.40	0.33	4.15×10^4	1.73	262,500

2.1. Numerical Methods

The volume of fluid (VOF) method, treating complicated free boundary configurations more flexible and efficient via volume fraction in stationary Euler mesh, was applied to track the air-water interface in the present work [28]. The differential equation in each control volume of the discrete region was integrated by the VOF method firstly and next the integral equation was linearized further to obtain the algebraic equation set of the corresponding parameters. Finally, all unknown quantities were solved. The solver employed pressure-based calculation, absolute velocity formulation and transient time and the SIMPLE algorithm was adopted for the pressure-velocity coupling. The computational domain was discretized using structured grid (shown in Figure 2) and the boundary layer cells were meshed in the vicinity of the wall and bed to ensure that they fell within the y$^+$ values required in the turbulence model. The number of the single trapezoidal U-shaped channel were $20 \times 30 \times 300$, whereas for the compound channel, were $25 \times 35 \times 300$. The number of volume cells for ten tests investigated in the paper were listed in Table 1.

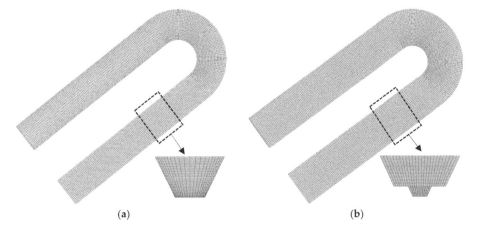

Figure 2. Meshing pattern of the U-shaped channel. (**a**) case3; (**b**) case10.

2.2. Governing Equations

The continuity equation, momentum equation, TKE budget [23] and the turbulent dissipation rate (ε) [29], adopted in the model, are shown below:

Continuity equation:

$$\frac{\partial \rho}{\partial t} + \frac{\partial \rho \tilde{u}_i}{\partial x_i} = 0 \tag{6}$$

Momentum equation:

$$\frac{\partial (\rho \tilde{u}_i)}{\partial t} + \frac{\partial (\rho \tilde{u}_i \tilde{u}_j)}{\partial x_j} = -\frac{\partial p}{\partial x_i} + \frac{\partial}{\partial x_j}\left[\mu\left(\frac{\partial \tilde{u}_i}{\partial x_j} + \frac{\partial \tilde{u}_j}{\partial x_i}\right)\right] + \frac{\partial}{\partial x_j}\left(-\rho\overline{u'_i u'_j}\right) \tag{7}$$

TKE budget:

$$
\begin{array}{ccccccccccccc}
\frac{\partial \overline{\rho K}}{\partial t} = & -C & + & P & + & T & + & \Pi & + & M & + & D & - & \overline{\varepsilon} \\
& \text{convection} & & \text{production} & & \text{turbulent} & & \text{pressure} & & \text{compressible} & & \text{viscous} & & \text{TKE} \\
& \text{term} & & \text{term} & & \text{transport} & & \text{dilatation} & & \text{mass flux} & & \text{diffusion} & & \text{dissipation} \\
& & & & & \text{term} & & \text{term} & & \text{contribution} & & \text{term} & & \text{rate term} \\
& & & & & & & & & \text{term} & & & &
\end{array} \tag{8}
$$

$$C = \frac{\partial \overline{\rho} \tilde{u}_j \overline{K}}{\partial x_j} \tag{9}$$

$$p = -\overline{\rho u'_i u'_j}\frac{\widetilde{\partial u_i}}{\partial x_j} \tag{10}$$

$$T = -\frac{\partial}{\partial x_j}\left[\overline{\rho}\frac{\overline{u'_i u'_i u'_j}}{2}\right] + \overline{p' u'_i \sigma_{ij}} \tag{11}$$

$$\Pi = -\overline{p'\frac{\partial u'_i}{\partial x_j}} \tag{12}$$

$$M = -\overline{\rho u'_i}\left(\frac{\partial \overline{\sigma}_{ij}}{\partial x_j} - \frac{\partial \overline{p}}{\partial x_i}\right) \tag{13}$$

$$D = \frac{\partial}{\partial x_j}\left(\overline{\sigma'_{ij} u'_i}\right) \tag{14}$$

k equation:

$$\frac{\partial\left(\rho \overline{k}\right)}{\partial t} + \frac{\partial\left(\rho \overline{k} \overline{u}_i\right)}{\partial x_i} = \frac{\partial}{\partial x_j}\left[\left(\mu + \frac{\mu_t}{\sigma_k}\right)\frac{\partial \overline{k}}{\partial x_j}\right] + G_k - \rho \varepsilon \tag{15}$$

ε equation:

$$\frac{\partial(\rho \overline{\varepsilon})}{\partial t} + \frac{\partial(\rho \overline{\varepsilon} \overline{u}_i)}{\partial x_i} = \frac{\partial}{\partial x_j}\left[\left(\mu + \frac{\mu_t}{\sigma_\varepsilon}\right)\frac{\partial \varepsilon}{\partial x_j}\right] + \frac{C^*_{1\varepsilon}\overline{\varepsilon}}{\overline{K}}G_k - C_{2\varepsilon}\rho\frac{\overline{\varepsilon}^2}{\overline{K}} \tag{16}$$

$$\mu_t = \rho C_\mu \frac{\overline{K}^2}{\overline{\varepsilon}} \tag{17}$$

$$G_k = \mu_t\left(\frac{\partial \overline{u}_i}{\partial x_j} + \frac{\partial \overline{u}_j}{\partial x_i}\right)\frac{\partial \overline{u}_i}{\partial x_j} \tag{18}$$

$$C^*_{1\varepsilon} = C_{1\varepsilon} - \frac{\eta(1 - \eta/\eta_0)}{1 + \beta\eta^3} \tag{19}$$

$$\eta = \sqrt{\frac{G_k}{\rho C_{\mu\varepsilon}}} \tag{20}$$

where K is the TKE, μ is the dynamic viscosity, μ_t is the turbulence viscosity, ρ is the corresponding density, u_i is the velocity component in the ith direction, t is the time and C_μ, C_μ, $C_{1\varepsilon}$, $C_{2\varepsilon}$, β and η_0 are empirical constants. In the RNG k-ε model, the empirical constants are given as $C_\mu = 0.0845$, $C_\mu = C_\varepsilon = 0.7179$, $C_{1\varepsilon} = 1.42$, $C_{2\varepsilon} = 1.68$, $\beta = 0.012$, and $\eta_0 = 4.38$ [29].

The compressible mass flux contribution term is zero for the incompressible fluid. Reynolds stress independent components can be solved via using Equations (9)–(14) above and $\varepsilon = \sigma'_{ik}\frac{\partial u'_i}{\partial x_k}$, however, supererogatory components, that is, the trible correlation matrix for fluctuating velocity $(\overline{u'_i u'_i u'_j})$ and correlation matrix for fluctuating velocity and pressure $(\overline{p' u'_i})$, are added in the turbulence kinetic energy budget, which made the budget is not closed. So K and ε equations are recommended to close the budget.

2.3. Boundary Conditions and Assumptions

In this study, without considering the dissolution of each other, the open channel flow was employed by VOF sub-models with air-water two phases. Pressure-inlet was adopted for the water surface and the pressure of the air-water interface was standard atmospheric pressure. Both sidewalls and bed of the channel, whose roughness height were 0.25 mm, were stationary wall and the fluent standard wall function approach was adopted. The velocity-inlet and pressure-outlet were adopted for the inlet and outlet of the U-shaped channel, respectively and the mean velocity of the inlet cross-section was 0.462 m/s.

2.4. Model Verification

It is necessary to validate the numerical model before conducting further research. Therefore, the mesh convergence study were performed via using grid convergence index (GCI) [30] with grid numbers of approximately 0.12 million, 0.20 million and 0.30 million. The effect of the grid sizes on the uncertainty of the computational velocity distribution in vertical direction was shown in Figure 3, which showed that the discretization uncertainties were little in most positions. The maximum of the GCI for the velocity profiles in concave bank, center line and convex bank were 6.3%, 2.4% and 8.2% and corresponding to 0.05 m/s, 0.06 m/s and 0.07 m/s, respectively. The grid number was set as 0.20 million to optimize the computational time.

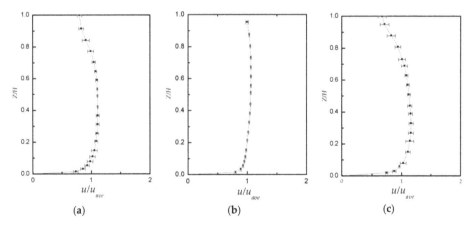

Figure 3. Discretization error bars computed using the GCI index for different grid densities, (a) the concave bank, (b)center line and (c)convex bank of the bend apex.

In addition, the results calculated by RNG k-ε turbulence model were compared with the experimental data obtained from Ma [24]. The geometry of the experimental curved channel was exactly the same as case5 in the present article, except the lengths of two straight cross-sections of the experiment were 6 m. The sidewalls and bed composed of cement had a low roughness and the velocity was obtained by ADV in experiment.

In the present work, the length of two straight cross-sections all were 4 m or 6 m in the U-shaped channel, which were compared with the experiment, respectively. The RNG k-ε model was adopted for simulation with the same experimental conditions and the data of simulation and experiment were dimensionless and compared. The longitudinal velocity in different position of cross-section of C0, C90 and C180, the structure of secondary flow of the cross-section in the bend apex and shear stress in the bend were analyzed to verify the numerical model. In Figure 4, the vertical coordinate is the relative depth, where Z is the instantaneous height and H is the water depth in corresponding position; the horizontal coordinate is the ratio of the longitudinal velocity (u) to the corresponding average velocity (u_{ave}).

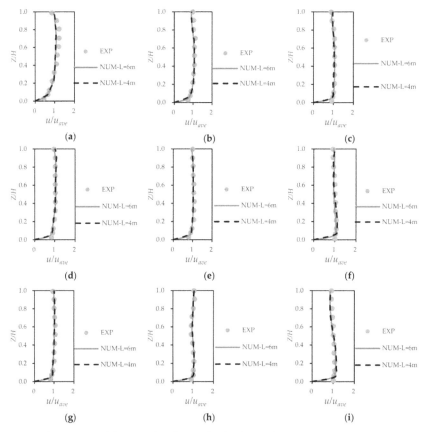

Figure 4. Comparison of dimensionless longitudinal velocity of various cross-sections between the numerical simulation and experiment, (**a–c**) the concave bank, center line and convex bank of the inlet of the bend, respectively; (**d–f**) the concave bank, center line and convex bank of the bend apex, respectively; and (**g–i**) the concave bank, center line and convex bank of the outlet of the bend, respectively.

It can be seen from Figure 4 that, in most locations, the longitudinal velocity distribution of the simulation exhibited a good agreement with the experimental in the vertical direction. The maximums of absolute error (the absolute value of the difference between experiment and numerical simulation) and relative error (the ratio of the absolute error to experiment) in different location were presented in Table 2. A well consistency was shown for the structure of secondary flow in Figure 5. Although the simulated shear stress in the bed was slightly larger than the experimental data, shown in Figure 6b, the commutated results were fairly consistent with Ma's investigation [24].

Table 2. Summary of the maximums of relative error and absolute error in different location. The locations of (a–i) were introduced in Figure 4.

		a	b	c	d	e	f	g	h	i
L = 6 m	Absolute error (m/s)	0.03	0.05	0.06	0.04	0.06	0.05	0.01	0.06	0.07
	Relative error (%)	12	10.8	13.9	15.4	12	12.5	4.6	13.3	14.6
L = 4 m	Absolute error (m/s)	0.03	0.04	0.05	0.04	0.07	0.06	0.02	0.06	0.06
	Relative error (%)	12	8.7	11.6	15.4	14	15	7.6	13.3	12.2

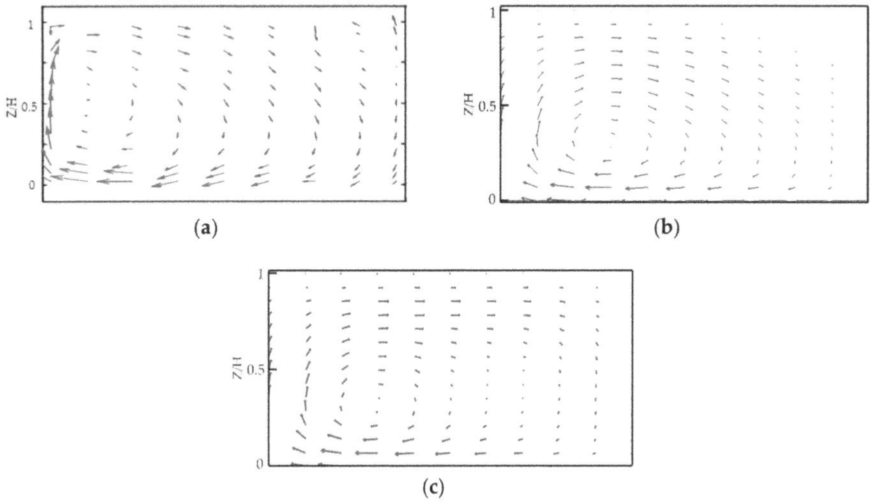

(a) (b)

(c)

Figure 5. Comparison of the secondary flow between the numerical simulation and experiment, (a) experiment [24], (b) numerical simulation L = 6 m, (c) numerical simulation L = 4 m.

Consequently, the RNG k-ε model could be used for the numerical investigation of the U-shaped channel and ensure the accuracy of the data. To optimize the computational time, the length of 4m were selected for the straight sections. Analyses were performed from the aspects of the WTS-BA, longitudinal velocity, secondary flow, shear stress and TKE in next section.

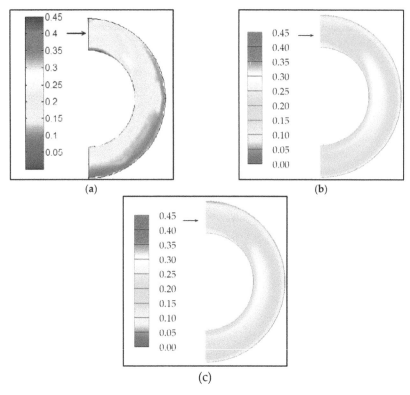

Figure 6. Comparison of the shear stress between the numerical simulation and experiment, (a) experiment [24], (b) numerical simulation L = 6 m, (c) numerical simulation L = 4 m.

3. Results and Discussion

3.1. Analysis of the Water Surface Transverse Slope in Bend Apex

The centrifugal motion of the water in bend is a result of centrifugal force, gravity and the friction with surrounding water. The transverse slope is presented on the water free surface, due to the descending water level in convex bank and rising water level in the concave bank [23]. The WTS varies with radius-to-width ratio, slope coefficient and cross-section shape [31]. The equation for calculating the WTS of cross-section of the bend apex (C90) is [31]:

$$J_r = \left(1 + \frac{g}{\kappa^2 C^2}\right)\frac{u^2}{gr} \tag{21}$$

where C is the Chezy coefficient, κ is the Carmen constant, g is the acceleration of gravity, r is the radius and u is the corresponding velocity. For this investigation, the empirical values $\frac{1}{\kappa^2}$ calculated by Zhang [32], Liu [31] and Rozovskii [32] were 5.75, 6.25 and 4, respectively. The cross-section in the bend apex (C90) was selected for further investigations, r and u were the corresponding centerline radius and the average longitudinal velocity of the cross-section of the bend apex, respectively.

The free surface level across the across-section in the bend apex with varying radius-to-width ratios and slope coefficients were shown in Figure 7a,b, respectively. The vertical coordinate is the water level and the horizontal coordinate is the relative distance from a point on the cross-section to the convex bank on the surface. The WTS-BA with varying radius-to-width ratios and slope coefficients were shown in Figure 8a,b, respectively. The vertical coordinate is a dimensionless WTS and the horizontal ordinate is radius-to-width ratio or slope coefficient.

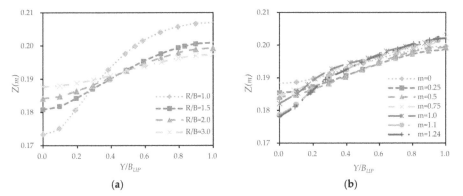

Figure 7. The free surface level across the across-section in the bend apex (C90), (**a**) the radius-to-width ratios (λ) of 1.0, 1.5, 2.0 and 3.0, (**b**) the slope coefficients (*m*) of 0, 0.25, 0.5, 0.75, 1.0, 1.10 and 1.24.

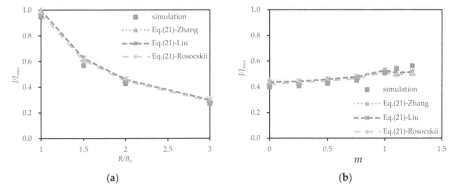

Figure 8. The WTS-BA (C90), (**a**) the radius-to-width ratios (λ) of 1.0, 1.5, 2.0 and 3.0; (**b**) the slope coefficients (*m*) of 0, 0.25, 0.5, 0.75, 1.0, 1.10 and 1.24.

The WTS decreased with increasing radius-to-width ratios, which can be explained by the decrease in the water level in the concave bank and the increased water level in the convex bank, shown in Figures 7a and 8a. It can be considered that the numerical simulation exhibited a good agreement with the theoretical Equation (21) for the maximum error is 8.2%, in spite of the fact that the data obtained via simulation was smaller than calculated by Equation (21) with the varying Carmen constants.

The WTS increased with the increasing slope coefficients, for the water level in the concave bank slightly decreased first and then increased, while decreased in a relative larger extent in the concave bank, shown in Figures 7b and 8b. As can be seen from Figure 8b, the data obtained via numerical simulation was smaller than calculated by Equation (21) except the compound cross-section, the maximum errors of which were 7.2% and 10.1%, respectively.

The reason was that Equation (21) was derived on the basis of a single trapezoidal cross-section, which could not be applicable to the compound cross-section. In the whole, the numerical simulation was consistent with the Equation (21) for the varying slope coefficients. As can be seen from Figures 7b and 8a, the effect of the radius-to-width ratio was greater than the slope coefficient on the WTS in the bend apex.

3.2. Analysis of the Longitudinal Velocity

Chang [33] believes that the vertical distribution of the longitudinal velocity in the bend conforms to the Equation (22):

$$\frac{u}{u_{ave}} = \left(1 + \frac{1}{n}\right)\left(\frac{Z}{H}\right)^{\frac{1}{n}} \tag{22}$$

where, n is obtained by equation $n = \kappa\sqrt{8/f}$, κ is the Carmen constant, f is the friction constant; Z is the instantaneous height and H is the corresponding water depth and u and u_{ave} are the longitudinal velocity and the corresponding average in the cross-section. The vertical distributions of the longitudinal velocity calculated by Equation (22) were shown in Figures 9 and 10.

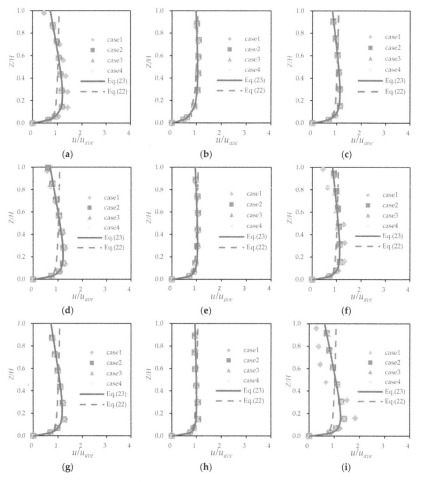

Figure 9. Vertical distribution of the dimensionless longitudinal velocity of the bend in different positions with varying radius-to-width ratios, (**a–c**) the concave bank, centerline and convex bank of the inlet of the bend, respectively; (**d–f**) the concave bank, centerline and convex bank of the bend apex, respectively; (**g–i**) the concave bank, centerline and convex bank of the outlet of the bend, respectively.

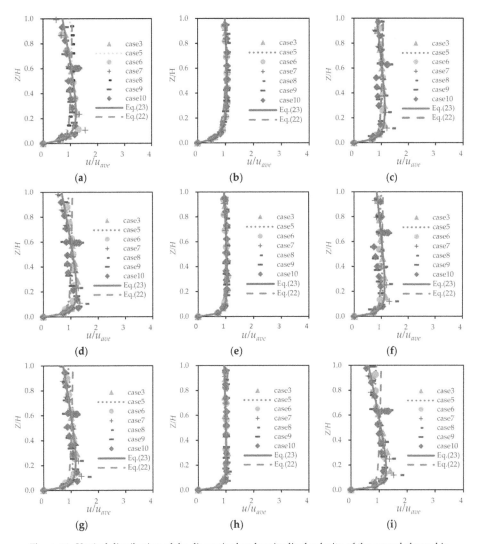

Figure 10. Vertical distribution of the dimensionless longitudinal velocity of the curved channel in different positions with varying slope coefficients, (**a–c**) the concave bank, center line and convex bank of the inlet of the bend, respectively; (**d–f**) the concave bank, center line and convex bank of the bend apex, respectively; (**g–i**) the concave bank, center line and convex bank of the outlet of the bend, respectively.

Figures 9 and 10 showed the vertical distribution of the dimensionless longitudinal velocity of the bend in different positions with the varying radius-to-width ratios and slope coefficients, respectively. In Figures 9 and 10, the vertical coordinate is the relative depth, where Z is instantaneous height, H is the corresponding water depth and the horizontal coordinate is the ratio of the longitudinal velocity (u) to the corresponding average velocity (u_{ave}). The velocity locating close to the sidewall at 5%B_0 was extracted to represent the velocity at the sidewall, as the velocity at the sidewall is zero.

As shown in Figure 10, the vertical distribution of longitudinal velocity calculated by Equation (22) was relatively consistent with the distribution of case5 (whose cross-section was rectangular).

The longitudinal velocity in the bend was affected by the shape of cross-section, radius-to-width ratio and other factors [34–37]. The differences of the velocity distribution in the both banks were bigger than the center line, which was shown in Figures 9 and 10. The longitudinal velocity extracted via the numerical simulation conformed to the Equation (13) in the center line, yet the obvious errors were exhibited in the both banks. In this study, Equation (23) was proposed, which was consistent with the vertical distribution of dimensionless longitudinal velocity for the varying operating conditions in any position:

$$\frac{u}{u_{ave}} = A\left(\frac{Z}{H}\right) + \frac{B}{\left(\frac{Z}{H}\right)} + C \tag{23}$$

where, A is equal to -0.09 or -0.03 for the sidewall or the center line, respectively, B is equal to -0.05 and the range of C is from 1 to 2.

Corresponding optimal fitting curves (the red continuous line) calculated by Equation (16) in different positions were shown in Figures 9 and 10. It can be seen that the vertical distribution of the longitudinal velocity in the bend with varying operating conditions were consistent with the optimal fitting curves in the center line but deviations are presented in the both banks. The deviations for the $\lambda = 1.0$ were conspicuous for the obvious gradient of the velocity in both banks and the vertical distribution for the $\lambda > 1.0$ were consistent with the optimal fitting curve. The deviations were lesser than others when the slope coefficient was 0.5. As shown in Figures 9 and 10, the influence of the slope coefficient on the distribution of longitudinal velocity was more conspicuous than the radius-to-width ratio.

The vertical distribution of the longitudinal velocity in both banks at the inlet of the bend were different for the varying operating conditions and demarcation were each $Z/H = 0.5$. As shown in Figure 9, the longitudinal velocity in the concave bank gradually decreased or increased for the $Z/H < 0.5$ or $Z/H > 0.5$ with increasing radius-to-width ratios but did not show the obvious trend for the varying radius-to-width ratios in the convex bank. As shown in Figure 10, the longitudinal velocity gradually increased or decreased for the $Z/H < 0.5$ or $Z/H > 0.5$ with increasing slope coefficients in the both banks. The difference of the velocity distribution between two types of compound cross-sections was slight.

The vertical distributions in the concave bank of the bend apex failed to show regular law for the varying operating conditions and were similar to the distributions in the convex bank of the inlet of the bend in the convex bank, whose demarcations were each $Z/H = 0.6$. The distributions of the longitudinal velocity of the outlet in the bend were similar to the distribution of the cross-section in the bend apex and the demarcations were $Z/H = 0.4$ and $Z/H = 0.5$.

3.3. Analysis of the Secondary Flow

The secondary flow is occurred for the momentum exchange [7]. The force of weeny water column are as follow [32]:

$$\Delta F = \underset{\substack{\text{gravity}\\\text{force}}}{g} + \underset{\substack{\text{pressure}\\\text{force}}}{\rho g h J_r} - \underset{\substack{\text{centrifugal}\\\text{force}}}{\alpha_0 \frac{u_r^2}{gr}} + \underset{\text{resistance}}{F_0} \tag{24}$$

where, g is the acceleration of gravity, r is the radius, h and u is the corresponding height and velocity, J_r is the transverse slope, α_0 is coefficient and F_0 is the summation of other resistance; in this investigation, gravity force is neglected in horizontal direction.

The structure of secondary flow of the cross-section in the bend apex for varying operating conditions was shown in Figure 11. In the present study, the clockwise vortex was defined as positive and counterclockwise vortex is reverse. As shown in Figure 11a–d, a reverse vortex is captured in case1 and the vortex structure do not change significantly from case2 to case4. The reason was that the WTS decreased sharply though the velocity increase in the concave bank with increasing radius-to-width ratios (shown in Figure 8a). The ΔF was positive for the convex bank and negative for the concave bank only in case1 in horizontal direction.

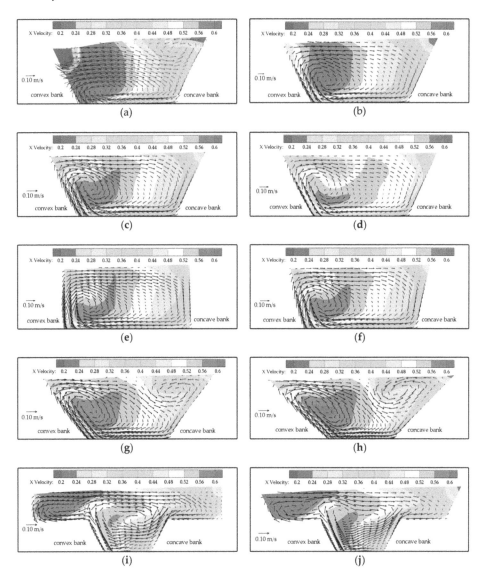

Figure 11. Structure of secondary flow of the cross-section in the bend apex for varying operating conditions, (**a–j**) case1–case10, respectively.

As shown in Figure 11c,e–h, a reverse vortex are gradually obvious in the cross-section of the bend apex with increasing slope coefficients. The obvious double vortexes were captured in Figure 11g,h, the positive vortex near the convex bank was close to the bed but the reverse vortex was generated near the water surface in the concave bank. The reason was that velocity decreased in the concave bank (shown in Figure 11), yet the transverse slope increased (shown in Figure 8b) with increasing slope coefficients and the reverse vortex occur for the ΔF was negative at a certain position. Obvious multi-vortex was captured in Figure 10i,j in the present study, due to the irregular velocity distribution caused by the compound section.

In this study, the equation of the intensity of secondary flow [38] is as follows:

$$\omega = \frac{u_w^2 + u_v^2}{u_l^2 + u_w^2 + u_v^2} \qquad (25)$$

where, ω is the intensity of secondary flow, u_w is the streamwise velocity, u_w is the spanwise velocity and u_V is the vertical velocity.

Figure 12a,b showed the graph of dimensionless intensity of secondary flow for varying radius-to-width ratios and slope coefficients. The vertical or horizontal coordinate were the dimensionless intensity of secondary flow or the angle of the bend, respectively. The distribution of the intensity of secondary flow for $m = 0$ is consistent with the investigation of Ma [24], and Vaghefi et al. [39] also presented that the maximum intensity of secondary flow appears at approximately 60°~80° via various analysis methods, which demonstrated good agreement with the above conclusion.

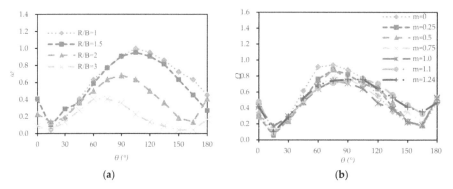

Figure 12. Graph of the dimensionless intensity of secondary flow for varying radius-to-width ratios and slope coefficients, (**a**) radius-to-width ratio, (**b**) slope coefficient.

The intensity of secondary flow decreased firstly, then increased and finally decreased with increasing slope coefficients. The reason why sudden increase was occurred was that the cross-section changes from trapezoid to compound cross-section. It can be considered that the intensity of secondary flow decreased with increasing slope coefficients but the intensity was the weakest when the slope coefficient was 1.0. As the radius-to-width ratio increased, the intensity of secondary flow decreased and the corresponding angle of the maximum intensity of secondary flow transforms from 105° to 60°.

3.4. Analysis of the Shear Stress

The higher the shear stress on the bed and sidewall generated by the water in the bend, the more obvious erosion or siltation in channel and shear stress was investigated by many scholars [8,9,38,39]. The shear stress cannot be measured directly in actual research and various calculation methods were adopted to calculate the shear stress via other physical parameters, for example, energy slope and hydraulic radius [40]. The average velocity and correlation coefficient were employed via linear calculation to determine the shear stress by Williams et al. but this calculation method was not accurate in local location [41] Fluctuating velocity [42] and TKE [43] can be used to calculate the shear stress. The TKE method less affected by local streamline is more suitable for complex areas in the bend [44]. In this study, TKE method was adopted to calculate the shear stress and the calculation equation is as follows:

$$\tau = C_1 \left[0.5\rho \left(\overline{u'^2} + \overline{v'^2} + \overline{\omega'^2} \right) \right] \qquad (26)$$

where C_1 is 0.19 [43], ρ is corresponding density and u', v' *and* ω' are the streamwise, spanwise and vertical fluctuating velocity, respectively.

The distributions of the shear stress in the bend were shown in Figure 13. The maximum shear stress and the corresponding angle for varying operating conditions were shown in Figures 14 and 15, respectively.

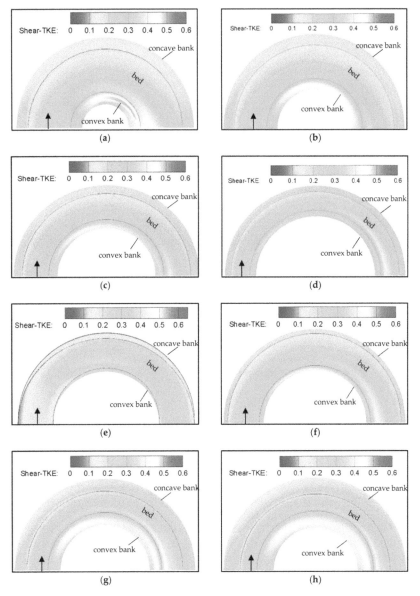

Figure 13. *Cont.*

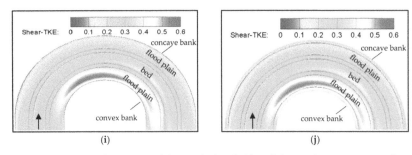

Figure 13. Shear stress distribution of the riverbed and sidewall for varying operating conditions, (**a**)–(**j**) case1–case10, respectively.

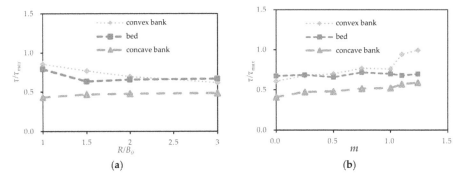

Figure 14. Maximum shear stress for varying operating conditions, (**a**) radius-to-width ratio, (**b**) slope coefficient.

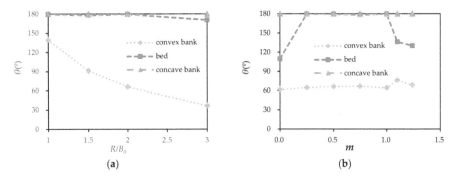

Figure 15. Corresponding angle of the maximum shear stress for varying operating conditions, (**a**) radius-to-width ratio, (**b**) slope coefficient.

According to Figure 13a–d, the transformation law of the shear stress in the convex bank, bed, concave bank decreased firstly and then increased with increasing radius-to-width ratios and the extent of maximum value variation were 27%, 15% and 12%, respectively (shown in Figure 14a). The position of the maximum shear stress moves to the upstream and the corresponding angle moved from 140° to 37° or 180° to 170° in the convex bank or bed (shown in Figure 15a) but did not move in the concave bank. Why shear stress was significantly higher than others for $\lambda = 1$ was that the momentum exchange and turbulence transfer violently for the dramatic change of the water in the bend.

The shear stress increased in the convex bank and concave bank with the increasing slope coefficients, which was consistent with the conclusion inferred by Ansar et al. [9], who revealed that the shear stress of sidewall increased with increasing slope coefficients in straight trapezoidal channel. Hence, the law of shear stress in both banks increased with increasing slope coefficients can be adopted in both straight and cured trapezoidal channels. The shear stress in the bed failed to present the regular variations with increasing slope coefficients and was higher than the trapezoidal cross-section in the convex bank of the compound cross-section in the same position. The position of maximum shear stress did not show the obvious law in both banks and bed with increasing slope coefficients.

3.5. Analysis of the TKE

Blanckaert K. [45] and Ma [24] investigated the distribution of TKE in the cross-section of the bend. In this study, the equation of TKE calculation is as follows:

$$K = \frac{1}{2}\left(u'^2 + v'^2 + \omega'^2\right) \tag{27}$$

where K presents the TKE, u', v', ω' are the streamwise, spanwise and vertical fluctuating velocity, respectively.

The graphs of the dimensionless TKE with the radius-to-width ratios and slope coefficients at various locations of the bend were shown in Figure 16. The relational graphs of the change extent of the dimensionless TKE with the radius-to-width ratios and the slope coefficients were shown in Figures 17 and 18, respectively. The TKE located at 5%B_0 was extracted to represent the sidewall. The turbulent kinetic energy was nondimensionalized via the mean turbulent kinetic energy (TKE$_{ave}$).

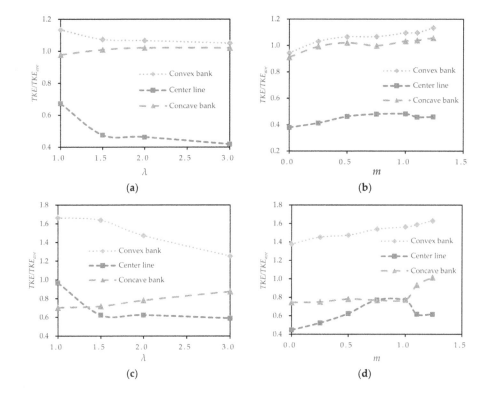

Figure 16. *Cont.*

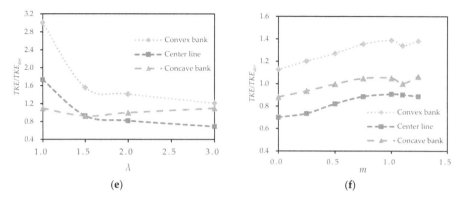

(e)   **(f)**

Figure 16. Graphs of the dimensionless TKE with the radius-to-width ratios and slope coefficients in different positions of the bend, (**a,b**) the inlet of the bend; (**c,d**) the apex of the bend; (**e,f**) the outlet of the bend.

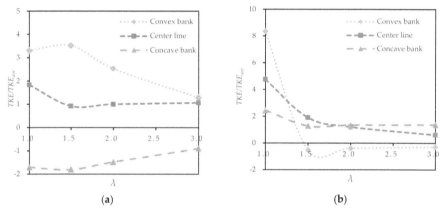

(a)   **(b)**

Figure 17. Relation graphs of the change extent of the dimensionless TKE with radius-to-width ratios, (**a**) the difference between inlet and apex of the bend, (**b**) the difference between apex and outlet of the bend.

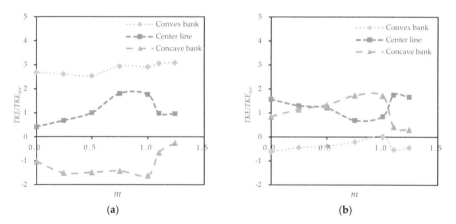

(a)   **(b)**

Figure 18. Relation graphs of the change extent of the dimensionless TKE with slope coefficients, (**a**) the difference between inlet and apex of the bend, (**b**) the difference between apex and outlet of the bend.

According to Figure 16, the TKE on both sidewalls was larger than the center line in the bend. The TKE in the convex bank and central line gradually decreased and increased slightly in the concave bank with increasing radius-to-width ratios. The reason was that the turbulence and pulsation were weaker in the convex bank and central line and more violent in the concave bank with the increasing radius-to-width ratios. The TKE of all single trapezoidal cross-section increased with increasing slope coefficients.

As the flow of water in the bend, the TKE obviously increased in the convex bank and center line and obviously reduced for the concave bank in the cross-section of the bend apex, as shown in Figure 16a–d. The amplitudes decreased with increasing radius-to-width ratios and increased with increasing slope coefficients, which were shown in Figures 17a and 18a.

In the outlet of the bend, the TKE significantly increased in the concave bank and center line and decreased in the convex bank, as shown in Figure 16c–f. The amplitudes decreased with increasing radius-to-width ratios and decreased in the convex bank and center line with increasing slope coefficients. An increase was observed in the concave bank for the single trapezoidal cross-section but an opposite trend was presented for the compound cross-section with the increasing slope coefficients.

4. Conclusions

In this study, the numerical model was verified via the existing experimental data and then numerical simulation was performed to investigate the hydraulic characteristics of the U-shaped channel with trapezoidal cross-section under various conditions, that is, four types of radius-to-width ratio and seven types of slope coefficient with the aspect ratio being constant. The conclusions are as follow:

(1) The transverse slope decreased with increasing radius-to-width ratios and increased with increasing slope coefficients. The effect of the radius-to-width ratio on the WTS in the bend apex was greater than that of the slope coefficient.

(2) A new equation concerning the vertical distribution of the longitudinal velocity in trapezoidal cross-section channel was proposed. The influence of the slope coefficient on the distribution of the longitudinal velocity was higher than that of the radius-to-width ratio.

(3) The slope coefficient had an effect on the structure of secondary flow in the bend. When the slope coefficient was large enough ($m = 0.75$ in the present study), obvious double vortexes were captured in the bend apex. The corresponding angle of the maximum intensity of secondary flow in the bend gradually moved to the upstream with increasing radius-to-width ratios. The intensity of secondary flow in the bend gradually decreased with the increase in slope coefficients.

(4) The maximum of the shear stress of the bend was observed in the convex bank, which decreased gradually and the corresponding angle decreased with increasing radius-to-width ratios. As the slope coefficient increased, the maximum of the shear stress in the convex bank gradually increased.

(5) The maximum TKE of the section in the bend decreased with increasing radius-to-width ratios and increased with increasing slope coefficients.

Author Contributions: Conceptualization and Methodology, J.Z.; Numerical Simulation, R.H.; Investigation, R.H.; Writing-Original Draft Preparation, R.H.; Writing-Review & Editing, J.Z.; Supervision, J.Z.; Project administration, J.Z.; Funding acquisition, J.Z.

Funding: This research was funded by the National Key Research and Development Program of China (No. 2016YFC0401707), National Science Fund for Distinguished Young Scholars (No. 51625901).

Conflicts of Interest: The authors declare no conflict of interest.

References

1. Blanckaert, K.; Graf, W.H. Momentum transport in sharp open-channel bends. *J. Hydraul. Eng.* **2004**, *130*, 186–198. [CrossRef]
2. Hey, R.D.; Thorne, C.R. Secondary flows in river channels. *Area* **1975**, *7*, 191–195.
3. Kabiri, S.A.; Farshi, F.; Chamani, M.R. Boundary shear stress in smooth trapezoidal open channel flows. *J. Hydraul. Eng.* **2013**, *139*, 205–212. [CrossRef]
4. Knight, D.W.; Omran, M.; Tang, X.N. Modeling depth-averaged velocity and boundary shear in trapezoidal channels with secondary flows. *J. Hydraul. Eng.* **2007**, *133*, 39–47. [CrossRef]
5. Shiono, K.; Muto, Y. Complex flow mechanisms in compound meandering channels with overbank flow. *J. Fluid Mech.* **1998**, *376*, 221–261. [CrossRef]
6. Knight, D.W.; Shiono, K. Turbulence measurements in a shear layer region of a compound channel. *J. Hydraul. Res.* **1990**, *28*, 175–196. [CrossRef]
7. Tominaga, A.; Nezu, L.; Ezaki, K.; Nakagawa, H. Three-dimensional turbulent structure in straight open channel flows. *J. Hydraul. Res.* **1989**, *27*, 149–173. [CrossRef]
8. Yang, S.Q.; Lim, S.Y. Boundary shear stress distributions in trapezoidal channels. *J. Hydraul. Res.* **2005**, *43*, 98–102. [CrossRef]
9. Ansar, K.; Morvan, H.P.; Hargreaves, D.M. Numerical Investigation into Secondary flow and Wall Shear in Trapezoidal Channels. *J. Hydraul. Eng.* **2011**, *137*, 432–440. [CrossRef]
10. Farhadi, A.; Sindelar, C.; Tritthart, M.; Glas, M.; Blanckaert, K.; Habersack, H. An investigation on the outer bank cell of secondary flow in channel bends. *J. Hydro-Envrion. Res.* **2018**, *18*, 1–11. [CrossRef]
11. Termini, D.; Piraino, M. Experimental analysis of cross-sectional flow motion in a large amplitude meandering bend. *Earth Surf. Process. Landf.* **2011**, *36*, 244–256. [CrossRef]
12. Shukla, D.R.; Shiono, K. CFD modelling of meandering channel during floods. *Water Manag.* **2008**, *161*, 1–12. [CrossRef]
13. Mosalman, A.; Mosalman, M.; Yazdi, H.M. Equations of unsteady flow in curved trapezoidal channels. *Int. J. Phys. Sci.* **2011**, *6*, 671–676. [CrossRef]
14. Shao, S.D. Simulation of breaking wave by SPH method coupled with k-epsilon model. *Int. J. Numer. Methods Fluids* **2006**, *44*, 338–349. [CrossRef]
15. Shao, S.D.; Ji, C.M. SPH computation of plunging waves using a 2-D sub-particle scale (SPS) turbulence model. *J. Hydraul. Res.* **2010**, *51*, 913–936. [CrossRef]
16. Peng, Y.; Zhou, J.G.; Burrows, R. Modelling the free surface flow in rectangular shallow basins by lattice Boltzmann method. *J. Hydraul. Eng.* **2011**, *137*, 1680–1685. [CrossRef]
17. Peng, Y.; Zhou, J.G.; Burrows, R. Modelling solute transport in shallow water with the lattice Boltzmann method. *Comput. Fluids* **2011**, *50*, 181–188. [CrossRef]
18. Peng, Y.; Mao, Y.F.; Wang, B.; Xie, B. Study on C-S and P-R EOS in pseudo-potential lattice Boltzmann model for two-phase flows. *Int. J. Mod. Phys. C* **2017**, *28*, 1750120. [CrossRef]
19. Bai, R.; Liu, S.; Tian, Z.; Wang, W.; Zhang, F. Experimental investigations of air-water flow properties of offse-aerator. *J. Hydraul. Eng.* **2017**, *144*, 04017059. [CrossRef]
20. Bai, R.; Zhang, F.; Liu, S.; Wang, W. Air concentration and bubble characteristics downstream of a chute aerator. *Int. J. Multiph. Flow* **2016**, *87*, 156–166. [CrossRef]
21. Ducoin, A.; Shadloo, M.S.; Roy, S. Direct Numerical Simulation of flow instabilities over Savonius style wind turbine blades. *Renew. Energy* **2017**, *105*, 374–385. [CrossRef]
22. Mendez, M.; Shadloo, M.S.; Hadjadj, A.; Ducoin, A. Boundary layer transition over a concave surface caused by centrifugal instabilities. *Comput. Fluids* **2018**, *171*, 135–153. [CrossRef]
23. Shadloo, M.S.; Hadjadj, A.; Hussain, F. Statistical behavior of supersonic turbulent boundary layers with heat transfer at $M_\infty = 2$. *Int. J. Heat Fluid Flow* **2015**, *53*, 113–134. [CrossRef]
24. Ma, M. Flow Characteristics and Influence of Geometric Factors in a Bend. Ph.D. Thesis, Xi'an University of Technology, Xi'an, China, 2017. (In Chinese)
25. Bai, Z.L.; Peng, Y.; Zhang, J.M. Three-Dimensional Turbulence Simulation of Flow in a V-Shaped Stepped Spillway. *J. Hydraul. Eng.* **2017**, *143*, 06017011. [CrossRef]
26. Sun, X.; Shiono, K.; Rameshwaran, P. Modelling vegetation effects in irregular meandering river. *J. Hydraul. Res.* **2010**, *48*, 775–783. [CrossRef]

27. Yakhot, V.; Orszag, S.A. Renormalization group analysis of turbulence: I. Basic Theory. *J. Sci. Comput.* **1986**, *1*, 3–51. [CrossRef]

28. Hirt, C.W.; Nichols, B.D. Volume of Fluid (VOF) Method for the Dynamics of Free Boundaries. *J. Comput. Phys.* **1981**, *39*, 201–225. [CrossRef]

29. Han, Z.; Reitz, R.D. Turbulence modeling of internal combustion engines using RNG k~ε models. *Combust. Sci. Technol.* **1995**, *106*, 267–295. [CrossRef]

30. Celik, I.B.; Ghia, U.; Roache, P.J.; Freitas, C.J. Procedure of estimation and reporting of uncertainty due to discretization in CFD applications. *J. Fluids Eng.* **2008**, *130*, 078001–078004. [CrossRef]

31. Liu, H.F. The study of water surface in curved channel. *J. Hydraul. Eng.* **1990**, *4*, 46–50. [CrossRef]

32. Zhang, H.W.; Lv, X. *Hydraulics in the Curved Channel*, 1st ed.; China Water Power Press: Beijing, China, 1993; pp. 48–49. ISBN 7–120-02333-0/TV•835. (In Chinese)

33. Chang, H.H. *Fluvial Processes in River Engineering*; John Wiley and Sons: New York, NY, USA, 1988.

34. Sarma, K.V.N.; Lakshminatayana, P.; Rao, N.S.L. Velocity distribution in smooth rectangular open channels. *J. Hydraul. Eng.* **1988**, *109*, 270–289. [CrossRef]

35. Yan, J.; Tang, H.W.; Xiao, Y.; Li, K.J.; Tian, Z.J. Experimental study on influence of boundary on location of maximum velocity in open channel flows. *Water Sci. Eng.* **2011**, *4*, 185–191. [CrossRef]

36. Song, T.; Graf, W.H. Velocity and turbulence distribution in unsteady open-channel flows. *J. Hydraul. Eng.* **1996**, *122*, 141–154. [CrossRef]

37. Kirkgoz, M.S. Turbulent Velocity Profiles for Smooth and Rough Open Channel Flow. *J. Hydraul. Eng.* **1989**, *115*, 1543–1561. [CrossRef]

38. Abhari, M.N.; Ghodsian, M.; Vaghefi, M.; Panahpur, N. Experimental and numerical simulation of flow in a 90° bend. *Flow Meas. Instrum.* **2010**, *21*, 292–298. [CrossRef]

39. Vaghefi, M.; Akbari, M.; Fiouz, A.R. An Experimental Study of Mean and Turbulent Flow in a 180 Degree Sharp Open Channel Bend: Secondary Flow and Bed Shear Stress. *KSCE J. Civ. Eng.* **2016**, *20*, 1582–1593. [CrossRef]

40. Koopaei, K.B.; Ervine, D.A.; Carling, P.A.; Cao, Z. Velocity and turbulence measurements for two overbank flow events in River Severn. *J. Hydraul. Eng.* **2002**, *128*, 891–900. [CrossRef]

41. Williams, J.J.; Rose, C.P.; Thorne, P.D.; O'connor, B.A.; Humphery, J.D.; Hardcastle, P.J.; Moores, S.P.; Cooke, J.A.; Wilson, D.J. Field observations and predictions of bed shear stresses and vertical suspended sediment concentration profiles in wave-current conditions. *Cont. Shelf Res.* **1999**, *19*, 507–536. [CrossRef]

42. Pope, S.B. *Turbulent Flow*, 1st ed.; Cambridge University Press: Cambridge, UK, 2000.

43. Kim, S.C.; Friedrichs, C.T.; Ma, J.P.Y.; Wright, L.D. Estimating bottom stress in tidal boundary layer from acoustic Doppler velocimeter data. *J. Hydraul. Eng.* **2000**, *126*, 399–406. [CrossRef]

44. Biron, P.M.; Robson, C.; Lapointe, M.F.; Gaskin, S.J. Comparing various methods of bed shear stress estimates in simple and complex flow fields. *Earth Surf. Process. Landf.* **2010**, *29*, 1403–1415. [CrossRef]

45. Blanckaert, K.; Vriend, H.J. Turbulence characteristics in sharp open-channel bends. *Phys. Fluids* **2005**, *17*, 717–731. [CrossRef]

Article

Three-Dimensional Turbulence Numerical Simulation of Flow in a Stepped Dropshaft

Yongfei Qi [1,2], Yurong Wang [1,*] and Jianmin Zhang [1,*]

[1]   State Key Laboratory of Hydraulics and Mountain River Engineering, Sichuan University,
     Chengdu 610065, China; xjqiyongfei@163.com
[2]   Department of Foundation Studies, Xinjiang Institute of Engineering, Urumqi 830023, China
*   Correspondence: wangyurong@scu.edu.cn (Y.W.); zhangjianmin@scu.edu.cn (J.Z.);
     Tel.: +86-139-8002-5149 (Y.W.); +86-139-8187-8609 (J.Z.)

Received: 18 November 2018; Accepted: 17 December 2018; Published: 24 December 2018

Abstract: The dropshaft structure is usually applied in an urban drainage system to connect the shallow pipe network and the deep tunnel. By using the renormalization group (RNG) $k\sim\varepsilon$ turbulence model with a volume of fluid method, the flow pattern and the maximum relative water depth over a stepped dropshaft with a different central angle of step were numerically investigated. The calculated results suggested that the flow in the stepped dropshaft was highly turbulent and characterized by deflection during the jet caused by the curvature of the sidewall. According to the pressure distribution on the horizontal step and the flow pattern above the step, the flow field was partitioned into the recirculating region, the wall-impinging region and the mixing region. In addition, with the increase in the central angle of step, the scope of the wall-impinging region and the mixing region increased and the scope of the recirculating region remained nearly unchanged. The maximum water depth increased with the increase in discharge. In the present work we have shown that, as the value of the central angle of step increased, the maximum water depth decreased initially and increased subsequently.

Keywords: stepped dropshaft; numerical simulation; flow region; central angle of step

1. Introduction

A dropshaft is a hydraulic structure that is installed in drainage systems and tunnel schemes to convey extreme rainfall so that it will not trigger urban flooding. The energy dissipation of a dropshaft is very notable. Studies by Rajaratnam et al., Chanson, and Camino et al. have shown that the energy dissipation rate of the dropshaft is normally more than 75% [1–4].

According to the flow features of a vertical dropshaft, the general dropshaft structure can be divided into four categories: plunge-flow dropshaft [5,6], vortex-flow dropshaft [7,8], baffle-flow dropshaft [9,10], and helicoidal-ramp dropshaft [11,12]. In recent years, vortex-flow dropshafts have become popular due to the excellent energy dissipation and air removal in the construction. The flow characteristic has been investigated by engineers and scientists through laboratory experimentation and numerical simulation. Jain and Kennedy [13] reported the application of vortex dropshafts in the Milwaukee Metropolitan Sewerage District in 1984. In 2010, Hager [14], combined with previous engineering experience, put forward a complete design plan for a vortex dropshaft. Zhao et al. [15] investigated the performance of a vortex drop structure with a relatively small height-to-diameter ratio and confirmed that the energy dissipation rate and air entrainment rate were very significant. Del Giudice and Gisonni [16] optimized the intake of a vortex drop structure in Naples and solved the problem of supercritical approach flow. Natarius [17] installed the Vortex Flow Insert (VFI) in a vortex drop which can effectively eliminate the public odor problem in that drop. Yu and Lee [18] investigated the hydraulics of tangential vortex intake and presented the design guideline that can convey the flow steadily without being influenced by the hydraulic jump.

In order to improve energy dissipation and reduce safety hazards, an innovative design of vortex-flow structures has been developed [19]. A series of steps was deployed on the dropshaft. The flow hit each step as an impinging jet, with the energy being dissipated through jet breakup in the air, collision and swirling on the step, and increased friction between jet and sidewall [20,21]. From the perspective of safety, the steps will significantly increase the air concentration and reduce the risks of cavitation damage [22]. The features of the standing waves concerning the relative height, the location, and the extent were studied theoretically and experimentally by Wu et al. [23].

Although researchers of existing studies have gained a comprehensive understanding of this particular type of dropshaft, little is known about the intricacies of the flow and influence of central angle of step on hydraulic properties. Therefore, in order to study the effect of central angle of step on the flow pattern and maximum water depth on the sidewall, numerical investigations of a stepped dropshaft with different central angles were carried out.

2. Numerical Simulation

2.1. Volume of Fluid Method

The volume of fluid [24] model, which relies on the fact that two or three phases are not interpenetrating, provides an efficient and economical way to track the volume fraction of each of the fluids throughout the domain. Because different fluid components are solved by a single set of momentum equations, the volume fraction of unit phase is defined in order to track the free surface of each computational cell. For an air–water two-phase flow model, in each computational cell, the sum of the volume fractions of the air and water is unity. Specifically, computational cells without water have a value of zero. Full cells are assigned a value of 1 and partially filled cells have a value between 0 and 1. Therefore, the volume fraction of air or water is defined as α_a and α_w, respectively. The relationship between α_a and α_w can be given as follows:

$$\alpha_a + \alpha_w = 1 \tag{1}$$

The variables and their attributes are shared by air and water and represent volume-averaged values if the volume fraction of air and water is obtained. Therefore, at any given control volume, the variables and their attributes represent either air or water, or a mixture of them. The tracking of the interface between air and water is accomplished by the solution of the continuity equation with the following form:

$$\frac{\partial \alpha_w}{\partial t} + u_i \frac{\partial \alpha_w}{\partial x_i} = 0 \tag{2}$$

where α_w is the volume fraction of water and u_i and x_i are the velocity components and coordinates ($i = 1, 2, 3$), respectively.

The density ρ and molecular viscosity μ, which are the volume-fraction-averaged properties and not constants can be expressed as follows:

$$\rho = \alpha_w \rho_w + (1 - \alpha_w)\rho_a \tag{3}$$

$$\mu = \alpha_w \mu_w + (1 - \alpha_w)\mu_a \tag{4}$$

where ρ_w and ρ_a are the density of water and air and μ_w and μ_a are the molecular viscosity of water and air. By the iterating solution of the volume fraction of water α_w, ρ and μ can be calculated.

2.2. Turbulence Model

With the rapid development of computer technology, numerical methods have been applied more and more to complex flows of industrial relevance [25–30]. The most widely used approach in both academia and industry for the modeling of turbulent flows is Reynolds-averaged Navier–Stokes (RANS) methods that solve additional transport equations for turbulence and introduce the turbulence eddy viscosity to simulations to mimic the effect of turbulence. In the RANS modeling framework, two distinct turbulence models were used, namely, k~ε and k~ω. The k~ε turbulent models have historically been used in flow simulations because of the better convergence and lower memory [31,32]. Desirable results can be achieved in various contexts by the k~ω turbulent models, such as boundary layers with adverse pressure gradients and flow separation [33,34]. There are many papers which simulated the complicated hydraulic properties in stepwise and vortex shaft spillway structures using the k~ε model [35–38], yielding reliable results compared with laboratory experiments. As an improvement of the standard k~ε model, the RNG k~ε model [39] is more accurate in simulating flows with high strained rates and streamline bending [40–42], such as the swirling flow. Therefore, the RNG k~ε turbulent model was adopted to model the intricate flow in the stepped dropshaft. The computational fluid dynamics module of the ANSYS 16.0 software (ANSYS®, Canonsburg, PA, USA), Fluent [43], was utilized to investigate the flow over the stepped dropshaft. The equations of the turbulent kinetic energy, k, and its dissipation rate, ε, are as follows:

k equation:

$$\frac{\partial}{\partial t}(\rho k) + \frac{\partial}{\partial x_i}(\rho k u_i) = \frac{\partial}{\partial x_j}\left[(\mu + \frac{\mu_t}{\sigma_k})\frac{\partial k}{\partial x_j}\right] + G_k - \rho\varepsilon, \tag{5}$$

ε equation:

$$\frac{\partial}{\partial t}(\rho\varepsilon) + \frac{\partial}{\partial x_i}(\rho\varepsilon u_i) = \frac{\partial}{\partial x_j}[(\mu + \frac{\mu_t}{\sigma_\varepsilon})\frac{\partial\varepsilon}{\partial x_j}] + C_{1\varepsilon}^* \frac{\varepsilon}{k}G_k - C_{2\varepsilon}\rho\frac{\varepsilon^2}{k}, \tag{6}$$

$$\mu_t = \rho C_\mu \frac{k^2}{\varepsilon} \tag{7}$$

$$G_k = \mu_t(\frac{\partial u_i}{\partial x_j} + \frac{\partial u_j}{\partial x_i})\frac{\partial u_i}{\partial x_j}, \tag{8}$$

$$C_{1\varepsilon}^* = C_{1\varepsilon} - \frac{\eta(1 - \eta/\eta_0)}{1 + \beta\eta^3}, \tag{9}$$

$$\eta = \sqrt{\frac{G_k}{\rho C_{\mu\varepsilon}}}, \tag{10}$$

where ρ is the corresponding density, μ is the dynamic viscosity, μ_t is the turbulence viscosity, u_i is the velocity component in the ith direction, t is the time, G_k is the generation of turbulent energy caused by the average velocity gradient, and C_μ, $C_{1\varepsilon}$, $C_{2\varepsilon}$, β, and η_0 are empirical constants. In the RNG k~ε model, the empirical constants are given as $C_\mu = 0.0845$, $\sigma_k = \sigma_\varepsilon = 0.7179$, $C_{1\varepsilon} = 1.42$, $C_{2\varepsilon} = 1.68$, $\beta = 0.012$, and $\eta_0 = 4.38$.

2.3. Numerical Algorithm

The Pressure-Implicit with Splitting of Operators (PISO) algorithm, based on the higher degree of the approximate relation between the corrections for pressure and velocity, was applied to implicitly couple the velocity and pressure. The implicit finite volume method was used to solve the iteration. The calculated domain was divided into discrete control volumes by the unstructured grid that has a high adaptability and is self-adjusting to the complicated geometry and boundary. The detailed location of the free surface is determined by the geometric reconstruction scheme, which is accurate and applicable for general unstructured meshes in Fluent. This scheme assumes that the interface between two fluids has a linear slope within each cell, and then uses this linear shape for calculation of the advective flux. Ultimately, the volume fraction of each cell can be calculated according to the advective flux balance of previous calculations. Refining the grid on interface between water and air is important for transient VOF calculations using the Geo-Reconstruct formulation. It was unfortunate that the water level in the dropshaft could not be measured beforehand because of the complexity and randomness in the multi-phase flow process. Consequently, in order to get the high-precision results, all the three calculated configurations were refined with total grid numbers of approximately 0.30 million, 0.46 million, and 0.65 million, respectively (the specific content will be introduced later).

The appropriate setting of time-dependent solution parameters is the magnitude guarantee of robustness and efficiency for calculation. For transient volume of fluid calculations that use the implicit scheme of VOF, the fixed times stepping method was selected. The max iterations per time step, a maximum for the number of iterations per time step, was set as 50. Since the ANSYS Fluent formulation is fully implicit, there is no stability criterion that must be met in determining the time step size. A good way to adjust the size of time step is to satisfy the ideal number of iterations per time step, which ranges from 5 to 10. To be specific, if the calculation needs only a few iterations per time step, the time step size should be increased; otherwise, it should be decreased. In this work, the range of the time step size was adjusted from 0.0001 s to 0.001 s and the iteration number was always less than 5.

2.4. Geometric Model

The calculated domain is shown in Figure 1. It has the same size as the physical model, given in the literature [19]. The numerical model consists of three main parts, namely, the inlet section, the vertical shaft section, and the outlet section. The inlet channel cross-section was rectangular, 0.15 m wide and 0.25 m deep. The length of the inlet channel was 1 m. For the vertical shaft section, the height was 1.834 m, the external radius (R) was 0.25 m, and the internal radius (r) was 0.1 m. The shaft vertical section consisted of 14 steps, which connected the inlet channel and the outlet channel. The width, b, and the height, h, of each step were 0.15 m and 0.131 m, respectively. They were numbered from S1 to S14. The central angle of step, one important structure feature of dropshaft, was defined as θ, and configurations with θ = 120°, 150°, or 180° were investigated. The dimensions of the cross-section of the outlet channel were $2 \times 0.15 \times 0.131$ m (length $\times$ width $\times$ height).

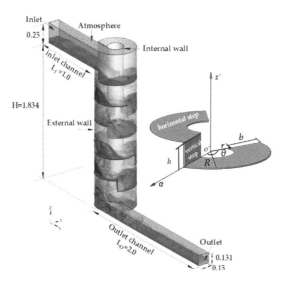

Figure 1. Computational model of the dropshaft.

In order to facilitate analysis, the cylindrical coordinate systems were established. The *z*-axis was the center axis of shaft. The relative polar coordinates were established on each step, in which the center of the circle was defined as the pole and the starting edge of the step was defined as the polar axis.

2.5. Boundary Conditions and Cases

The boundary conditions were as follows:

(1) Inlet boundary: the velocity inlet was used for the intake, which was set at 0.89–2.69 m/s;
(2) Outlet boundary: the outlet boundary was set as pressure outlet and the normal gradient of all variables was equal to 0;
(3) Free surface: the free surface of water was assumed to be the pressure inlet and the pressure value was $P = 0$; and
(4) Wall boundary: no-slip velocity boundary condition; the near-wall regions of the flow were analyzed using the method of standard wall function.

In this study, nine cases with different flow rates and different central angles of step were investigated. The calculation program is shown in Table 1. The design condition of verified cases was set up the same way as the experiment from the literature [19]: *h* is the step height; *D* is the width-radius ratio, which is the ratio of step width to external radius; *i* is the slope coefficient, which is spreading out from the external wall; and *Fr* is the Froude number of approach flow.

Table 1. Summary of the operating conditions for the simulations.

h (m)	*D*	θ (°)	*i*	*Q* (m³/s)	*Fr*	Case
0.131	0.6	150	0.20	12.75	0.58	test1
				41.50	1.90	test2
		120	0.25	80.00	3.67	1
				48.00	2.20	2
				26.50	1.22	3
		150	0.20	80.00	3.67	4
				48.00	2.20	5
				26.50	1.22	6
		180	0.17	80.00	3.67	7
				48.00	2.20	8
				26.50	1.22	9

2.6. Verification

2.6.1. Grid Testing

It is very important to use the grid reasonably for the accuracy of calculation results. Therefore, the effects of different grid densities on the uncertainty of the calculation results were tested using the grid convergence index (GCI) [44] with grid numbers of approximately 0.65 million (grid 1), 0.46 million (grid 2), and 0.30 million (grid 3). The GCI is defined as follows:

$$GCI = \frac{1.25|(\phi_1 - \phi_2)/\phi_1|}{(k_2 - k_1)^p - 1}, \tag{11}$$

in which ϕ_i represents the variable of the *i*th calculation, k_i represents the representative grid size for the *i*th calculation, and let $k_1 < k_2 < k_3$. For three-dimensional calculations, *k* is given by the following:

$$k = \left(\left[\frac{1}{N} \sum_{i=1}^{N} (\Delta V_i) \right] \right)^{1/3} \tag{12}$$

The order of p in Equation (11) can be estimated by the following:

$$p = \frac{1}{\ln(k_2/k_1)} \left| \ln|(\phi_3 - \phi_2)/(\phi_2 - \phi_1)| + \left| \ln\left(\frac{(k_2/k_1)^P - \text{sgn}[(\phi_3 - \phi_2)/(\phi_2 - \phi_1)]}{(k_3/k_2)^P - \text{sgn}[(\phi_3 - \phi_2)/(\phi_2 - \phi_1)]} \right) \right| \right|. \tag{13}$$

In this paper, the pressure and tangential velocity that, respectively, were written in dimensionless form as p/h and $V_\tau/V_{\tau max}$, were variables. Figure 2b,d presents pressure and tangential velocity along a radial line that was in the *yoz* plane and paralleled to the *y*-axis (as depicted in Figure 2a) for different grid densities with $Q = 41.25$ m^3/s. The data obtained from grid 2 were close to the data obtained with the grid 1 and the computational efficiency decreased by 20%. In Figure 2c,e, the maximum uncertainties in pressure and tangential velocity were approximately 3.12% and 4.17%, respectively, implying that the discretization uncertainties were little in most locations. For the consideration of computational efficiency and accuracy, the grid number of 0.46 million (grid 2) was adopted in the present study.

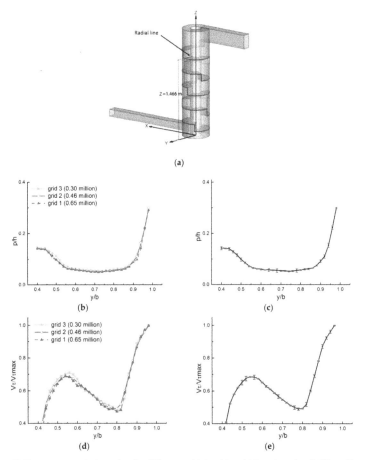

Figure 2. Grid convergence index value for different grid densities: (**a**) location of radial line; (**b**) pressure profile in models with different mesh numbers; (**c**) fine-grid solution, with discretization error bars computed using Equation (11); (**d**) tangential velocity profile in models with different mesh numbers; (**e**) fine-grid solution, with discretization error bars computed using Equation (11).

2.6.2. Model Verification

The calculated flow pattern was fairly consistent with the experimental results under different discharges, as is shown in Figure 3. When transferring from the upper steps to the lower steps, and owing to the collision between the flow and the steps, part of the flow changed direction with the mainstream travelling downstream. The flow that changed direction impacted the vertical surface of the step and formed a reflux. In addition, the cavities near the vertical step were obvious. As shown in Figure 3a,c, the main differences are as follows: (1) the recirculation zone for $Q = 41.5$ m^3/s was smaller than that of $Q = 12.75$ m^3/s; and (2) for $Q = 12.75$ m^3/s, there was an increase in the water level when flow entered the next step, but for $Q = 41.5$ m^3/s, the increase in water level occurred after the nappe flow impacted the horizontal step. As can be seen from Figure 3b,d, the simulated flow patterns were in good agreement with experimental results.

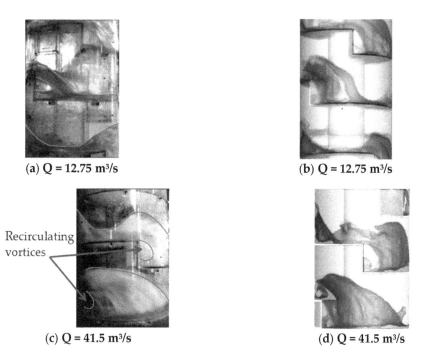

(a) Q = 12.75 m³/s   (b) Q = 12.75 m³/s

Recirculating vortices

(c) Q = 41.5 m³/s   (d) Q = 41.5 m³/s

Figure 3. Comparison of experimental and calculated flow pattern: (**a,c**) flow pattern in experiment; (**b,d**) flow pattern in simulation.

In addition, experimental data obtained from Wu et al. [19] were used to verify the numerical results by comparing the stream-wise and radial pressure at S9, as is shown in Figure 4. In Figure 4a, four measuring points (P1~P4) were set on the center line of the bottom step along the flow direction. P3, and the other three measuring points (P5, P6, P7) were on a straight line that was paralleled to the y-axis in the radial direction. It can be seen from Figure 4b,c that, at most locations, the calculated results of pressure distribution were fairly consistent with that of laboratory tests. The maximums of the relative error for $Q = 12.75$ m^3/s and $Q = 41.5$ m^3/s were 7.5% and 10%, respectively, indicating that the numerical simulations produced reliable and acceptable results.

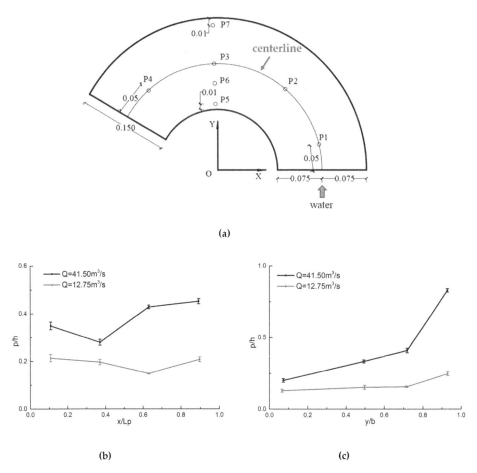

Figure 4. Comparison of numerical and experimental data: (**a**) location of the measuring points (m); (**b**) pressure distribution along the flow direction; (**c**) pressure distribution along the radical direction. Notes: L_p represents the arc length of the center line; b is the width of the horizontal step.

2.6.3. Fluctuation of Calculation Results

There existed instabilities and uncertainties in the computed results because of the complexity of the flow. In order to improve the reliability of results, the pressure for R = 0.25 m at S6 and the water surface on outer wall at S6 were monitored to analyze the fluctuation of pressure and water level in the calculation process. After a long time calculation, when the average relative errors of the mass flow rate were below 10%, the flow variables were considered relatively constant in this work. Five relative locations, namely, 10°, 30°, 60°, 90°, and 120°, were selected from each of the nine cases for determining the dispersion of pressure and the water surface at the same time period. The data in a duration of 10 s with an interval of 1 s were selected for analysis. The fluctuation of the data was calculated in terms of the standard deviation ($\sigma_{p/h}$ and $\sigma_{h_w/h}$). Table 2 shows the standard deviation of the variables for each case. In Table 2, the standard deviations of pressure and water surface were small, ranging from 0.0565 to 0.1008, indicating the reliability of the simulated results.

Table 2. Summary of the standard deviations of pressure and water surface in different locations at the same time period, where Er_{ave} represents the average relative errors of the mass flow rate of inlet and outlet; $\sigma_{p/h}$ and $\sigma_{hw/h}$ are the standard deviation of pressure and water surface.

	Er_{ave} (%)		10°	30°	60°	90°	120°
case1	8.25	$\sigma_{p/h}$	0.0707	0.0692	0.0723	0.0818	0.0726
		$\sigma_{hw/h}$	0.0852	0.0912	0.0823	0.0797	0.0906
case2	7.18	$\sigma_{p/h}$	0.0675	0.0704	0.0681	0.0619	0.0621
		$\sigma_{hw/h}$	0.0823	0.0834	0.0885	0.0818	0.0911
case3	4.58	$\sigma_{p/h}$	0.0754	0.0652	0.0801	0.0781	0.0702
		$\sigma_{hw/h}$	0.0925	0.0922	0.0879	0.0921	0.0942
case4	7.32	$\sigma_{p/h}$	0.0583	0.0621	0.0612	0.0565	0.0669
		$\sigma_{hw/h}$	0.0725	0.0822	0.0850	0.0818	0.0861
case5	7.07	$\sigma_{p/h}$	0.0754	0.0689	0.0692	0.0717	0.0722
		$\sigma_{hw/h}$	0.0628	0.0587	0.0603	0.0614	0.0592
case6	5.68	$\sigma_{p/h}$	0.0718	0.0782	0.0811	0.0777	0.0798
		$\sigma_{hw/h}$	0.0823	0.0898	0.0912	0.0884	0.0879
case7	8.12	$\sigma_{p/h}$	0.0905	0.0972	0.0883	0.0878	0.0928
		$\sigma_{hw/h}$	0.0923	0.0985	0.0984	0.1008	0.0954
case8	7.22	$\sigma_{p/h}$	0.0661	0.0622	0.0704	0.0683	0.0688
		$\sigma_{hw/h}$	0.0921	0.0918	0.0856	0.0885	0.0892
case9	6.64	$\sigma_{p/h}$	0.0775	0.0721	0.0605	0.0644	0.0786
		$\sigma_{hw/h}$	0.0858	0.0805	0.0734	0.0713	0.0728

All verifications for computational models, including the grid sensitivity study, the model verification, and the fluctuation of the calculated results, showed some errors and uncertainties due to the use of the eddy viscosity-based turbulence models. The errors of such eddy viscosity-based turbulence models in flow separation and recirculation zones have been documented in prior studies [45,46]. Nevertheless, using RNG $k\sim\varepsilon$ to reveal the general flow patterns could satisfy the research requirements at the present stage.

3. Results and Analysis

3.1. Region Division in the Flow

As the simulation reveals, the flow transfers from one step to another, in which the flow regime has similar features on different steps. First, the successive impinging produces a bifurcation, where part of the flow moves backwards and forms a recirculation zone, while the main flow rotates along the sidewall with the vortex intensity decaying. Second, the flow undergoes a significant mixing and jumping process at the end of step. In order to present the complicated flow pattern of the stepped dropshaft accurately, the flow field is divided into the recirculating region (I), the wall-impinging region (II) and the mixing region (III), according to the flow characteristics. The relative ranges of these regions are expressed in terms of angles, namely, α_1, α_2, and α_3, respectively. In the subsequent sections, flow regions are discussed qualitatively according to individual streamlines and analyzed quantitatively on the basis of the pressure distribution for $R = 0.25$ m.

Figure 5 highlights the streamlines on the first step in each configuration. Although the central angle of each configuration was different, they shared similar flow features. In particular, the flow pattern had similar characteristics as well on other steps having the same central angle. Hence, only the flow on S6 in the different configuration under $Q = 80$ m^3/s was analyzed.

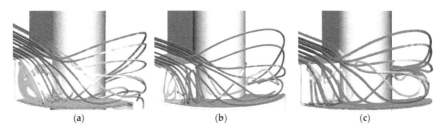

(a) (b) (c)

Figure 5. Streamlines above the S2 in different configurations: (**a**) 120°; (**b**) 150°; and (**c**) 180°.

(1) Recirculating region

The recirculating region was the zone where there was insignificant change of pressure gradient in the radial direction and streamline deflection in the opposite direction. Figure 6 illustrates the flow pattern along the flow direction in the cross-section for different configurations. Recirculating vortices can be found on the step due to the pressure difference between the upper and lower surfaces of the nappe flow during the falling process. Combined with the pressure difference and resistance caused by the nappe flow colliding with the horizontal step surface, the nappe flow departed from its original direction and formed a reflux. In addition, the location of recirculating moved from the external wall to the internal wall because of the curvature of the sidewall along the downstream direction. Figure 7 presents the pressure distributions on the horizontal step surfaces and the range of flow regimes. Most of the falling points were located at the intersection line of the horizontal step and the external wall. The most remarkable feature was that it was the minimum value of relative pressure that could be detected in the recirculating region.

(2) Wall-impinging region

The definition for wall-impinging region was the same as the recirculating region, which was mainly based on pressure variations and streamline features. Therefore, the range of the wall-impinging region was from the location where the minimum relative pressure occurred to the location where the local minimum pressure took place.

In the wall-impinging region, the flow was subjected to resistance and centrifugal force. However, the location and time that the inner and outer fluid were impacting the step or sidewall were different because of the curvature. First, the tangential velocity changed abruptly because of the impact of the surface flow near the external wall on that of the lateral one, leading it to increase rapidly and then to decline gradually on the water surface. The flow near the internal wall jetted along the tangent of that under pressure difference and gravity, and the maximum distance of the falling point of flow was in the circumferential direction, whereas the minimum was in the radial direction. For the lower level fluid, the flow quickly dived to the bottom of the step and traveled forward along the sidewall under the influence of gravity. However, the falling points of inside flow were close to the axis of the horizontal step surface. Deflected by the step, the flow deviated to the internal wall, resulting in it hitting the internal wall and attaching to it.

Therefore, the water surface was higher on the external wall than it was on the internal wall (Figure 5). Moreover, maximum pressure occurred and the pressure gradient showed drastic change at the start of the wall-impinging region (Figure 6).

(3) Mixing region

The range of the mixing region was from the location where the local minimum pressure occurred to the end of the step. The possibility of a relatively steady flow in the mixing region may be determined by the declining intensity of the swirling flow. The flow near the external wall moved inward from the bottom and the one close to the internal wall moved to the opposite from the surface due to the inward pressure gradient near the horizontal step which prevailed over the centrifugal force. As a result, there were less variations in the water surface in the transverse direction. Flow from the internal and the external walls mixed and interacted with each other near the axis of the step, causing the fluctuation of water surface along the circumference. At the end of the step, the elevation of the water surface caused by the transverse gradient in the outer wall was higher than that in the inner wall. The pressure distribution in the mixing region exhibited no obvious change.

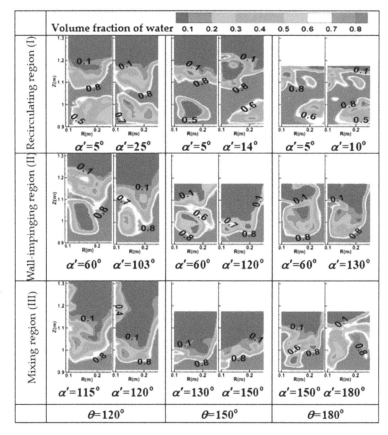

Figure 6. Flow pattern in the cross-section for different configurations, where α' is the relative location of the cross-section.

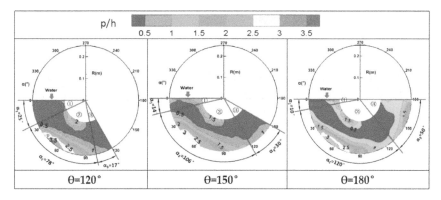

Figure 7. Pressure distribution and flow regions at S6.

3.2. Regional Scope

Figure 8 highlights the pressure distributions on the horizontal step for $R = 0.25$ m. As can be seen, the impact pressure fluctuated greatly with the maximum and minimum occurring in the recirculating region and the wall-impinging region, respectively, under all conditions. In addition, the pressure in the wall-impinging region showed greater gradient than that in the mixing region and the recirculating region. As the θ increased, α_1 decreased, while α_2 and α_3 increased. The cause of this phenomenon was attributed to the increase in the circumferential length of the step. These changes were able to bring the falling point of the jet flow to the vertical step closer and to make the recirculating region smaller. On the contrary, the ranges of the wall-impinging region and the mixing region were enhanced. Flow regime had little impact on the spectrum of flow regions in most cases with $\theta = 120°$ and $\theta = 150°$, but a different phenomenon could be observed when $\theta = 180°$. When flow discharge increased, α_2 increased, whereas α_3 decreased.

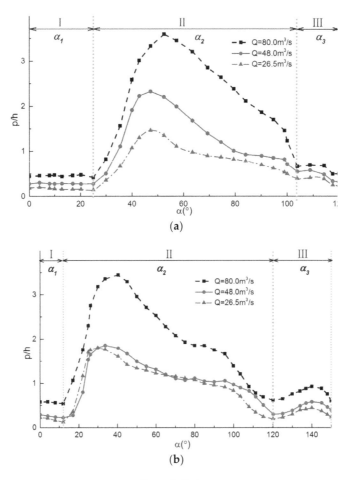

Figure 8. *Cont.*

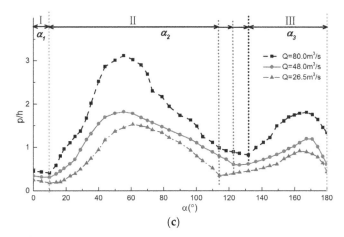

Figure 8. Pressure distribution for R = 0.25 m in different θ: (**a**) 120°; (**b**) 150°; and (**c**) 180°.

Figure 9a displays the average range of flow region on each step. Although some slight variations can be found, it is noticeable that the ranges of the regions were approximately the same under different flow conditions, indicating that the scope was mainly affected by the step angle rather than the flow regime. The chart compares the average range of different flow regions in all configurations.

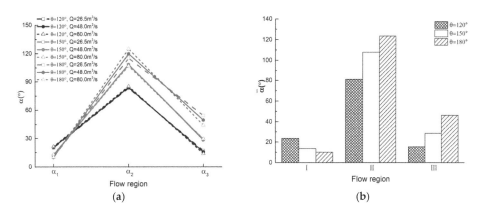

Figure 9. Flow region: (**a**) range of flow regions and (**b**) range of flow regions under different configurations.

3.3. Water Depth

For the traditional dropshaft, standing waves can only occur in a supercritical approach flow [47–49] channel due to sidewall deflection or curvature; therefore, the height of sidewalls must be carefully designed in a prototype environment. In addition, these standing waves can spread to downstream channels, causing fluctuations in the water surface. Nevertheless, for the dropshaft in the present work, sidewall standing waves were generated on each step, although the standing wave was quite insignificant under a small flow rate. Therefore, research studies should be undertaken to gain insights into the typical properties of standing waves. A sketch of the wave along the external wall is shown in Figure 10. h_{max} is the maximum water depth and α_{max} is the relative position of the maximum water depth. The distance between the steps is h_s, and the relative position of the vertical plane of the upper step is α_s. A dimensionless height, h_s/h, for $\theta = 120°$ and $\theta = 180°$ is, respectively,

3 and 2, and the location, α_s, for both $\theta = 120°$ and $\theta = 180°$ is 0. However, for $\theta = 150°$, h_s/h and α_s vary significantly with h_s/h ranging from 3 to 2 and α_s ranging from 0° to 60°.

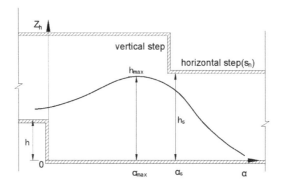

Figure 10. Water depth along the external wall.

Figure 11 demonstrates the flow profile along the external wall. The flow-depth curves indicated that the water surface along the outer wall presented a similar wavy pattern under different conditions.

(1) With the increase in the central angles of step, drastic fluctuation began to occur. For $\theta = 120°$, the water-surface profile was steady and the water depth decreased gradually along the flow direction in which the maximum relative water depth occurred in the recirculating region. For $\theta = 150°$, the variation of free surface became remarkable. When the central angle of step increased to 180°, the oscillation of water surface was more violent than that of the other cases.

For the small center angle, as the rotational flow approached the mixing region, the flow fell into the next step suddenly owing to the small range of the mixing region, which could lead to a dramatic change in axial velocity gradient and water depth. However, the larger the angle, the greater the diffusion in the mixing region. Water surface fluctuated greatly due to the resistance of the sidewall and centrifugal force.

(2) The influence of flow rate on water depth was relatively noteworthy: the water level rose with the increase in discharge under different conditions. For $\theta = 120°$ and $\theta = 150°$, the free surface changed without distinct oscillation under all circumstances. For $\theta = 180°$, the water level oscillated more vigorously with a further increase in discharge, causing an increase in the water surface, and hit the vertical surface of the next step in the mixing region.

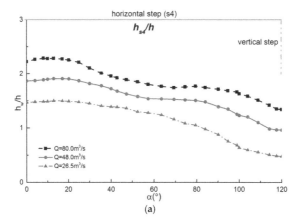

Figure 11. *Cont.*

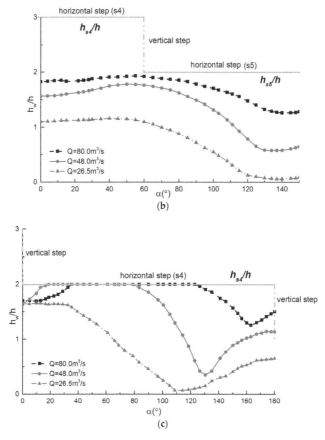

Figure 11. Water surface curves: (**a**) 120°; (**b**) 150°; and (**c**) 180°.

Figure 12 illustrates the change of relative maximum water depth in all cases and yields the following conclusions:

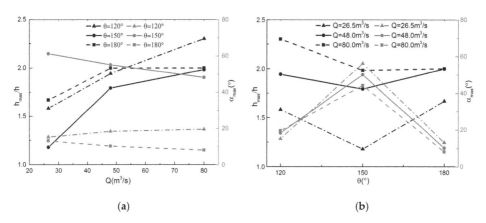

Figure 12. Water depth characteristics: (**a**) the relationship between h_{max}/h, α_{max} and Q and (**b**) the relationship between h_{max}/h, α_{max} and θ.

(1) For a fixed central angle of step θ, h_{max}/h and α_{max} increased as Q increased.

(2) For a fixed discharge Q, h_{max}/h and α_{max} showed no obvious trend as θ increased.

(3) With the increment of θ, h_{max}/h decreased initially and then increased, whereas α_{max} exhibited the opposite phenomenon.

(4) With the increase in Q, h_{max}/h increased and α_{max} presented no evident change.

Therefore, it seemed that the range of h_{max}/h depended on the discharge and the central angle of step, whereas the range of α_{max} was associated with the central angle of step. For $\theta = 180°$, when $h_{max}/h = 2$ ($Q = 48$ m^3/s, 80 m^3/s), the flow impacted the bottom of the upper step, which can only occur on some steps when $\theta = 150°$ for $Q = 80$ m^3/s. For $\theta = 120°$, h_{max}/h was considerably lower than the height of the upper step in all discharge.

4. Conclusions

For all the cases considered above in this paper, the RNG $k\sim\varepsilon$ turbulence flow model is capable of predicting the main characteristics of the flow. This included the definition of the flow regime and the determination of the maximum water depth. The following conclusions can be drawn:

1. Because of the complexity of the flow regime, the flow was partitioned into the recirculating region, the wall-impinging region, and the mixing region.

2. The range of the three regions was affected by the central angle of step. With the increase in θ, the range of the wall-impinging region and the mixing region increased, whereas the range of the recirculating region had no significant change.

3. The maximum water depth increased with the increase in the flow rate, whereas the flow rate showed little effect on α_{max}. Moreover, with the increase in θ, the maximum water depth initially decreased and then increased; however, α_{max} displayed the opposite trend.

Although the results of this study are encouraging, same solution processes can be adopted in future work to model the energy dissipation performance and air entertainment characteristics of stepped dropshafts.

Author Contributions: Conceptualization and Methodology, Y.W. and J.Z.; Numerical Simulation, Y.Q.; Investigation, Y.Q.; Writing—Original Draft Preparation, Y.Q.; Writing—Review and Editing, Y.W. and J.Z.; Supervision, Y.W. and J.Z.; Project Administration, Y.W. and J.Z.; Funding Acquisition, Y.W. and J.Z.

Funding: This research was funded by the National Key Research and Development Program of China (No. 2016YFC0401707), the National Natural Science Foundation of China (No. 51579165), and the National Science Fund for Distinguished Young Scholars (No. 51625901).

Conflicts of Interest: The authors declare no conflict of interest.

References

1. Rajaratnam, N.; Mainali, A.; Hsung, C. Observations on flow in vertical dropshafts in urban drainage systems. *J. Environ. Eng.* **1997**, *123*, 486–491. [CrossRef]

2. Chanson, H. Hydraulics of rectangular dropshafts. *J. Irrig. Drain. Eng.* **2004**, *130*, 523–529. [CrossRef]

3. Chanson, H. Understanding air–water mass transfer in rectangular dropshafts. *J. Environ. Eng. Sci.* **2004**, *3*, 319–330. [CrossRef]

4. Adriana Camino, G.; Zhu, D.Z.; Rajaratnam, N. Flow observations in tall plunging flow dropshafts. *J. Hydraul. Eng.* **2014**, *141*, 06014020. [CrossRef]

5. Granata, F.; de Marinis, G.; Gargano, R.; Hager, W.H. Hydraulics of circular drop manholes. *J. Irrig. Drain. Eng.* **2010**, *137*, 102–111. [CrossRef]

6. Ma, Y.; Zhu, D.Z.; Rajaratnam, N.; van Duin, B. Energy dissipation in circular drop manholes. *J. Irrig. Drain. Eng.* **2017**, *143*, 04017047. [CrossRef]

7. Jain, S.C. Air transport in vortex-flow drop shafts. *J. Hydraul. Eng.* **1988**, *114*, 1485–1497. [CrossRef]

8. Vischer, D.; Hager, W. Vortex drops. In *Energy Dissipators*; Routledge: Abingdon, UK, 2018; pp. 167–181. ISBN 9781351451345.

9. Odgaard, A.J.; Lyons, T.C.; Craig, A.J. Baffle-drop structure design relationships. *J. Hydraul. Eng.* **2013**, *139*, 995–1002. [CrossRef]

10. Stirrup, M.; Margevicius, T.; Hrkac, T.; Baca, A. A baffling solution to Sewage Conveyance In York Region, Ontario. *Proc. Water Environ. Fed.* **2012**, *2012*, 74–90. [CrossRef]

11. Kennedy, J.F.; Jain, S.C.; Quinones, R.R. Helicoidal-ramp dropshaft. *J. Hydraul. Eng.* **1988**, *114*, 315–325. [CrossRef]

12. Tamura, S.; Matsushima, O.; Yoshikawa, S. Helicoidal-ramp type drop shaft to deal with high head drop works in manholes. *Proc. Water Environ. Fed.* **2010**, *2010*, 4991–5002. [CrossRef]

13. Jain, S.C.; Kennedy, J.F. *Vortex-Flow Drop Structures for the Milwaukee Metropolitan Sewerage District Inline Storage System*; Iowa Institute of Hydraulic Research, The University of Iowa: Iowa City, IA, USA, 1983.

14. Hager, W.H. *Wastewater Hydraulics: Theory and Practice*; Springer Science & Business Media: Berlin, Germany, 2010.

15. Zhao, C.H.; Zhu, D.Z.; Sun, S.K.; Liu, Z.P. Experimental study of flow in a vortex drop shaft. *J. Hydraul. Eng.* **2006**, *132*, 61–68. [CrossRef]

16. Del Giudice, G.; Gisonni, C. Vortex dropshaft retrofitting: Case of Naples city (Italy). *J. Hydraul. Res.* **2011**, *49*, 804–808. [CrossRef]

17. Natarius, E.M. Aeration performance of vortex flow insert assemblies in sewer drop structures. *Proc. Water Environ. Fed.* **2008**, *2008*, 842–851. [CrossRef]

18. Yu, D.; Lee, J.H. Hydraulics of tangential vortex intake for urban drainage. *J. Hydraul. Eng.* **2009**, *135*, 164–174. [CrossRef]

19. Wu, J.H.; Yang, T.; Sheng, J.Y.; Ren, W.C.; Fei, M.A. Hydraulic characteristics of stepped spillway dropshafts with large angle. *Chin. J. Hydrodyn.* **2018**, *33*, 176–180. (In Chinese) [CrossRef]

20. Christodoulou, G.C. Energy dissipation on stepped spillways. *J. Hydraul. Eng.* **1993**, *119*, 644–650. [CrossRef]

21. Sorensen, R.M. Stepped spillway hydraulic model investigation. *J. Hydraul. Eng.* **1985**, *111*, 1461–1472. [CrossRef]

22. Frizell, K.W.; Renna, F.M.; Matos, J. Cavitation potential of flow on stepped spillways. *J. Hydraul. Eng.* **2012**, *139*, 630–636. [CrossRef]

23. Wu, J.H.; Ren, W.C.; Ma, F. Standing wave at dropshaft inlets. *J. Hydrodyn. Ser. B* **2017**, *29*, 524–527. [CrossRef]

24. Hirt, C.W.; Nichols, B.D. Volume of fluid (VOF) method for the dynamics of free boundaries. *J. Comput. Phys.* **1981**, *39*, 201–225. [CrossRef]

25. Peng, Y.; Zhou, J.G.; Burrows, R. Modelling the free surface flow in rectangular shallow basins by lattice Boltzmann method. *J. Hydrau. Eng.* **2011**, *137*, 1680–1685. [CrossRef]

26. Peng, Y.; Zhou, J.G.; Burrows, R. Modelling solute transport in shallow water with the lattice Boltzmann method. *Comput. Fluids.* **2011**, *50*, 181–188. [CrossRef]

27. Peng, Y.; Zhang, J.M.; Zhou, J.G. Lattice Boltzmann Model Using Two-Relaxation-Time for Shallow Water Equations. *J. Hydrau. Eng.* **2016**, *142*, 06015017. [CrossRef]

28. Peng, Y.; Zhang, J.M.; Meng, J.P. Second order force scheme for lattice Boltzmann model of shallow water flows. *J. Hydraul. Res.* **2017**, *55*, 592–597. [CrossRef]

29. Peng, Y.; Mao, Y.F.; Wang, B.; Xie, B. Study on C-S and P-R EOS in pseudo-potential lattice Boltzmann model for two-phase flows. *Int. J. Mod. Phys. C* **2017**, *28*, 1750120. [CrossRef]

30. Peng, Y.; Wang, B.; Mao, Y.F. Study on force schemes in pseudopotential lattice Boltzmann model for two-phase flows. *Math. Probl. Eng.* **2018**. [CrossRef]

31. Galván, S.; Reggio, M.; Guibault, F. Assessment study of k-ε turbulence models and near-wall modeling for steady state swirling flow analysis in draft tube using fluent. *Eng. Appl. Comput. Fluid* **2011**, *5*, 459–478. [CrossRef]

32. Morovati, K.; Eghbalzadeh, A.; Javan, M. Numerical investigation of the configuration of the pools on the flow pattern passing over pooled stepped spillway in skimming flow regime. *Acta Mech.* **2015**, *227*, 1–14. [CrossRef]

33. Devolder, B.; Troch, P.; Rauwoens, P. Performance of a buoyancy-modified k-ω and k-ω SST turbulence model for simulating wave breaking under regular waves using OpenFOAM®. *Coast. Eng.* **2018**, *138*, 49–65. [CrossRef]

34. Fuhrman, D.R.; Dixen, M.; Jacobsen, N.G. Physically-consistent wall boundary conditions for the k-ω turbulence model. *J. Hydraul. Res.* **2010**, *48*, 793–800. [CrossRef]

35. Bai, Z.L.; Zhang, J.M. Comparison of different turbulence models for numerical simulation of pressure distribution in V-shaped stepped spillway. *Math. Probl. Eng.* **2017**, *2017*, 1–9. [CrossRef]

36. Li, S.; Zhang, J. Numerical investigation on the hydraulic properties of the skimming flow over pooled stepped spillway. *Water* **2018**, *10*, 1478. [CrossRef]

37. Chan, S.; Qiao, Q.; Lee, J.H. On the three-dimensional flow of a stable tangential vortex intake. *J. Hydro-Environ. Res.* **2018**, *21*, 29–42. [CrossRef]

38. Gao, X.P.; Zhang, H.; Liu, J.J.; Sun, B.; Tian, Y. Numerical investigation of flow in a vertical pipe inlet/outlet with a horizontal anti-vortex plate: Effect of diversion orifices height and divergence angle. *Eng. Appl. Comput. Fluid Mech.* **2018**, *12*, 182–194. [CrossRef]

39. Yakhot, V.; Orszag, S.A. Renormalization group analysis of turbulence. I. Basic theory. *J. Sci. Comput.* **1986**, *57*, 1722. [CrossRef]

40. Liu, Z.P.; Guo, X.L.; Xia, Q.F.; Fu, H.; Wang, T.; Dong, X.L. Experimental and numerical investigation of flow in a newly developed vortex drop shaft spillway. *J. Hydraul. Eng.* **2018**, *144*, 04018014. [CrossRef]

41. Zhang, W.; Wang, J.; Zhou, C.; Dong, Z.; Zhou, Z. Numerical simulation of hydraulic characteristics in a vortex drop shaft. *Water* **2018**, *10*, 1393. [CrossRef]

42. Guo, X.L.; Xia, Q.F.; Fu, H.; Yang, K.L.; Li, S.J. Numerical study on flow of newly vortex drop shaft spillway. *J. Hydraul. Eng.* **2016**, *47*, 733–741. (In Chinese) [CrossRef]

43. *ANSYS Fluent Theory Guide*; Release 16.0; ANSYS Inc.: Canonsburg, PA, USA, 2015.

44. Celik, I.B.; Ghia, U.; Roache, P.J. Procedure for estimation and reporting of uncertainty due to discretization in CFD applications. *J. Fluids Eng.* **2008**, *130*. [CrossRef]

45. Iaccarino, G.; Mishra, A.A.; Ghili, S. Eigenspace perturbations for uncertainty estimation of single-point turbulence closures. *Phys. Rev. Fluids* **2017**, *2*, 024605. [CrossRef]

46. Mishra, A.A.; Iaccarino, G. Uncertainty Estimation for Reynolds-Averaged Navier–Stokes Predictions of High-Speed Aircraft Nozzle Jets. *AIAA J.* **2017**, 3999–4004. [CrossRef]

47. Del Giudice, G.; Gisonni, C.; Rasulo, G. Design of a scroll vortex inlet for supercritical approach flow. *J. Hydraul. Eng.* **2010**, *136*, 837–841. [CrossRef]

48. Hager, W.H. Vortex drop inlet for supercritical approaching flow. *J. Hydraul. Eng.* **1990**, *116*, 1048–1054. [CrossRef]

49. Kawagoshi, N.; Hager, W. Wave type flow at abrupt drops Wave type flow at abrupt drops: I. Flow geometry. *J. Hydraul. Res.* **1990**, *28*, 235–252. [CrossRef]

Article

Investigation of Free Surface Turbulence Damping in RANS Simulations for Complex Free Surface Flows

Arun Kamath[1,*], Gábor Fleit[2] and Hans Bihs[1]

[1] Department of Civil and Environmental Engineering, Norwegian University of Science of Technology, NTNU, 7491 Trondheim, Norway; hans.bihs@ntnu.no

[2] Department of Hydraulic and Water Resources Engineering, Budapest University of Technology and Economics, Moegyetem rkp. 3, 1111 Budapest, Hungary; fleitg@gmail.com

[*] Correspondence: arun.kamath@ntnu.no

Received: 1 February 2019; Accepted: 25 February 2019; Published: 4 March 2019

Abstract: The modelling of complex free surface flows over weirs and in the vicinity of bridge piers is presented in a numerical model emulating open channel flow based on the Reynolds Averaged Navier-Stokes (RANS) equations. The importance of handling the turbulence at the free surface in the case of different flow regimes using an immiscible two-phase RANS Computational Fluid Dynamics (CFD) model is demonstrated. The free surface restricts the length scales of turbulence and this is generally not accounted for in standard two-equation turbulence modelling approaches. With the two-phase flow approach, large-velocity gradients across the free surface due to the large difference in the density of the fluids can lead to over-production of turbulence. In this paper, turbulence at the free surface is restricted with an additional boundary condition for the turbulent dissipation. The resulting difference in the free surface features and the consequences for the solution of the flow problem is discussed for different flow conditions. The numerical results for the free surface and stream-wise velocity gradients are compared to experimental data to show that turbulence damping at the free surface provides a better representation of the flow features in all the flow regimes and especially in cases with rapidly varying flow conditions.

Keywords: free surface flow; embankment weir; bridge piers; hydraulic jump; turbulence; CFD

---

1. Introduction

The interface between air and water or the free water surface is one of the most significant features of open channel flow. The motion of the free surface and the flow properties such as velocities and the pressure in the flow are strongly related. A change in the geometry of the channel, such as in the case of flow around hydraulic structures, steep pressure gradients occur due to flow constriction. According to Bernoulli's equation, under ideal conditions the total energy in the flow is conserved through transfer of energy between potential and kinetic energy. In addition, energy is lost to fluid-fluid and fluid-solid interface friction. This results in a change in the free surface location and the flow features, and governs the design of hydraulic structures.

The free surface features resulting from different flow conditions such as hydraulic jumps, shock waves, and standing waves can have negative consequences on the hydraulic structure and should be considered during design. Several researchers have proposed different empirical formulae to estimate different features such as the height of a shock front or deflection angles [1–4]. Physical model tests investigating different flow conditions and the resulting flow features can provide more details regarding the flow phenomena [5–8].

Numerical modelling of the open channel flow phenomena can yield further detail into the flow physics. Due to the strong relationship between the flow hydrodynamics and the free surface, the correct representation of the free surface along with the boundary conditions are essential to accurately represent the flow in a numerical model. Over recent decades many different methods for the prediction of the free water surface elevation have been developed. They reach from simple empirical estimations with 1D numerical tools, e.g., Brunner [9] to more advanced 3D models, e.g., Olsen [10].

The application of Computational Fluid Dynamics (CFD) methods to study free surface flows is an interesting avenue where different flow features can be calculated in detail with the solution of the Navier-Stokes equations, where the Reynolds Averaged Navier-Stokes (RANS) framework is employed [11,12]. Further several recent studies in literature present eddy-resolving frameworks such as Kirkil et al. [13], Kara et al. [14], Kang and Sotiropoulos [15] and McSherry et al. [16]. Eddy-resolving simulations provide a higher level of detail regarding the coherent structures in a turbulent flow and a better resolution of the flow compared to the RANS models [17]. On the other hand, the potential of the RANS models to evaluate complex free surface flows without resolving details regarding the turbulent eddies is not yet fully used. Unsteady RANS simulations offer a balance between computational efficiency and resolution of the flow, which can be used for many engineering applications. A challenge with two-phase RANS simulations is that the effect of the free surface is not accounted for in the standard RANS turbulence models. The turbulence production is determined by the strain rate in fluids and at the interface, the abrupt large difference in the fluid density leads to a high turbulence production. In reality, the presence of the free surface results in the elongation of the turbulent eddies parallel to the flow and dissipation of the turbulent energy in the direction normal to the free surface. This turbulence dissipation at the free surface must be included using additional boundary conditions in RANS turbulence models.

Several authors have presented different approaches to overcome this limitation in the RANS equations for two-phase flow. Jacobsen et al. [18] proposed an alternative method to calculate the turbulence production in the two-equation turbulence models for two-phase CFD models using the vorticity in the flow instead of the strain rate to avoid the over-production of turbulence around the free surface. Devolder et al. [19] demonstrated excessive damping of the free surface due to the over-production of turbulence around the free surface with the $k - \omega$ SST model in OpenFOAM and proposed a buoyancy modification term to account for this problem. Furthermore, Larsen and Fuhrman [20] showed that the overestimation of turbulence around the free surface is inherent in the classical RANS turbulence modelling approach and proposed a new closure term to avoid the exponential growth of the eddy viscosity.

In the current study, the relationship shown by Naot and Rodi [21], following the concept of wall function dissipation adjusted to free surface conditions based on experimental observations is implemented to address turbulence over-production around the free surface. This provides a boundary condition for the turbulence dissipation around the free surface based on the flow physics. The boundary condition is implemented in the open-source CFD model, REEF3D [22] and used to study complex flow features downstream of different hydraulic structures for different flow conditions. The model has been previously applied to various problems in which the representation of the complex deformation of the free surface is essential such as wave trapping between cylinders [23], breaking wave forces [24,25], breaking wave kinematics [26] and floating bodies on waves [27].

The influence of the free surface turbulence boundary condition is investigated for various flow cases in this paper to demonstrate the applicability of the method in a universal manner to complex free surface flows. Numerical simulations are carried out for controlled and free stream outflow conditions to obtain different flow conditions such as sub-critical flow throughout the domain, change from sub-critical to super-critical flow regime, transitional flow, and change from sub-critical to super-critical and back to sub-critical. These changing flow conditions are investigated for free surface flow over hydraulic structures such an embankment weir, a broad-crested weir and in the vicinity of bridge piers with focus on the boundary condition for the turbulence dissipation at the free surface. The numerical results for the free surface and the stream-wise velocity profiles over the water depth are compared to experimental data. The change in the distribution of the eddy viscosity around the free surface and the stream-wise velocity profiles in the presence and absence of the free surface turbulence boundary condition is also analyzed to investigate the influence and importance of turbulence damping at the free surface in the different cases.

2. Numerical Model

REEF3D uses the incompressible RANS equations together with the continuity equation to solve the immiscible two-phase fluid flow problem:

$$\frac{\partial u_i}{\partial x_i} = 0 \tag{1}$$

$$\frac{\partial u_i}{\partial t} + u_j \frac{\partial u_i}{\partial x_j} = -\frac{1}{\rho}\frac{\partial p}{\partial x_i} + \frac{\partial}{\partial x_j}\left[(\nu + \nu_t)\left(\frac{\partial u_i}{\partial x_j} + \frac{\partial u_j}{\partial x_i}\right)\right] + g_i \tag{2}$$

where u is the time averaged velocity, ρ is the density of the fluid, p is the pressure, ν is the kinematic viscosity, ν_t is the eddy viscosity and g the acceleration due to gravity. The equations are solved simultaneously for the two phases and the location of the free surface determines the material properties of the fluid cell such as density and viscosity.

The projection method [28] is used for pressure treatment and the resulting Poisson pressure equation is solved with a geometric multigrid preconditioned BiCGStab solver [29] provided by the high-performance solver library, HYPRE [30]. Turbulence modelling is carried out using the two-equation $k - \omega$ model [31]:

$$\frac{\partial k}{\partial t} + u_j \frac{\partial k}{\partial x_j} = \frac{\partial}{\partial x_j}\left[\left(\nu + \frac{\nu_t}{\sigma_k}\right)\frac{\partial k}{\partial x_j}\right] + P_k - \beta_k k\omega \tag{3}$$

$$\frac{\partial \omega}{\partial t} + u_j \frac{\partial \omega}{\partial x_j} = \frac{\partial}{\partial x_j}\left[\left(\nu + \frac{\nu_t}{\sigma_\omega}\right)\frac{\partial \omega}{\partial x_j}\right] + \frac{\omega}{k}\alpha P_k - \beta\omega^2 \tag{4}$$

where k is the turbulent kinetic energy, ω is the specific turbulent dissipation rate, P_k is the turbulent production rate, the coefficients have the values $\alpha = \frac{5}{9}$, $\beta_k = \frac{9}{100}$, $\beta = \frac{3}{40}$, $\sigma_k = 2$ and $\sigma_\omega = 2$.

This turbulence model is chosen as it is the most suitable for two-phase flow calculations with unsteady flow situations as in the cases presented in this study. This is because $k - \omega$ model tends to be numerically more stable due to the linear relationship between k and ω, $\nu_t = k/\omega$. In comparison, the $k - \epsilon$ model defines this relationship with a quadratic term as in $\nu_t = \rho\,C\,k^2/\epsilon$, which can lead to issues with numerical stability for areas of low levels of turbulence intensity.

To avoid the over-production of eddy viscosity outside the boundary layer in simulations of highly strained flow, eddy-viscosity limiters proposed by [32] are used:

$$\nu_t = \min\left(\frac{k}{\omega}, \sqrt{\frac{2}{3}}\frac{k}{|S|}\right) \tag{5}$$

where $|S|$ is the magnitude of the strain tensor.

The free surface influences the turbulence in such a way that the components normal to the surface are damped whereas the components parallel to the surface are enhanced. This is included in the turbulence model through an additional boundary condition at the free surface as shown by [21]. Here, the specific turbulence dissipation term at the free surface ω_s is defined as:

$$\omega_s = \frac{c_\mu^{-\frac{1}{4}}}{\kappa}k^{\frac{1}{2}} \cdot \left(\frac{1}{y'} + \frac{1}{y^*}\right) \tag{6}$$

where $c_\mu = 0.07$ and $\kappa = 0.4$. The variable y' is the virtual origin of the turbulent length scale, and was empirically found to be 0.07 times the mean water depth [33]. Including the distance y^* from the nearest wall gives a smooth transition from the free surface value to the wall boundary value of ω. The damping is carried out only around the interface using the Dirac delta function $\delta(\phi)$:

$$\delta(\phi) = \begin{cases} \frac{1}{2\epsilon}\left(1 + \cos\left(\frac{\pi\phi}{\epsilon}\right)\right) & if \ |\phi| < \epsilon \\ 0 & else \end{cases} \tag{7}$$

where $\epsilon = 1.6dx$ is the width over which the delta function is applied and dx is the grid size. The value is chosen to ensure that at least one grid cell is included around the interface in each direction and the boundary condition is applied just around the interface.

2.1. Free Surface

The free surface is determined with the level set method. The zero-level set of a signed distance function $\phi(\vec{x}, t)$ is used to represent the interface between air and water [34]. Moving away from the interface, the level set function gives the shortest distance from the interface. The sign of the function distinguishes between the two fluids across the interface as shown in Equation (8):

$$\phi(\vec{x}, t) \begin{cases} > 0 & \text{if } \vec{x} \text{ is in the air phase} \\ = 0 & \text{if } \vec{x} \text{ is at the interface} \\ < 0 & \text{if } \vec{x} \text{ is in the water phase} \end{cases} \tag{8}$$

The level set function is moved under the influence of the velocity field u_j obtained from the solution of the RANS equations to represent the change in the free surface as the simulation progresses, with the convection equation in Equation (9):

$$\frac{\partial\phi}{\partial t} + u_j\frac{\partial\phi}{\partial x_j} = 0 \tag{9}$$

The kinematic and dynamic boundary conditions at the free surface are implicitly enforced by the level set function. The discontinuity in the material properties such as the density and viscosity are handled using an interpolation with the Heaviside function around the interface over a distance of $2.1dx$ [22]. The level set function loses its signed distance property on convection and is reinitialized after every iteration using a partial differential equation-based reinitialization procedure [35] to regain its signed distance property.

2.2. Discretization Schemes

The fifth-order conservative finite difference Weighted Essentially Non-Oscillatory (WENO) scheme [36] is applied to discretize the convective terms of the RANS equations. The level set function, turbulent kinetic energy and the specific turbulent dissipation rate are discretized using the Hamilton-Jacobi formulation of the WENO scheme [37]. The WENO scheme is a minimum third-order accurate and numerically stable even in the presence of large gradients. Time advancement for the momentum and level set equations is carried out using a Total Variation Diminishing (TVD) third-order Runge-Kutta explicit time scheme [38]. Adaptive time stepping is employed to satisfy the CFL criterion based on the maximum velocity in the domain. This ensures numerical stability throughout the simulation with an optimal value of time step size. An implicit scheme is used for the time advancement of the turbulent kinetic energy and the specific turbulent dissipation, as these variables are mostly source term driven with a low influence of the convective terms. Diffusion terms of the velocities are also subjected to implicit treatment to remove them from the CFL criterion. The numerical model has been validated in detail by Bihs et al. [22] and the convergence and order of the model presented. The model is found to be monotonically convergent with the convergence ratio $R \in [0, 1]$. The combination of the higher-order schemes results in a combined order of about 2.7 for the model.

The numerical model uses a Cartesian grid for the spatial discretization together with the Immersed Boundary Method (IBM) to represent the irregular boundaries in the domain. Berthelsen and Faltinsen [39] developed the local directional ghost cell IBM to extend the solution smoothly in the same direction as the discretization, which is adapted to three dimensions in the current model. REEF3D is fully parallelized using the domain decomposition strategy and Message Passing Interface (MPI).

3. Numerical Setup

The free surface flow features in open channel flow vary with the flow conditions. Depending on the boundary conditions, the flow around hydraulic structures can stay sub-critical throughout, change from a sub-critical to a super-critical flow or change from sub-critical to super-critical and then back to a sub-critical flow. An overview of the flow conditions and cases investigated is provided in Table 1. The details regarding numerical setup and boundary conditions are provided in this section.

Table 1. Overview of the flow conditions and cases investigated in the study.

No.	Flow Condition	Case	Re	Fr_{max}
1.	sub-critical flow	embankment weir	4.5×10^4	0.33
		bridge piers	2.0×10^5	0.40
2.	sub-critical to super-critical flow	broad-crested weir	4.6×10^4	3.98
3.	transitional flow	embankment weir	6.4×10^4	2.63
4.	sub-critical to super-critical to sub-critical flow	embankment weir	7.8×10^4	4.16
		bridge piers	1.95×10^5	2.09

3.1. General Boundary Conditions

In all the cases presented in this study, a logarithmic velocity profile is prescribed at the inflow for the water phase. The air velocities are not prescribed and are a part of the solution. The outflow boundary condition for the simulations with an embankment weir is a zero-gradient or free stream outflow condition, where no resistance is offered to the flow. The required tailwater level in the different cases is maintained using a rectangular obstacle placed towards the end of the channel to represent the outflow gate in the experiments. The free stream outflow boundary condition is also applied in the simulation of the rectangular weir where the flow transitions from sub-critical to super-critical. In the case of the bridge piers, the tailwater level is prescribed at the outflow. The side walls in the two-dimensional numerical channel are treated with a zero-gradient Neumann condition, whereas a no-slip wall boundary condition is used in the three-dimensional channel. A no-slip wall boundary condition is enforced at the bottom of the channel and a zero-gradient Neumann condition is applied to the top of the numerical channel in all the simulations. An initial water depth is prescribed to initialize the free surface in the channel at $t = 0$ and a discharge given in the experiments is also prescribed. A Cartesian grid is employed in all the simulations with $dx = dy = dz$ for all the cases.

3.2. Sub-Critical Flow

The first case for this flow condition is the experiment carried out for submerged overflow on an embankment weir by Fritz and Hager [5]. The experiments are carried out in a channel 7 m long, 0.499 m wide and 0.700 m high. The embankment weir is 0.3 m high and 1.5 m long, with a slope of 1V: 2H on both the upstream and downstream faces. The two-dimensional numerical channel setup is similar to the experimental setup and is illustrated in Figure 1 showing the water and air phases within the outline of the channel. The channel has a length of 12 m and a height of 0.7 m, and the water depth is initialized as 0.454 m and a discharge of 0.038 m³/s/m is specified. The tailwater level in the channel is maintained at $d_t = 0.450$ m using a rectangular obstacle of height 0.355 m and length 0.5 m placed at 11.0 m in the channel. Stream-wise velocity profiles are calculated at four locations in the channel, upstream of the weir at 1.94 m, on the upstream face at 2.54 m, over the downstream toe at 3.54 m and downstream of the weir at 3.94 m.

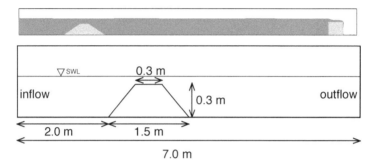

Figure 1. Numerical channel setup for the cases with an embankment weir following Fritz and Hager [5].

The second case for this flow condition is the flow around a pair of bridge piers presented by Szydłowski [40]. The channel in the experiments is 15 m long, 0.62 m wide and a pair of bridge piers of diameter 0.11 m are placed 2 m downstream of the inflow section equidistant from each other and the channel walls. The numerical setup is 4.0 m long, 0.62 m wide and 1.0 m high. The water depth is initialized to 0.48 m at $t = 0$ s and the tailwater level is prescribed to be 0.48 m along with a discharge of 0.114 m³/s in the channel. The piers are placed 1.30 m from the inflow region, equidistant from the channel walls and from each other with a distance of 0.134 m. Stream-wise velocity profiles are calculated at four locations, upstream of the piers at 0.8 m, in between the piers at 1.30 m, downstream of the piers at 1.63 m and 3.0 m. The numerical setup is illustrated in Figure 2 showing the water and air phases within the outline of the numerical channel and the piers.

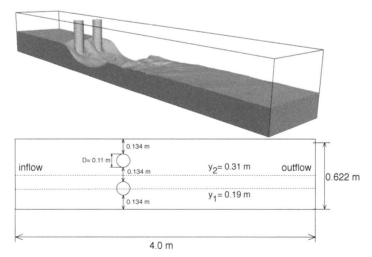

Figure 2. Numerical channel setup with bridge piers following Szydłowski [40].

3.3. Sub-Critical to Super-Critical Flow

The experiment presented by Sarker and Rhodes [41] for flow over a rectangular broad-crested weir is simulated for this flow condition. The experimental channel is 3.162 m long and 0.105 m wide. A rectangular weir with sharp edges 0.4 m long and 0.1 m high is placed in the channel with a discharge of 0.004684 m^3/s. The numerical setup is illustrated in Figure 3 showing the outline of the numerical channel, the water, and air phases and the weir. The numerical domain is 3.16 m long, 0.105 m wide and 0.4 m high. The broad-crested weir is 0.4 m long and 0.1 m high as in the experiments and placed at 1.20 m from the inlet. The free surface elevation at the inflow and the discharge are prescribed to be 0.194 m and 0.004684 m^3/s respectively, using the values from the experimental data.

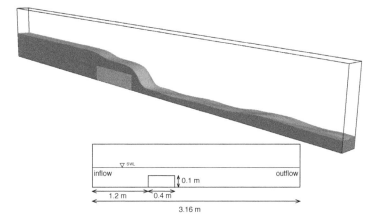

Figure 3. Numerical channel setup for broad-crested weir following Sarker and Rhodes [41].

3.4. Transitional Flow

For this flow condition, the plunging jet flow resulting from a 1% submergence of the embankment weir presented by Fritz and Hager [5] is presented. The numerical setup is similar to that illustrated in Figure 1, with the height of the rectangular obstacle set to 0.19 m to maintain a tailwater level of $d_t = 0.301$ m. A discharge of 0.054 m^3/s/m is specified in the two-dimensional channel. The water depth is initialized at 0.40 m at $t = 0$ s in the simulation. Stream-wise velocity profiles are calculated at four locations in the channel, upstream of the weir at 1.94 m, on the upstream face at 2.54 m, over the downstream toe at 3.54 m and downstream of the weir at 3.94 m.

3.5. Sub-Critical to Super-Critical to Sub-Critical Flow

The first case for this flow condition is the free overflow on an embankment weir presented by Fritz and Hager [5]. The two-dimensional numerical channel is the same as that illustrated in Figure 1. The tailwater level is maintained at $d_t = 0.159$ m using a rectangular obstacle of height 0.06 m. A discharge of 0.054 m^3/s/m is prescribed in the two-dimensional channel and the water depth is initialized at 0.40 m at $t = 0$ s. Stream-wise velocity profiles are calculated at four locations in the channel, upstream of the weir at 1.95 m, on the upstream face at 2.54 m, over the downstream toe at 3.54 m and downstream of the weir at 3.94 m.

The second case for this flow condition is the flow around a pair of bridge piers presented by Szydłowski [40], using the numerical setup illustrated in Figure 2, but with a height of 0.5 m. A tailwater level of 0.19 m, a discharge of 0.114 m^3/s is prescribed and the water depth initialized at 0.19 m at $t = 0$ s. Stream-wise velocity profiles are calculated at four locations, upstream of the piers at 0.8 m, in between the piers at 1.30 m, downstream of the piers at 1.63 m and 3.0 m.

4. Results and Discussion

The different flow conditions analyzed in this study have special free surface features associated with them. The change of the flow conditions from a super-critical condition to a sub-critical condition involves major energy losses, formation of special free surface features such as a hydraulic jump and can be especially challenging to represent correctly in numerical simulations. In the following sections, the numerical results for different scenarios with different flow conditions around an embankment weir, a broad-crested weir, and a pair of bridge piers are presented. The numerical results for the free surface elevation in all the cases are compared to experimental data. The velocity profiles at various locations in the numerical channel and the difference in the eddy viscosity in the numerical channel with and without the application of the free surface turbulence damping (FSTD) treatment is presented. In addition, the extent of the recirculation zones from Fritz and Hager [5] is presented along with the numerical results for the embankment weirs.

4.1. Sub-Critical Flow

4.1.1. Flow over an Embankment Weir

In the first case, submerged overflow on an embankment weir with a surface jet flow and a nearly horizontal free surface is studied. A grid convergence study is carried out for this case with FSTD using $dx = 0.02$ m, 0.01 m and 0.005 m. The grid convergence for the free surface elevation is presented in Figure 4 and for the stream-wise velocity profiles in Figure 5. The numerical results for the free surface and the velocity profiles are seen to be similar for $dx = 0.01$ m and $dx = 0.005$ m and the simulation with a grid size of $dx = 0.01$ m is used for further analysis of this flow condition.

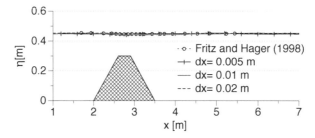

Figure 4. Grid convergence study with FSTD for the free surface elevation in the case of submerged flow with a surface jet over an embankment weir.

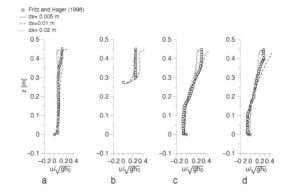

Figure 5. Grid convergence study with FSTD for the horizontal velocity profiles over the water depth at various locations for submerged flow over an embankment weir: (**a**) $x = 1.95$ m, (**b**) $x = 2.54$ m, (**c**) $x = 3.54$ m, (**d**) $x = 4.04$ m.

The numerical results for the free surface elevation with and without FSTD are compared to the experimental data in Figure 6. A good agreement is seen between the numerical result with FSTD and the experimental data. The numerical free surface elevation without FSTD in Figure 6 is seen to be similar but slightly lower just downstream of the weir between 2.6 m and 4.5 m. Due to the sub-critical flow condition throughout the channel, the free surface has a relatively simple, almost horizontal form and as a result the two simulations show similar results for the free surface elevation. The stream-wise velocity profiles in the simulations with and without FSTD are presented in Figure 7. A large deviation is seen in the velocities closer to the free surface in Figure 7b–d, indicating that the recirculation zones in the flow are calculated differently in the two simulations. The velocity profiles with FSTD show a reasonable agreement with the experimental data.

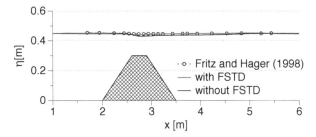

Figure 6. Comparison of the experimental data [5] and the numerical results with and without FSTD for the free surface elevation in the case of submerged flow with a surface jet over an embankment weir.

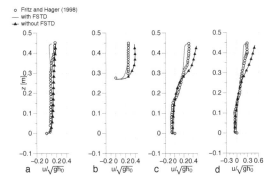

Figure 7. Horizontal velocity (u) profiles over the water depth at various locations along the channel with and without FSTD for submerged flow with a surface jet over an embankment weir: (**a**) $x = 1.95$ m, (**b**) $x = 2.54$ m, (**c**) $x = 3.54$ m, (**d**) $x = 4.04$ m.

The flow field in the numerical model for this scenario with and without FSTD is presented in Figure 8. The lower limit of the recirculation zone presented by Fritz and Hager [5] is overlaid as white squares on the streamlines to provide a comparison of the recirculation zone calculated in the simulations and the experiments. The streamlines in the water phase in Figure 8a show that while the surface jet is represented, the recirculation zone is over-estimated in the simulation without FSTD. In the simulation with FSTD, the surface jet and the extent of the recirculation zone below the free surface downstream of the embankment weir is correctly represented, as seen from Figure 8b. Thus, despite the numerical free surface elevations in the two simulations being similar, the flow features agree better with the experimental observation in the simulation with FSTD.

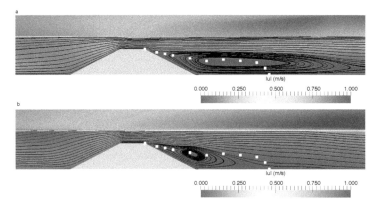

Figure 8. Free surface features for submerged flow over an embankment weir without and with FSTD, the white dots denote the lower limit of the mixing layer presented in Fritz and Hager [5]: (**a**) without FSTD, (**b**) with FSTD.

Furthermore, the eddy viscosity in the numerical channel is presented in Figure 9. In the simulation without FSTD, Figure 9a shows that the eddy viscosity has large values around the free surface downstream of the weir, both in the air and water phases. Whereas, in Figure 9b with FSTD, the eddy-viscosity values are much lower around the free surface and higher around the recirculation zone under the free surface and further downstream. Thus, it is seen that even in the simplest case of sub-critical flow throughout the channel, the damping of the turbulence at the free surface significantly aids in the better representation of the flow, even though no significant changes are seen in the calculation of the free surface elevation.

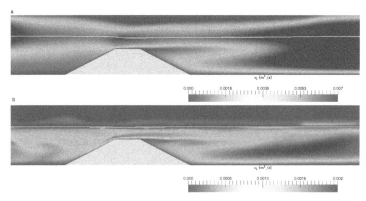

Figure 9. Eddy-viscosity contours for submerged flow over an embankment weir: (**a**) without FSTD, high values of eddy viscosity around the free surface both in air and water phases, (**b**) with FSTD, the reduction in eddy viscosity around the free surface is clearly seen.

4.1.2. Flow around a Pair of Bridge Piers

In the second case with sub-critical flow throughout the domain, the flow around a pair of bridge piers is presented. The grid convergence study for the free surface elevation along a line through a pier and along the center of the channel is shown in Figure 10a,b and is seen that the numerical results are similar for all the grid sizes. The results from the simulation with $dx = 0.01$ m are used for further analysis. The free surface calculated in simulations with and without FSTD treatment in Figure 11 show that the results are similar. There is a small deflection in the free surface downstream of the piers along the center of the channel which is slightly over-estimated in the simulation without

FSTD treatment. The velocity profiles upstream of the piers in Figure 12a at 0.8 m and Figure 12b at 1.30 are identical, indicating that the FSTD treatment does not influence the flow upstream of the cylinder. The velocity profile in the vicinity of the deflection at 1.63 m in Figure 12c shows a difference in the flow field corresponding to the small difference in the location of the free surface at this location. Further downstream from the piers, the velocity profiles at 3.0 m in Figure 12d, are identical again for the simulations with and without FSTD treatment. The results show that the FSTD treatment has only minor implications for the flow in this case as well.

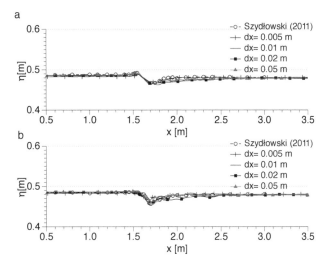

Figure 10. Grid convergence study with FSTD for the numerical free surface elevation for flow around a pair of bridge piers: (**a**) along the center line through a pier $y_1 = 0.19$ m, (**b**) along the center of the channel $y_2 = 0.31$ m.

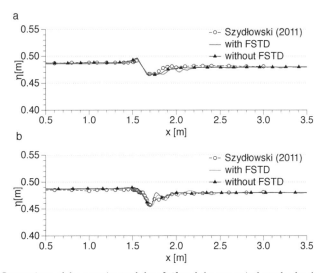

Figure 11. Comparison of the experimental data [40] and the numerical results for the free surface elevation with and without FSTD: (**a**) along the center line through a pier $y_1 = 0.19$ m, (**b**) along the center of the channel $y_2 = 0.31$ m.

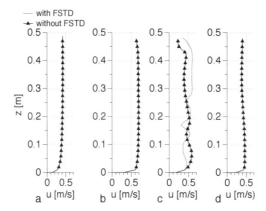

Figure 12. Horizontal velocity (u) profiles over the water depth at various locations along the channel with and without FSTD: (**a**) $x = 0.80$ m, (**b**) $x = 1.30$ m, (**c**) $x = 1.63$ m, (**d**) $x = 3.00$ m.

From the results presented for sub-critical flow throughout the channel with an embankment weir and a pair of bridge piers, it is seen that the representation of the free surface is similar in simulations with and without FSTD treatment. The flow pattern in the vicinity of the structures, on the other hand, can be influenced to some extent. The velocity profiles in the case of the embankment weir with a shallower water depth show a significant difference and consequently the recirculation pattern is not represented correctly without FSTD treatment. In the case of the bridge piers with a higher water depth, the velocity profiles show only a minor difference in the vicinity of the piers.

4.2. Sub-Critical to Super-Critical

In this scenario, the flow condition changes from sub-critical to super-critical after passing the hydraulic structure and stays super-critical until the outflow region. A grid convergence study is carried out for this case with FSTD using $dx = 0.025$ m, 0.01 m, 0.005 m and 0.0025 m. Figure 13 shows that the most coarse grid with $dx = 0.025$ m results in a spurious wave-like flow, which disappears on refining the grid to $dx = 0.01$ m. The results for the free surface elevation presented in Figure 13 show that the solution converges at grid size $dx = 0.005$ m and this is selected for further analysis of this flow condition.

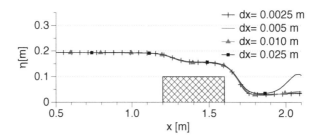

Figure 13. Grid convergence study with FSTD for the free surface elevation in the case of flow over a broad-crested weir with a free stream outflow condition.

The numerical results with and without FSTD for the free surface elevation are compared to the experimental data from Sarker and Rhodes [41] in Figure 14. The numerical result with FSTD agrees well with the experimental data. Also, the difference in the numerical results for the free surface elevation with and without FSTD is not very significant.

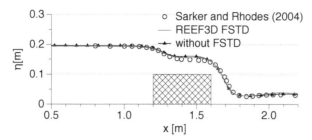

Figure 14. Comparison of the experimental data [41] and the numerical results with and without FSTD for the free surface elevation over a broad-crested weir with free stream outflow condition.

The streamlines and the velocity magnitude contours in the two simulations are shown in Figure 15. It is seen that the extent of the recirculation zone is similar in the two simulations whereas the maximum velocities in the numerical channel are higher in the simulation without FSTD. The distribution of the eddy viscosity along the center of the numerical channel is presented in Figure 16. The higher values of eddy viscosity are distributed around the free surface, in the region with super-critical flow downstream of the weir in Figure 16a for the simulation without FSTD. Whereas in the simulation with FSTD shown in Figure 16b, the higher values of eddy viscosity are found below the free surface and in the recirculation zone downstream of the weir and a region of reduced eddy viscosity around the free surface is clearly seen.

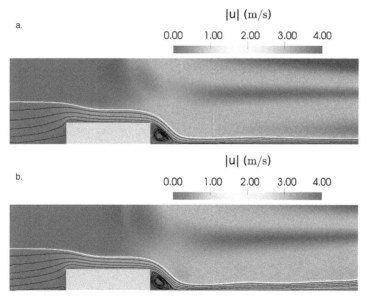

Figure 15. Free surface features around the broad-crested weir with and without FSTD showing streamlines and velocity magnitude contours: (**a**) without FSTD, (**b**) with FSTD.

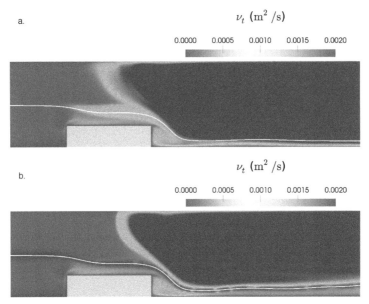

Figure 16. Eddy viscosity around the broad-crested weir: (**a**) without FSTD, high values of eddy viscosity around the free surface in both air and water phases, (**b**) with FSTD, the reduction in the eddy viscosity around the free surface is clearly seen.

Thus, in the scenario where the flow condition changes from sub-critical to super-critical and stays super-critical until the outlet, the free surface elevation is not changed significantly by FSTD. Some changes are seen in the distribution of the eddy viscosity and the maximum velocity in the channel.

4.3. Transitional Flow

In this section, the flow condition with a plunging jet flow on an embankment weir is presented. The grid convergence study for the free surface elevation in Figure 17 shows that for a coarse grid with $dx = 0.02$ m, the plunging jet flow is not represented. With finer grids of $dx = 0.01$ m and $dx = 0.005$ m the numerical free surface elevation agrees with the experimental data. This is also reflected in the grid convergence study for the stream-wise velocity profiles presented in Figure 18, where the results for $dx = 0.02$ m does not agree with the experimental observation downstream of the weir. The velocity profiles for $dx = 0.01$ m and $dx = 0.005$ m are similar and the results from $dx = 0.01$ m are used for further analysis of the flow situation.

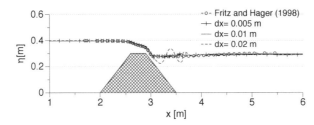

Figure 17. Grid convergence study with FSTD for the free surface elevation in the case of plunging jet flow on an embankment weir.

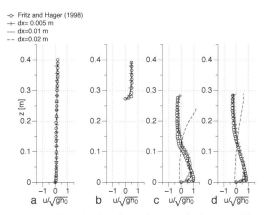

Figure 18. Grid convergence study with FSTD for the horizontal velocity profiles over the water depth at various locations for plunging jet flow on an embankment weir: (**a**) $x = 1.94$ m, (**b**) $x = 2.54$ m, (**c**) $x = 3.54$ m, (**d**) $x = 3.94$ m.

The free surface elevations in the simulations with and without FSTD treatment are presented in Figure 19. In the simulation without FSTD, the deflection in the free surface at the location of the plunging jet is over-estimated, whereas in the simulation with FSTD, the free surface agrees with the experimental data. In addition, the free surface further downstream of the weir is underestimated in the simulation without FSTD treatment. The stream-wise velocity profiles in Figure 20a,b show that the calculated flow field is identical upstream of the weir at $x = 1.94$ m and on the upstream face of the weir at $x = 2.54$ m is identical and in agreement with the experimental data. For the velocity profiles on the downstream face of the weir at $x = 3.54$ m and further downstream at $x = 3.94$ m in Figure 20c,d respectively, the numerical results with FSTD agree with the experimental data, whereas without FSTD, the velocities close to the free surface are over-estimated.

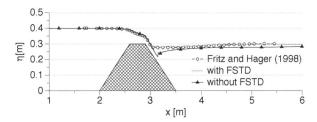

Figure 19. Comparison of the experimental data [5] and the numerical results with and without FSTD for the free surface elevation in the case of plunging jet flow on an embankment weir.

The flow in the numerical channel with and without FSTD treatment is investigated using streamlines and the extent of the recirculation zone is compared with the observations in the experiments by Fritz and Hager [5] in Figure 21. While the plunging jet flow is seen in both simulations from the velocity magnitude contours, the recirculation pattern is different in the two simulations. In the simulation without FSTD treatment in Figure 21a, the extent of the recirculation zone is smaller than in the experiments but is correctly represented in the simulation with FSTD as seen in Figure 21b. The eddy viscosity in the simulation without FSTD treatment is seen to have very high values around the free surface in Figure 22a. The FSTD treatment results in lower turbulent production around the interface as seen in Figure 22b.

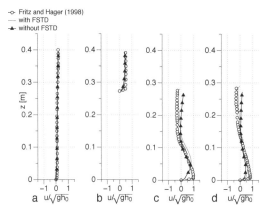

Figure 20. Horizontal velocity (u) profiles over the water depth for plunging jet flow at various locations along the channel with and without FSTD (FSTD): (**a**) $x = 1.94$ m, (**b**) $x = 2.54$ m, (**c**) $x = 3.54$ m, (**d**) $x = 3.94$ m.

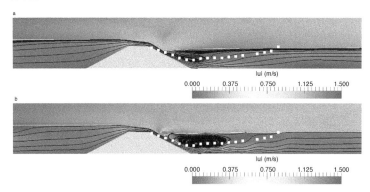

Figure 21. Free surface features for plunging jet flow on an embankment weir without and with FSTD, the white dots denote the lower limit of the mixing layer presented in Fritz and Hager [5]: (**a**) without FSTD, (**b**) with FSTD.

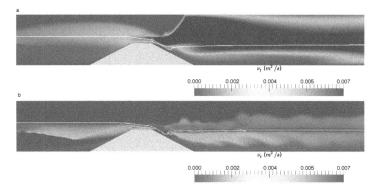

Figure 22. Eddy-viscosity contours for plunging jet flow on an embankment weir: (**a**) without FSTD, high values of eddy viscosity are noticed around the interface in both air and water phases, (**b**) with FSTD, the reduction in the eddy viscosity around the free surface is clearly seen.

From the results for the transitional flow regime with a plunging jet flow on an embankment weir, it is seen that the difference in the calculation of the free surface and the recirculation pattern is significantly influenced by the over-production of turbulence around the free surface in the absence of the FSTD algorithm. This contrasts with the flow scenarios with sub-critical flow throughout the channel and sub-critical to super-critical flow transition presented in the previous sections, where the difference in the numerical results are generally smaller with and without the application of FSTD.

4.4. Sub-Critical to Super-Critical to Sub-Critical

This is the most complex flow scenario simulated in this study where the flow conditions change from sub-critical to super-critical and back to sub-critical in the vicinity of the hydraulic structure. Two cases are presented for this scenario, the first case is the free overflow over an embankment weir presented by Fritz and Hager [5] and the second case is the flow around a pair of bridge piers presented by Szydłowski [40].

4.4.1. Flow over an Embankment Weir

A grid convergence study is carried out with FSTD using $dx = 0.03$ m, 0.02 m, 0.01 m and 0.005 m. The results for the free surface elevation shown in Figure 23. For the coarse grid with $dx = 0.03$ m, the hydraulic jump is not represented, and additional waves are seen downstream of the weir for $dx = 0.02$ m. The numerical results improve with grid refinement and the free surface elevation with $dx = 0.01$ m and $dx = 0.005$ m are seen to be similar and agree with the experimental data. The grid convergence study for the stream-wise velocity profile is presented in Figure 24. The velocity profiles are identical upstream of the weir at $x = 2.94$ m, and on the upstream face of the weir at $x = 2.54$ m for all the grid sizes in Figure 24a,b, respectively. At the location of the hydraulic jump at $x = 3.54$ m and further downstream at $x = 3.94$ m in Figure 24c,d respectively, the velocity profiles for $dx = 0.01$ m and $dx = 0.005$ m are similar. The results from $dx = 0.01$ m are used for further analysis of the case.

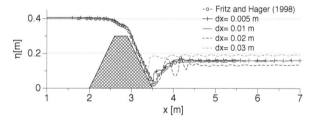

Figure 23. Grid convergence study with FSTD for the free surface elevation in the case of free overflow on an embankment weir.

The numerical result for the free surface elevation with and without FSTD is compared to the experimental data in Figure 25. The numerical result with FSTD agrees with the experimental data, including the representation of the hydraulic jump. The difference between the calculated free surface elevations in the simulations with and without FSTD is clearly seen. The free surface elevation returns to the tailwater level sharply downstream of the weir and has a constant free surface in the simulation without FSTD. The stream-wise velocity profiles show that the flow field is similar upstream of the weir in Figure 26a,b respectively with and without FSTD. The velocity profile at the location of the hydraulic jump in Figure 26c shows a significant difference between the two simulations corresponding to the difference in the free surface location. The velocity profile in the simulation with FSTD follows the observations from the experiments, though a slight overestimation of the free surface location by about 0.01 m and a corresponding deviation from the experimental observation is seen in Figure 26c. Further downstream of the weir at $x = 3.94$ m in Figure 26d, the over-estimated horizontal free surface in the simulation without FSTD corresponds to a different velocity profile, while the result with FSTD treatment is closer to the experimental observation.

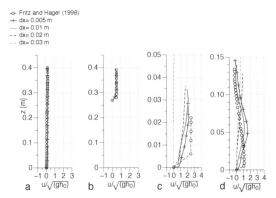

Figure 24. Grid convergence study with FSTD for the horizontal velocity profiles over the water depth at various locations for free overflow on an embankment weir: (**a**) $x = 1.94$ m, (**b**) $x = 2.54$ m, (**c**) $x = 3.54$ m, (**d**) $x = 3.94$ m.

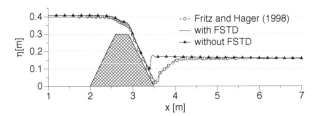

Figure 25. Comparison of the experimental data [5] and the numerical results with and without FSTD for the free surface elevation in the case of free overflow on an embankment weir.

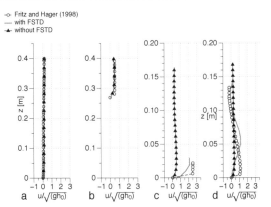

Figure 26. Horizontal velocity (u) profiles over the water depth for free overflow on an embankment weir at various locations along the channel with and without FSTD: (**a**) $x = 1.94$ m, (**b**) $x = 2.54$ m, (**c**) $x = 3.54$ m, (**d**) $x = 3.94$ m.

The lower limit of the recirculation zone presented by Fritz and Hager [5] is overlaid as white squares on the streamlines to provide a comparison of the recirculation zone calculated in the simulations in Figure 27. The streamlines and the velocity magnitude contours in Figure 27a show that the recirculation zone and the submerged flow are not calculated in the simulation without FSTD. The simulation with FSTD in Figure 27b shows that the calculated recirculation region agrees well with the lower limit of the recirculation zone in the experiments. The distribution of the eddy viscosity in the channel for the two simulations is presented in Figure 28. In the simulation without FSTD, the higher values of eddy viscosity are located around the free surface, in the region of the hydraulic jump. This has a damping effect on the flow in the region and the hydraulic jump is not represented correctly as seen in Figure 28a. In the simulation with FSTD in Figure 28b, the values of eddy viscosity are the lowest just around the free surface.

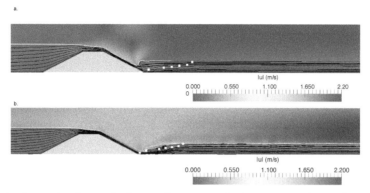

Figure 27. Free surface features for free overflow on an embankment weir without and with FSTD, the white dots denote the lower limit of the mixing layer presented in Fritz and Hager [5]: (**a**) without FSTD, (**b**) with FSTD.

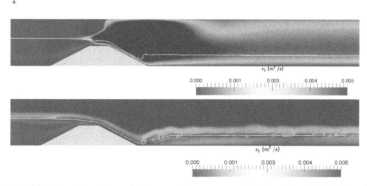

Figure 28. Eddy-viscosity contours for free overflow on an embankment weir: (**a**) without FSTD, high values of eddy viscosity are noticed around the interface in both air and water phases, (**b**) with FSTD, the reduction in the eddy viscosity around the free surface is clearly seen.

Thus, it is clearly seen from this case that the FSTD treatment has a significant effect on the calculation of the free surface elevation and the flow features involved in a scenario where the flow conditions undergo rapid changes from sub-critical to super-critical and back to sub-critical. The representation of the free surface elevation and the recirculation zone is more realistic in the simulations with FSTD as seen from the comparisons with the experimental results.

4.4.2. Flow around a Pair of Bridge Piers

A grid convergence study is carried out in this case using grid sizes $dx = 0.00625$ m, 0.01 m, 0.025 m, 0.05 m and 0.10 m using FSTD in the simulations. The results for the free surface along the center of the channel ($y_2 = 0.31$m) and along the center line through a pier ($y_1 = 0.19$ m) are presented in Figure 29. The free surface elevations agree with the experimental data for $dx = 0.01$ m and $dx = 0.00625$ m with the hydraulic jump correctly represented in the numerical result. The grid convergence study for the stream-wise velocity profiles in Figure 30 shows that the numerical results improve with grid refinement and similar values are obtained with $dx = 0.01$ m and $dx = 0.00625$ m. The results obtained with the grid size of $dx = 0.01$ m are used in the analysis of this flow condition.

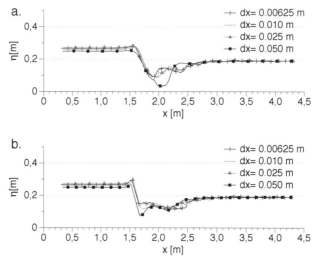

Figure 29. Grid convergence study with FSTD for the numerical free surface elevation for flow around a pair of bridge piers: (**a**) along the center line through a pier $y_1 = 0.19$ m, (**b**) along the center of the channel $y_2 = 0.31$ m.

The obstruction due to the bridge piers leads to super-critical flow downstream of the piers and the flow regime changes back to sub-critical near the outflow. The comparison of the numerical results with experimental data in this case with and without FSTD along the center line through a pier ($y_1 = 0.19$ m) is shown in Figure 31a and along the center of the channel ($y_2 = 0.31$ m) is presented in Figure 31b. These results show the importance of FSTD in complex flow situations. The free surface downstream of the piers is elevated and the free surface in the super-critical region is not represented correctly both along the center of the channel and along the center line through one of the piers without FSTD. The free surface downstream of the piers slopes downwards towards the outlet to meet the controlled outflow boundary condition. This together with the elevated free surface upstream of the piers indicates an unphysical energy loss in the numerical channel due to the over-production of eddy viscosity around the free surface. On the other hand, in the numerical results with FSTD follow the experimental data all along the channel including the hydraulic jump and the slope of the downstream free surface does not have a distinct slope towards the outlet.

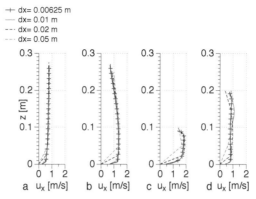

Figure 30. Grid convergence study with FSTD for the horizontal velocity profiles over the water depth at various locations in the numerical channel: (**a**) $x = 0.8$ m, (**b**) $x = 1.30$ m, (**c**) $x = 1.63$ m, (**d**) $x = 3.0$ m.

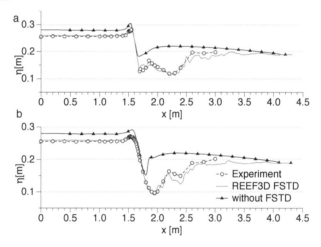

Figure 31. Comparison of the experimental data [40] and the numerical results for the free surface elevation with and without FSTD: (**a**) along the center line through a pier $y_1 = 0.19$ m, (**b**) along the center of the channel $y_2 = 0.31$ m.

The stream-wise velocity profiles with and without FSTD at various locations in the channel show that the flow field is similar upstream of the piers at $x = 0.8$ m and between the piers at $x = 1.30$ m in Figure 32a,b, respectively. The major difference is seen in Figure 32c at the trough of the hydraulic jump at $x = 1.63$ m. The application of FSTD results in a more physical distribution of the horizontal velocity along the water depth and the maximum horizontal velocity is seen just under the free surface. In the absence of FSTD, the free surface is higher, and the maximum horizontal velocity occurs much below the free surface. Further downstream at $x = 3.0$ m in Figure 32, the velocity profiles differ only in the vicinity of the free surface, corresponding to the small difference in the free surface elevation.

The difference between the flow regimes in the two simulations is further demonstrated with the eddy viscosity in the channel. Figure 33 shows the eddy-viscosity distribution around the free surface without and with FSTD. Without FSTD, high values are seen for the eddy viscosity around the free surface especially in the region of the hydraulic jump behind the piers in Figure 33a. The strain in the flow at this location is the highest in the channel due to the occurrence of the hydraulic jump. The highly strained flow results in an unphysical high value of eddy viscosity in the region, resulting in a damped out free surface. On the other hand, with FSTD, the eddy viscosity around the interface is reduced in Figure 33b and the free surface resembles the experimental observations. The flow in the channel and the free surface features in the three-dimensional numerical channel without FSTD is shown in Figure 34a shows the abrupt jump in the free surface downstream of the piers and the horizontal free surface further downstream. In the simulation with FSTD in Figure 34b, the region downstream of the piers shows a rapid decrease in free surface elevation with a trans-critical flow, the formation of a hydraulic jump and the flow returns to a sub-critical flow further downstream.

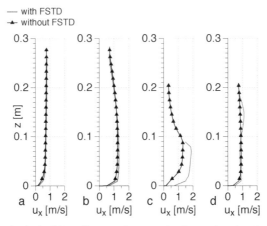

Figure 32. Horizontal velocity (u) profiles over the water depth at various locations along the center of the channel in simulations with and without FSTD: (**a**) $x = 0.80$ m, (**b**) $x = 1.30$ m, (**c**) $x = 1.63$ m, (**d**) $x = 3.00$ m.

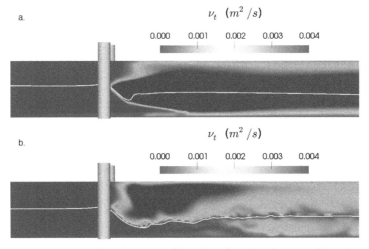

Figure 33. Eddy viscosity along the center of the channel with bridge piers: (**a**) without FSTD, high values of eddy viscosity are seen around the free surface in both air and water phases, (**b**) with FSTD, reduction of the eddy viscosity around the interface is clearly seen.

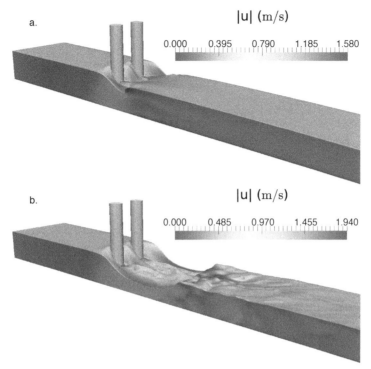

Figure 34. Free surface features in the simulations with and without FSTD along with the velocity magnitude contours throughout the channel: (**a**) without FSTD, (**b**) with FSTD.

The results presented above for a rapidly varying flow situation from sub-critical to super-critical and back to sub-critical with the free surface elevations, velocity profiles, the recirculation zone and the distribution of the eddy viscosity show that the application of a FSTD algorithm is essential to obtain realistic and physical results in such cases using immiscible two-phase RANS equations.

5. Conclusions

The open-source CFD model REEF3D is used to simulate free surface flows around different hydraulic structures with different flow conditions using immiscible two-phase RANS equations. The distribution of the eddy viscosity, the velocity magnitude, and the recirculation zones in the numerical channel are analyzed for the different cases and compared to experimental data and correlated with the effect of FSTD treatment. From the investigation of the different flow scenarios presented in this study, the following conclusions can be drawn:

- When the flow conditions remain sub-critical throughout the domain, the calculated free surface elevations are similar with and without FSTD but the submerged recirculation zone is better represented when FSTD is used.
- When the flow conditions change from sub-critical to super-critical, the calculated free surface elevation and the flow features such as the recirculation zone are similar with and without FSTD.
- In the case of transitional flow, the differences in the calculation of the free surface and the recirculation pattern become more apparent.
- When the flow conditions change from sub-critical to super-critical and back to sub-critical, the calculated free surface elevation and the flow features such as the hydraulic jump and the recirculation zone are represented correctly only in the simulation with FSTD.

- In general, simulations with FSTD provide a more physical representation of the recirculation zone. The calculation of free surface elevation is significantly affected in the complex case of flow conditions changing from sub-critical to super-critical to sub-critical.

The results demonstrate the importance of FSTD treatment in immiscible two-phase RANS simulations of flow around hydraulic structures. The implementation also addresses an important issue of over-production of turbulence at the free surface in two-phase RANS modelling during complex free surface flow using an empirical approach that is based on the flow physics. The proposed boundary condition is shown to be numerically stable in all the flow situations studied and has varying degree of influence on the results, depending in the flow situation. It is also demonstrated that two-phase RANS modelling that does not resolve the detailed eddies in turbulent flow can be used to obtain reasonably accurate representation of complex flow conditions. In the case of flows with hydraulic jumps, the internal structures of the flow and the air entrainment are not presented, but the bulk quantities that influence engineering design of hydraulic structures such as the free surface and the velocity profile over depth are well represented. This provides an alternative tool to analyze engineering design of hydraulic structures at a computational expense that is much lower than that required by eddy-resolving models.

Author Contributions: Lead Author A.K. was responsible for the design of the simulations, compilation of the results and drafting the manuscript. G.F. carried out initial testing and validation. Concept development, coding and academic supervision by H.B.

Funding: This research received no external funding.

Acknowledgments: The authors thank Prof Michał Szydłowski for providing the experimental data for the free surface elevations for the flow scenario with the bridge piers. GF was supported through the ÚNKP-18-3-I New National Excellence Program of the Ministry of Human Capacities. This research was supported in part with computational resources at the Norwegian University of Science and Technology (NTNU) provided by NOTUR (No. NN2620K), http://www.notur.no.

Conflicts of Interest: The authors declare no conflict of interest.

References

1. Rouse, H.; Bhoota, B.; Hsu, E.Y. Design of Channel Expansion. *Proc. Am. Soc. Civ. Eng. (ASCE)* **1951**, *75*, 1369–1385.
2. Ippen, A.T.; Dawson, J.H. Design of channel contractions. *Proc. Am. Soc. Civ. Eng. (ASCE)* **1951**, *75*, 1348–1368.
3. Hager, W.H. Supercritical flow in channel junctions. *J. Hydraul. Eng.* **1989**, *115*, 595–616. [CrossRef]
4. Mazumder, S.K.; Hager, W.H. Supercritical expansion flow in Rouse modified and reversed transitions. *J. Hydraul. Eng.* **1993**, *119*, 201–219. [CrossRef]
5. Fritz, H.M.; Hager, W.H. Hydraulics of embankment weirs. *J. Hydraul. Eng.* **1998**, *124*, 963–971. [CrossRef]
6. Tullis, B.; Neilson, J. Performance of submerged ogee-crest weir head-discharge relationships. *J. Hydraul. Eng.* **2008**, *134*, 486–491. [CrossRef]
7. Murzyn, F.; Chanson, H. Free-surface fluctuations in hydraulic jumps: Experimental observations. *Exp. Therm. Fluid Sci.* **2009**, *33*, 1055–1064. [CrossRef]
8. Chachereau, Y.; Chanson, H. Free-surface fluctuations and turbulence in hydraulic jumps. *Exp. Therm. Fluid Sci.* **2011**, *35*, 896–909. [CrossRef]
9. Brunner, G.W. HEC-RAS (River Analysis System). In Proceedings of the North American Water and Environment Congress & Destructive Water, Anaheim, CA, USA, 22–28 June 1996; pp. 3782–3787.
10. Olsen, N. *A Three Dimensional Numerical Model for Simulation of Sediment Movements in Water Intakes with Multiblock Option: Version 1.1 and 2.0 User's Manual;* Department of Hydraulic and Environmental Engineering, The Norwegian University of Science and Technology: Trondheim, Norway, 2003; p. 138.
11. Haun, S.; Olsen, N.R.B.; Feurich, R. Numerical modeling of flow over trapezoidal broad-crested weir. *Eng. Appl. Comput. Fluid Mech.* **2011**, *5*, 397–405. [CrossRef]
12. Fleit, G.; Baranya, S.; Bihs, H. CFD Modeling of Varied Flow Conditions Over an Ogee-Weir. *Period. Polytech. Civ. Eng.* **2017**, 1–7. [CrossRef]

13. Kirkil, G.; Constantinescu, S.G.; Ettema, R. Coherent structures in the flow field around a circular cylinder with scour hole. *J. Hydraul. Eng.* **2008**, *134*, 572–587. [CrossRef]

14. Kara, S.; Kara, M.C.; Stoesser, T.; Sturm, T.W. Free-surface versus rigid-lid LES computations for bridge-abutment flow. *J. Hydraul. Eng.* **2015**, *141*, 04015019. [CrossRef]

15. Kang, S.; Sotiropoulos, F. Large-eddy simulation of three-dimensional turbulent free surface flow past a complex stream restoration structure. *J. Hydraul. Eng.* **2015**, *141*, 04015022. [CrossRef]

16. McSherry, R.J.; Chua, K.V.; Stoesser, T. Large eddy simulation of free-surface flows. *J. Hydrodyn. Ser. B* **2017**, *29*, 1–12. [CrossRef]

17. Viti, N.; Valero, D.; Gualtieri, C. Numerical Simulation of Hydraulic Jumps. Part 2: Recent Results and Future Outlook. *Water* **2019**, *11*, 28. [CrossRef]

18. Jacobsen, N.G.; Fuhrman, D.R.; Fredsøe, J. A wave generation toolbox for the open-source CFD library: OpenFOAM. *Int. J. Numer. Methods Fluids* **2012**, *70*, 1073–1088. [CrossRef]

19. Devolder, B.; Rauwoens, P.; Troch, P. Application of a buoyancy-modified $k - \omega$ SST turbulence model to simulate wave run-up around a monopile subjected to regular waves using OpenFOAM®. *Coast. Eng.* **2017**, *125*, 81–94. [CrossRef]

20. Larsen, B.E.; Fuhrman, D.R. On the over-production of turbulence beneath surface waves in Reynolds-averaged Navier-Stokes models. *J. Fluid Mech.* **2018**, *853*, 419–460. [CrossRef]

21. Naot, D.; Rodi, W. Calculation of secondary currents in channel flow. *J. Hydraul. Div. ASCE* **1982**, *108*, 948–968.

22. Bihs, H.; Kamath, A.; Alagan Chella, M.; Aggarwal, A.; Arntsen, Ø.A. A new level set numerical wave tank with improved density interpolation for complex wave hydrodynamics. *Comput. Fluids* **2016**, *140*, 191–208. [CrossRef]

23. Kamath, A.; Bihs, H.; Alagan Chella, M.; Arntsen, Ø.A. Upstream-Cylinder and Downstream-Cylinder Influence on the Hydrodynamics of a Four-Cylinder Group. *J. Waterw. Port Coast. Ocean Eng.* **2016**, *142*, 04016002. [CrossRef]

24. Kamath, A.; Alagan Chella, M.; Bihs, H.; Arntsen, Ø.A. Breaking wave interaction with a vertical cylinder and the effect of breaker location. *Ocean Eng.* **2016**, *128*, 105–115. [CrossRef]

25. Bihs, H.; Kamath, A.; Alagan Chella, M.; Arntsen, Ø.A. Breaking-Wave Interaction with Tandem Cylinders under Different Impact Scenarios. *J. Waterw. Port Coast. Ocean Eng.* **2016**, *142*, 04016005. [CrossRef]

26. Alagan Chella, M.; Bihs, H.; Myrhaug, D.; Muskulus, M. Breaking solitary waves and breaking wave forces on a vertically mounted slender cylinder over an impermeable sloping seabed. *J. Ocean Eng. Mar. Energy* **2017**, *3*, 1–19. [CrossRef]

27. Bihs, H.; Kamath, A. A combined level set/ghost cell immersed boundary representation for floating body simulations. *Int. J. Numer. Methods Fluids* **2017**, *83*, 905–916. [CrossRef]

28. Chorin, A. Numerical solution of the Navier-Stokes equations. *Math. Comput.* **1968**, *22*, 745–762. [CrossRef]

29. Ashby, S.F.; Falgout, R.D. A Parallel Mulitgrid Preconditioned Conjugate Gradient Algorithm for Groundwater Flow Simulations. *Nucl. Sci. Eng.* **1996**, *124*, 145–159. [CrossRef]

30. Center for Applied Scientific Computing. *HYPRE High Performance Preconditioners—User's Manual*; Lawrence Livermore National Laboratory: Livermore, CA, USA 2006.

31. Wilcox, D.C. *Turbulence Modeling for CFD*; DCW Industries Inc.: La Canada, CA, USA, 1994.

32. Durbin, P.A. Limiters and wall treatments in applied turbulence modeling. *Fluid Dyn. Res.* **2009**, *41*, 1–18. [CrossRef]

33. Hossain, M.S.; Rodi, W. Mathematical modeling of vertical mixing in stratified channel flow. In Proceedings of the 2nd Symposium on Stratified Flows, Trondheim, Norway, 24–27 June 1980.

34. Osher, S.; Sethian, J.A. Fronts propagating with curvature-dependent speed: Algorithms based on Hamilton-Jacobi formulations. *J. Comput. Phys.* **1988**, *79*, 12–49. [CrossRef]

35. Peng, D.; Merriman, B.; Osher, S.; Zhao, H.; Kang, M. A PDE-based fast local level set method. *J. Comput. Phys.* **1999**, *155*, 410–438. [CrossRef]

36. Jiang, G.S.; Shu, C.W. Efficient implementation of weighted ENO schemes. *J. Comput. Phys.* **1996**, *126*, 202–228. [CrossRef]

37. Jiang, G.S.; Peng, D. Weighted ENO schemes for Hamilton-Jacobi equations. *SIAM J. Sci. Comput.* **2000**, *21*, 2126–2143. [CrossRef]

38. Shu, C.W.; Osher, S. Efficient implementation of essentially non-oscillatory shock capturing schemes. *J. Comput. Phys.* **1988**, *77*, 439–471. [CrossRef]
39. Berthelsen, P.A.; Faltinsen, O.M. A local directional ghost cell approach for incompressible viscous flow problems with irregular boundaries. *J. Comput. Phys.* **2008**, *227*, 4354–4397. [CrossRef]
40. Szydłowski, M. Numerical Simulation of Open Channel Flow between Bridge Piers. *TASK Q.* **2011**, *15*, 271–282.
41. Sarker, M.A.; Rhodes, D.G. Calculation of free-surface profile over a rectangular broad-crested weir. *Flow Meas. Instrum.* **2004**, *15*, 215–219. [CrossRef]

Article

Consistent Particle Method Simulation of Solitary Wave Interaction with a Submerged Breakwater

Yaru Ren [1], Min Luo [2,*] and Pengzhi Lin [1]

[1] State Key Laboratory of Hydraulics and Mountain River Engineering, Sichuan University, Chengdu 610065, China; 2017223060054@stu.scu.edu.cn (Y.R.); cvelinpz@126.com (P.L.)

[2] Zienkiewicz Centre for Computational Engineering, College of Engineering, Swansea University, Swansea SA1 8EN, UK

* Correspondence: min.luo@swansea.ac.uk; Tel.: +44(0)-1792-604391

Received: 3 January 2019; Accepted: 26 January 2019; Published: 2 February 2019

Abstract: This paper presents a numerical study of the solitary wave interaction with a submerged breakwater using the Consistent Particle Method (CPM). The distinct feature of CPM is that it computes the spatial derivatives by using the Taylor series expansion directly and without the use of the kernel or weighting functions. This achieves good numerical consistency and hence better accuracy. Validated by published experiment data, the CPM model is shown to be able to predict the wave elevations, profiles and velocities when a solitary wave interacts with a submerged breakwater. Using the validated model, the detailed physics of the wave breaking process, the vortex generation and evolution and the water particle trajectories are investigated. The influence of the breakwater dimension on the wave characteristics is parametrically studied.

Keywords: consistent particle method; solitary wave; submerged breakwater; breaking wave; vortex

1. Introduction

Tsunamis possess tremendous destructive power and are among the most horrible natural hazards in coastal regions. For example, the catastrophic tsunami disaster in Indonesia in September 2018 caused more than 2100 deaths, more than 4400 people seriously wounded and more than 67,000 houses destroyed or damaged [1]. The most recent tsunami disaster triggered by the collapse of the volcano in Indonesia has also caused heavy casualties and damaged thousands of properties [2]. To protect the coastal communities, breakwaters are extensively constructed along the coast for reflecting waves and/or dissipating wave energy. Among the various types of breakwaters, the submerged breakwater has been frequently used because it has less effect on the ecosystem and is cheaper to construct [3]. However, the wave-structure interaction induces complex flows, which can be dangerous for underwater craft navigations and swimmers as well as transport excessive sediments and hence cause the erosion of the structure foundation. In this context, the present study aims to investigate the detailed physics of the tsunami wave-submerged breakwater interaction. The wave breaking process, vortex generation and evolution and fluid particle trajectory will be focused.

Solitary waves have resemblances to tsunami waves and hence are frequently used as the substitute in tsunami studies [4,5], although some studies looked into the differences between a solitary wave and a tsunami wave [6,7]. The review here discusses the works on both solitary and tsunami waves without emphasizing their differences. An early study of the solitary wave-structure interaction is Grilli et al. [8] who investigated the breaking characteristics of solitary waves over submerged and emerged trapezoidal breakwaters by experimental measurements of the wave profiles and a fully nonlinear potential model. Lin et al [9] studied the solitary wave runup and rundown on sloping beaches by using a RANS (Reynolds Average Navier-Stokes) model that was able to reproduce the breaking wave and the associated turbulence. Chang et al [10] studied the solitary wave interaction with a submerged rectangular obstacle and particularly measured the velocity field near the obstacle by the Particle Image Velocimetry (PIV). Huang and Dong [11] numerically studied the viscous interaction between a solitary wave and a submerged rectangular dike by a finite-analytical method and found that the primary vortex generated at the leeward of the dike and the secondary vortex at the right toe of the dike may scour the bottom of the dike. Based on laboratory experiments and the numerical model developed in Lin et al [9], Liu and Al-Banaa [12] studied the action of non-breaking solitary waves on a vertical surface-piercing barrier with the emphasis on wave runup and forces. Lin [13] systematically explored the wave reflection, transmission and dissipation of a solitary wave interacting with submerged rectangular obstacles using his RANS model. More recently, Hsiao and Lin [14] experimentally and numerically studied the solitary wave impinging and overtopping on a seawall, and Wu et al [15] investigated the propagation of solitary waves over a rectangular breakwater. In both studies, the characteristics of breaking waves and the accompanied vortexes were analyzed numerically based on the COBRAS model (COrnell BReaking And Structure), which is a cognate of Lin's numerical model.

All the numerical studies mentioned above are based on the mesh-based methods. In recent years, another category of numerical method has obtained significant developments, i.e., the particle method. Two typical methods belonging to this category are the Smoothed Particle Hydrodynamics (SPH) [16,17] and Moving Particle Semi-implicit (MPS) method [18]. Without predefined meshes, the particle methods are advantageous in dealing with the large deformations (e.g., fluid merging and splitting) and tracking the free surface or fluid interface, and hence have been extensively applied to study free surface flows [19,20] and wave-structure interaction problems [21–23]. The most up-to-date developments of particle methods are reviewed in [24]. The pioneer particle method study of solitary wave-structure interaction is Monaghan and Kos [25], who studied the run-up and return of a solitary wave on a beach. Shao [26] is the pioneer of using the incompressible SPH (ISPH) for solitary wave interaction with structure. Specifically, this work investigated the characteristics of wave reflection, transmission, and dissipation of solitary waves impinging on a vertical surface-piercing barrier. Zhang and Tang [27] employed the MPS method to study the interaction between the solitary wave and a flat plate. Huang and Zhu [28] investigated the tsunami actions on a seawall.

Although a large number of researches have investigated the solitary wave interaction with submerged structures, some key phenomena associated with the breaking wave generation and evolution in the wave-structure interaction process are not fully understood. The present study aims to reveal the physical process of these phenomena. Particularly for the particle methods such as SPH and MPS, although having the advantage of being able to capture large deformations, they suffer from spurious pressure fluctuations [29]. One major cause is that the derivative approximation schemes introduce numerical errors particularly for irregular particle distributions [30]. To overcome this issue, the Consistent Particle Method (CPM) has been developed [30]. Different from the kernel approximation schemes in SPH and the weighted-average particle interaction models in MPS, the CPM computes the first- and second-order derivatives simultaneously based on the Taylor series expansion [30]. In this way, the numerical consistency that is a key issue in the derivative approximation of particle methods can be achieved and hence the numerical accuracy is improved. The performance of CPM in alleviating spurious pressure fluctuation has been well demonstrated by the cases of violent free surface flow [30,31] and fluid-structure interaction [32]. Another issue that has restricted the application of particle methods in large-scale problems is the relatively low computational efficiency. CPM adopts the OpenMP parallel computing to accelerate the computation [33], and the Graphics Processing Unit parallel computing is ongoing.

In this paper, the CPM is used to study the solitary wave interaction with a submerged breakwater. The numerical accuracy is validated by the published experimental results of wave elevation, wave profile and wave velocity distributions at three sections. Using the validated model, the characteristics of the breaking wave process and the associated vortex structure will be discussed in detail. The water particle trajectories will be analyzed to get insights into the possible sediment transport. And a parametric study will be conducted to explore the influences of the breakwater dimension on the fluid fields. It is expected that the research findings can provide some guidance for the engineering design.

2. CPM Methodology

2.1. Governing Equations

The governing equations of CPM are the conservations of mass and momentum, i.e., the Navier-Stokes equations, as follows:

$$\frac{1}{\rho}\frac{D\rho}{Dt} + \nabla \cdot \mathbf{v} = 0 \tag{1}$$

$$\frac{D\mathbf{v}}{Dt} = -\frac{1}{\rho}\nabla p + \nu\nabla^2\mathbf{v} + \mathbf{g}, \tag{2}$$

where ρ is the density of a fluid, $\mathbf{v}$ the particle velocity vector, p the fluid pressure, $\mathbf{v}$ the kinematic viscosity of a fluid, $\mathbf{g}$ the gravity acceleration and t the time. The fluid domain is discretized by non-connecting particles. Each particle has a fixed mass and moves under the external forces arising from the gravity, pressure difference and viscosity as in Equation (2).

2.2. Two-Step Projection Method

CPM solves the governing equations by a two-step projection method [34]. In the predictor step, the intermediate particle velocities and positions are computed by neglecting the pressure gradient term as follows:

$$\mathbf{v}^* = \mathbf{v}^{(k)} + \left[\nu\nabla^2\mathbf{v}^{(k)} + \mathbf{g}\right]\Delta t \tag{3}$$

$$\mathbf{r}^* = \mathbf{r}^{(k)} + \mathbf{v}^*\Delta t, \tag{4}$$

where $\mathbf{v}^{(k)}$ and $\mathbf{r}^{(k)}$ are the particle velocity and position in the k-th (previous) time step respectively, $\mathbf{v}^*$ and $\mathbf{r}^*$ are the temporary particle velocity and position, respectively, and Δt is the time step size.

In the corrector step, a pressure Poisson equation (PPE) is derived as:

$$\nabla \cdot \left(\frac{1}{\rho^*}\nabla p^{(k+1)}\right) = \frac{1}{\Delta t^2}\frac{\rho^{(k+1)} - \rho^*}{\rho^{(k+1)}}. \tag{5}$$

The fluid incompressibility condition is enforced by setting the fluid density at the current time step ($\rho^{(k+1)}$) to the initial value (ρ_0). The intermediate fluid density ($\rho*$) of a particle is computed by evaluating the relative distances of this particle to its neighbor particles within the influence domain (the influence radius $r_e = 2.1L_0$, where L_0 the initial particle spacing) [31].

Applying a proper derivative computation scheme to the left-hand side of Equation (5), a linear equation system with sparse and non-symmetric coefficients can be obtained and solved efficiently. With the solved pressure, particle velocities and positions in the entire computational domain are corrected as:

$$\mathbf{v}^{(k+1)} = \mathbf{v}^* - \left(\frac{\nabla p}{\rho}\right)^{(k+1)}\Delta t \tag{6}$$

and

$$\mathbf{r}^{(k+1)} = \mathbf{r}^{(k)} + 0.5\left(\mathbf{v}^{(k)} + \mathbf{v}^{(k+1)}\right)\Delta t. \tag{7}$$

The computational time step size Δt is governed by the Courant condition as

$$\frac{v_{max}\Delta t}{L_0} \leq C_{max}, \tag{8}$$

where v_{max} is the maximum particle velocity at the k-th time step and the coefficient C_{max} is selected to be 0.25.

This above is a standard solving scheme that has been used in other particle methods such as MPS [18] and ISPH [26]. The distinct feature of CPM compared to other particle methods lies in the computation of the spatial derivatives in the governing equations, as elaborated in the following section.

2.3. Derivative Computation Based on Taylor Series Expansion

The Taylor series expansion for a smooth function $f(x, y)$ in the vicinity of a reference particle (x_0, y_0) can be expressed as:

$$f(x,y) = f_0 + hf_{,x0} + kf_{,y0} + \frac{1}{2}h^2 f_{,xx0} + hk f_{,xy0} + \frac{1}{2}k^2 f_{,yy0} + O(r^3), \tag{9}$$

where $h = x - x_0$, $k = y - y_0$, $f_0 = f(x_0, y_0)$, $f_{,x0}$ is the first order derivative of function f with respect to x at (x_0, y_0), and $f_{,xy0}$ the second-order derivative of function f with respect to x and y at (x_0, y_0). By writing Equation (9) for each of the neighboring particles, the following equation system can be obtained:

$$[A]\{Df\} - \{f\} = 0, \tag{10}$$

where $[A]$ is a function of relative particle positions (i.e., h and k), $\{f\}$ is a combination of the variable differences between the reference particle and its neighboring particles, (i.e., $f - f_0$), and $\{Df\}$ is a vector including the five derivatives in Equation (9). Theoretically, five neighboring particles are enough to solve Equation (10). In practice, more particles, (typically 12–18 for 2D problems), are involved and hence Equation (10). is over-determined. The over-determined equation system is solved by the weighted-least-squares scheme [35], giving the first- and second-order derivatives simultaneously as follows [30]:

$$\frac{\partial p_i}{\partial x} = \sum_{j \neq i} \left[w_j^2 \left(a_1 h_j + a_2 k_j + 0.5 a_3 h_j^2 + a_4 h_j k_j + 0.5 a_5 k_j^2 \right) (p_j - p_i) \right] \tag{11}$$

and

$$\frac{\partial^2 p_i}{\partial x^2} = \sum_{j \neq i} \left[w_j^2 \left(c_1 h_j + c_2 k_j + 0.5 c_3 h_j^2 + c_4 h_j k_j + 0.5 c_5 k_j^2 \right) (p_j - p_i) \right], \tag{12}$$

where w_j is the weighting function used in the weighted-least-squares approximation to solve an overdetermined equation system. Note that this weighting function is essentially different from the kernel in SPH and the weighting function in the particle interaction model of MPS, both of which serve as the weighting in the weighted average calculation of function value or derivatives. p_i in Equations (11) and (12) is the pressure on particle i, and a and c the coefficients generated by the weighted-least-squares scheme (refer to Equation (21) in [30]).

2.4. Free Surface and Solid Boundaries

In CPM, the free surface particles are recognized by the "arc" method [35]. If any arcs of a circle around a reference particle are not covered by the circles of its neighbors, this reference particle is a free surface particle, (the reader is referred to [30] for more details). The essential boundary condition, i.e., $p = 0$, is enforced on the free surface particles. The impermeable walls are modeled by the mirror particle approach [26,36], i.e., generating fictitious particles outside the physical boundary by mirroring real fluid particles along the boundary line. Particularly, a moving wall can be simulated by updating the axis (in accordance with the physical wall motion) along which mirror particles are generated. This is an intrinsic advantage compared to the mesh-based methods. The submerged structure is simulated by fixed particles. On all the solid boundaries, the Neumann boundary condition $\partial p / \partial n = -\rho g \times n$ is applied, where n is the outward unit vector of the solid boundary.

3. Parameters of the Studied Problem

The CPM model is used to study the experimental case of solitary wave interaction with a submerged bottom-mounted breakwater in Wu et al. [15]. Figure 1 is the schematic view of this example, in which the length of the wave flume is 6.5 m, the water depth $h = 0.14$ m and the rectangular structure (dimension 0.1 m × 0.02 m) located 2.5 m from the home position of the wave maker. Wu's experimental study was conducted in the wave flume at Tainan Hydraulics Laboratory, National Cheng Kung University. Wave heights at E1 ($x = -0.657$ m), E2 ($x = 0.010$ m) and E3 ($x = 0.357$ m) were measured by wave gauges. The wave profile and velocity filed in a certain area at the leeward of the breakwater were measured by a PIV system. The solitary wave was produced by a translational wave maker, whose motion was generated by the procedure presented in Goring [37]. In CPM simulations, stationary particles of initial particle distance $L_0 = 0.0025$ m are generated in the whole domain. Using the approach introduced in Section 2.3 to move the wave maker in the described way, a solitary wave can be produced. A fixed time step $\Delta t = 0.0005$ s is adopted to satisfy the Courant condition. The CPM results of wave elevations at E1, E2 and E3, wave profiles and wave velocities at three sections (V1-$x = 0.06$ m, V2-$x = 0.10$ m and V3-$x = 0.14$ m, as shown in Figure 1) are compared with the experimental results of Wu et al. [15].

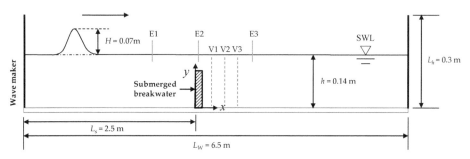

E1: $x = -0.657$ m E2: $x = 0.010$ m E3: $x = 0.357$ m V1: $x = 0.060$ m V2: $x = 0.100$ m V3: $x = 0.140$ m

Figure 1. Schematic view of the solitary wave propagating over a submerged breakwater.

4. Results and Discussions

4.1. Wave Transmission and Reflection

The wavemaker generates a solitary wave of wave height of $H = 0.07$ m, which propagates forward. The wave elevations just in front of (E1), at the location of (E2) and just behind of (E3) the breakwater is presented in Figure 2. The time when the solitary wave crest occurs at E1 is $t = 0$ s. In general, the CPM results and the COBRAS numerical results of Wu et al. [15] match quite well and are in a good agreement with Wu's experimental data [15]. The relative differences of these two sets of numerical wave crests against the experimental data are presented in Table 1. The maximum difference is 3.1% for COBRAS and 0.4% for CPM. Both numerical models are very accurate in predicting the wave elevations. From the next section on, the capability of CPM in reproducing the wave breaking process and tracking the particle trajectories will be demonstrated.

The solitary wave-breakwater interaction happens when the wave approaches the structure. The wave crests at E1 and E3 are treated as the amplitudes of the incident and transmitted waves. Since most of the energy of a solitary wave concentrates on the bulge part of the wave, the ratio of the transmitted and incident wave heights, i.e., Ht/Hi, is used as an indicator of the wave transmission. Based on this, the wave transmission coefficient of this case is 92.6%. The breakwater also reflects part of the wave, which is manifested in the second crest of wave elevation at E1 that occurs at around $t = 1$ s. With the same argument for the wave transmission, the wave reflection coefficient is evaluated by the ratio of the reflected and incident wave heights, i.e., Hr/Hi. It is computed to be 17.7% for this case. Analytical solutions have been derived for estimating the wave transmission (Equation 4.8 in [38]) and reflection (Equation (11) in [13]) coefficients of a solitary wave propagating rectangular submerged obstacles. Based on these equations, the wave transmission and reflection coefficients of this case are 95.3% and 18.1%, the relative differences between which and the CPM solutions are 2.8% and 2.2%, respectively.

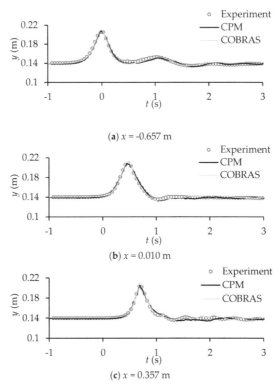

(a) x = -0.657 m

(b) x = 0.010 m

(c) x = 0.357 m

Figure 2. Wave elevations at E1, E2 and E3: **(a)** x = −0.657 m; **(b)** x = 0.010 m; **(c)** x = 0.357 m.

Table 1. Relative differences between the numerical and experimental results of wave crests at E1, E2 and E3.

Numerical models	Relative Errors Against Experimental Data in [15]		
	E1	E2	E3
COBRAS in [15]	2.0%	1.4%	3.1%
CPM	0.2%	0.4%	0.2%

4.2. Fluid Characteristics of Wave-Structure Interaction

When the wave approaches the breakwater, the wave accelerates due to the blockage of part of the flume section. This leads to a jet flow toward the leeward side of the breakwater. Accompanied is a clockwise vortex behind and at an elevation near the crest of the breakwater. These phenomena can be clearly seen from the velocity profile at t = 0.46 s as shown in Figure 3a. The jet flow impinges the main water body behind the breakwater. Due to the velocity of the main water body being smaller than that of the impinging flow, the elevation of the water body is raised (see Figure 3b). The raised water body and the crest of the incoming jet flow form the double crests. This is the so-called crest-crest exchange phenomenon [15,39]. As the main crest of the solitary wave propagates downstream, the jet flow impingement increases the steepness of the tail of the transmitted wave (see the wave profile at *t* = 0.74 s). The impinging flow continuously reduces the water elevation of the impinging location and causes the tail surface to break in the direction opposite to that of the main wave direction (Figure 3d). In this process, the vortex develops in size and moves downstream with the wave. Specifically, the vortex occupies the area of 0.2 m × 0.1 m behind the breakwater at *t* = 0.88 s. It is noteworthy that the CPM model has accurately reproduced the experimental wave profiles at the time instants discussed above. And the predicted velocity distributions at sections V1, V2 and V3 are in a reasonably good agreement with the experimental results as shown in Figure 4. Although some discrepancies exist near the center of the vortex region, the numerical model is able to capture the key phenomena during this process. This shows the satisfactory accuracy of CPM.

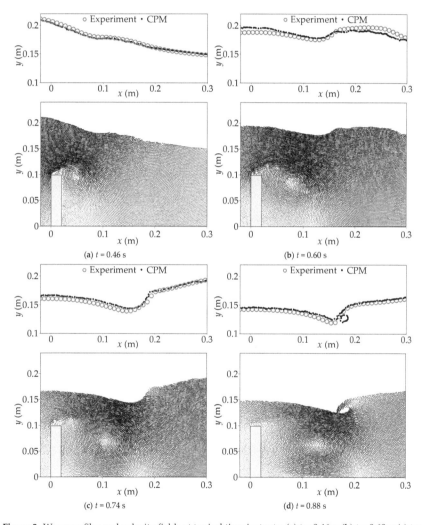

Figure 3. Wave profiles and velocity fields at typical time instants: (**a**) *t* = 0.46 s; (**b**) *t* = 0.60 s; (**c**) *t* = 0.74 s; (**d**) *t* = 0.88 s.

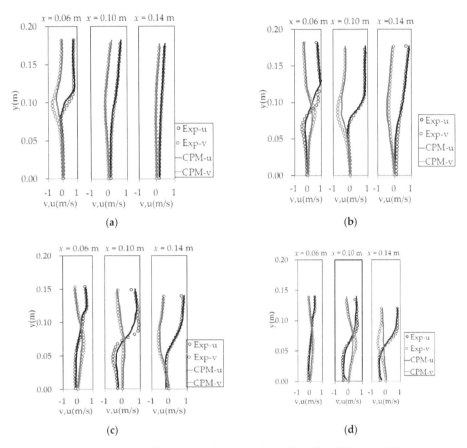

Figure 4. Horizontal and vertical velocity distributions at the sections of $x = 0.06$ m, $x = 0.10$ m and $x = 0.14$ m at typical time instants: (**a**) $t = 0.46$ s; (**b**) $t = 0.60$ s; (**c**) $t = 0.74$ s; (**d**) $t = 0.88$ s.

The backward (in terms of the main wave direction) plunging breaker interacts with the forward main jet flow, inducing energetic and complicated local fluid motion. Specifically, the impingement of the plunging breaker on the main jet causes a velocity bifurcation, as shown by the bent arrows in Figure 5a. The fluid with an upward velocity component forms a second jet, while the majority of the main jet maintains the original moving direction and merges into the main water body. The second jet rises to an elevation higher than the crest of the plunging wave because the wave kinetic energy is converted to the potential energy, and then splits into two parts (see Figure 5b). The leftward part overturns and impinges on the main jet flow again, and the rightward part impinges on the plunging breaker. All violent wave-breaking scenarios during this process dissipate wave energy. Another interesting phenomenon accompanying this process is the air entrapment as shown in Figure 5a. Since the present CPM model does not include the air phase, the effects of the air entrapment on the local fluid characteristics are not simulated. To capture this in detail, a two-phase simulation that allows for the compressible air necessitates and is left for the future work. The air entrapment region is occupied by water very quickly because of gravity. The fluid that surrounds this region forms an anti-clockwise vortex as shown in Figure 5b.

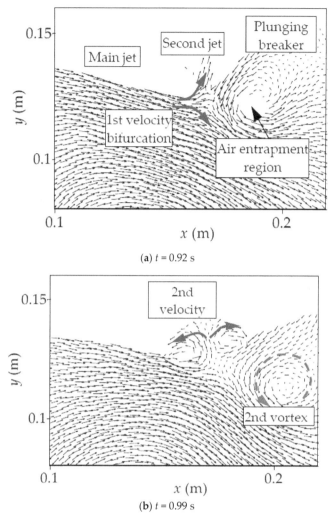

Figure 5. The interaction between the backward plunging breaker and the main jet flow: (**a**) *t* = 0.92 s; (**b**) *t* = 0.99 s.

The main vortex induced by the main jet flow and the second vortex initiated by the plunging breaker keep on developing after the main body of the solitary wave passes by this region. While the second vortex is rotating itself, it rotates around the main vortex, liking the movement relationship between a satellite and a planet. When the water at the leeward starts to flow back to the seaward at around *t* = 1.2 s (manifested in the wave elevation at E2 in Figure 2), a third vortex is initiated near the leeward toe of the breakwater (see Figure 6a). By rotating around the main vortex, this vortex subsequently goes up to the free surface (Figure 6b,c). Due to the energy dissipation induced by fluid viscosity, the third and second vortex disappear at *t* = 2.55 s (Figure 6d) and *t* = 3.59 s (Figure 6e), respectively. The backflow goes to the wave paddle and is reflected again. This new incident wave that has a much smaller amplitude pushes the remaining main vortex onshore and generates a small vortex near the crest of the breakwater at the leeward (Figure 6f). These vortexes will disappear eventually under the action of fluid viscous force.

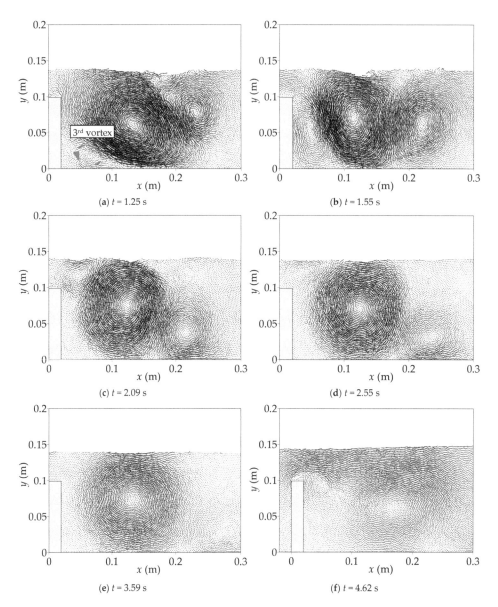

Figure 6. Snapshots of backward plunging breaker and forward jet flow interaction: (**a**) $t = 1.25$ s; (**b**) $t = 1.55$ s; (**c**) $t = 2.09$ s; (**d**) $t = 2.55$ s; (**e**) $t = 3.59$ s; (**f**) $t = 4.62$ s.

4.3. Trajectories of the Fluid Particles

The vortex transports sediment particles and can cause the failure of structure foundations or excessive sediment deposition, both of which will affect the proper functioning of a submerged breakwater. Since the sediment motion is closely related to fluid motion, this section discusses the trajectories of fluid particles at typical locations, to provide some insights into the possible sediment transportations. Note that, because of the Lagrangian description, the particle method offers convenience for this analysis.

Eight particles (the initial positions shown in Figure 7) at the seaward and leeward sides of the breakwater are traced and the trajectories are shown in Figure 8. Particles 1 and 2, which are initially located at the seaward and at an elevation of half of the breakwater height, go along the onshore direction with the main jet flow because of the vortex and their trajectories display spirals, as shown in Figure 8a,b. Particle 3 moves from a position near the flume bed to an elevation more than half of the breakwater (Figure 8c). The seaward bottom corner of the structure is a stagnation point. Hence, particle 4 that is near to structure toe has a small velocity during this period. However, this particle does move up a distance (Figure 8d). It can be anticipated that, with the actions of some more waves, particles 3 and 4 will go up further and eventually move to the leeward of the breakwater. The trajectories of particles 1 to 4 imply that the running-up component of the wave upon the interaction with the breakwater has the power to move the sediment particles near the seaward toe to the leeward. This will induce the erosion of the breakwater foundation.

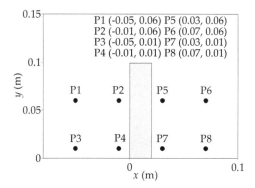

Figure 7. Initial positions of the traced particles.

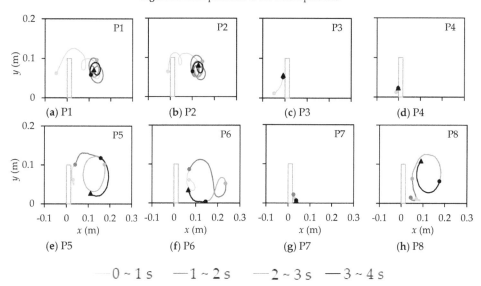

Figure 8. Trajectories of eight particles from 0 ~ 4 s: Blue line (0 ~ 1 s), red line (1 ~ 2 s), green line (2 ~ 3 s) and black line (3 ~ 4 s)

The particles originally located at the leeward and at an elevation of half of the breakwater height (i.e., particles 5 and 6) move with the main vortex, and the trajectories are spirals (Figure 8e,f). Particle

8 that initially locates near the flume bed is easily "transported" by the main vortex (Figure 8h). Similar to particle 4, particle 7 (near to the leeward toe of the breakwater) does not have a long trajectory as shown in Figure 8g. The main feature of the water particle trajectories at the lee side of the breakwater is that the vortex flow initiates the sediment particles. With the combined actions of vortexes and incident waves, the sediment particles will be transported onshore and the foundation of the breakwater will be eroded. Therefore, these factors should be considered carefully in practical designs.

5. Influence of Breakwater Dimension on Fluid Characteristics

In practical applications, the dimension of a breakwater and the wave conditions are important factors that affect the effectiveness of the breakwater in preventing incident waves and dissipating wave energy. To provide some insights, this section conducts a parametric study by analyzing the wave characteristics of the abovementioned solitary wave interacting with six rectangular submerged breakwaters, whose dimensions are shown in Table 2. The wave transmission (K_T) and reflection coefficients (K_R) with breakwaters of different dimensions are shown in Figure 9. As can be seen, increasing the breakwater height increases the wave reflection and reduces the wave transmission, while the breakwater width has little effect on wave reflection and transmission.

Table 2. Dimensions of six rectangular breakwaters.

Dimension	Case 1	Case 2	Case 3	Case 4	Case 5	Case 6
Width a (m)	0.02	0.02	0.02	0.1	0.1	0.1
Height b (m)	0.08	0.1	0.12	0.08	0.1	0.12

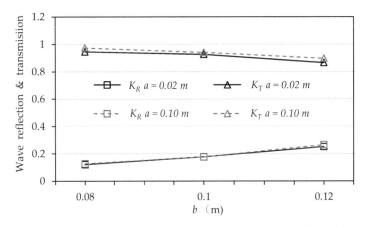

Figure 9. Wave transmission and reflection coefficients with breakwaters of difference dimensions.

When the breakwater width is fixed, a higher breakwater causes more wave reflections and hence less transmitted waves. However, because of the reduction of the section that the wave goes through, the jet flow that impinges towards the lee side has a larger velocity. This induces breaking waves and vortexes of higher intensity. Besides, the main vortex is further away from the structure because the jet flow ejects further from the crest of the breakwater. These phenomena are manifested by the velocity fields of cases 1 (Figure 10a,b), 2 (Figure 3b,d) and 3 (Figure 10c,d) at $t = 0.60$ s and 0.88 s. The strong vortex induces more sediment transportations and hence erosions. In practical designs, therefore, one needs to consider the balance between reflecting more waves and the severity of the erosion at the lee side of the breakwater. The effect of the breakwater width on the flow field is demonstrated by comparing cases 2 and 5 (Figure 10e,f). The key difference is that the main vortex in the case with a wider breakwater is closer to the lee side toe of the breakwater. The practical implementation is that more sediment particles near the structure toe will be initiated and transported away, causing the attached scour as termed in Young and Testik [40]. For a narrow breakwater, since the main vortexes are relatively far from the toe of the breakwater, the sediment particles underneath of the main vortexes are initiated. Some sediments are transported closer to the breakwater and deposited at the lee side toe. This causes the detached scour [40].

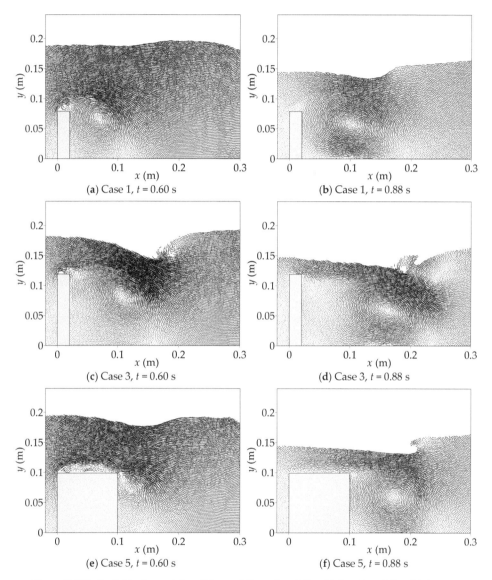

Figure 10. Velocity fields of cases 1, 3 and 5 at $t = 0.60$ s and $t = 0.88$ s: (**a**) Case 1, $t = 0.60$ s; (**b**) Case 1, $t = 0.88$ s; (**c**) Case 3, $t = 0.60$ s; (**d**) Case 3, $t = 0.88$ s; (**e**) Case 5, $t = 0.60$ s; (**f**) Case 5, $t = 0.88$ s.

6. Conclusions

In this work, the CPM is used to study the solitary wave interaction with rectangular submerged breakwaters. The distinct feature of CPM is that it computes the gradient and Laplacian operators in the governing equations in a way consistent with the Taylor series expansion. This addresses the issue of numerical consistency that exists in some other particle methods and hence enhances the numerical accuracy. The CPM model is used to study a documented experimental case. The wave elevation, wave profile and velocity distributions are in good agreement with the experimental data.

Using the validated numerical model, the process of solitary wave-breakwater interaction is studied in detail. The incident solitary wave turns into a jet flow when passing through the breakwater. The jet flow moves onshore and impinges on the water body at the lee side of the breakwater. Subsequently, a backward plunging wave is generated, the interaction between which and the jet flow induces a second jet. The second jet goes up and then splits into two parts. One part re-impinges on the main jet flow and the other part impinges on the plunging breaker, both of which induce splashes. The wave splitting and merging during the wave breaking process cause the dissipation of wave energy.

The main jet flow induces a big vortex behind the breakwater and the plunging breaker induces a secondary vortex. While both vortexes cause the spin of fluid particles, the second vortex, as a whole, rotates around the main vortex. The vortexes play key roles in transporting sediment particles and dissipating wave energy. The potential or capability of the wave flows around the breakwater to transport sediment particles are further studied by analyzing the trajectories of fluid particles. In general, the waves tend to move the seaward sediment particles near the breakwater to the leeward, and transport the sediment leeward particles further onshore.

The influence of the breakwater dimension on the fluid flow around the breakwater is studied. It is found that a higher breakwater causes more wave reflection and hence fewer waves go through the structure. Although the breakwater width has a negligible effect on wave reflection and transmission, the vortex distribution and the associated toe scour pattern are closely related to the width of a breakwater. Specifically, the detached scour is more likely to happen for a narrow breakwater and the attached scour for a wide breakwater.

Author Contributions: Conceptualization, Y.R., M.L. and P.L.; methodology, Y.R., M.L. and P.L.; simulation and data analysis, Y.R. and M.L.; writing—original draft preparation, Y.R. and M.L.; writing—review and editing, Y.R. and M.L.

Funding: This research work is supported by the Open Funding SKHL1710 and SKHL1712 from the State Key Laboratory of Hydraulics and Mountain River Engineering in Sichuan University, China.

Acknowledgments: The authors thank Yun-ta Wu at Tamkang University and Shih-Chun Hsiao in National Cheng Kung University for providing the experimental data as presented in Section 4.

Conflicts of Interest: The authors declare no conflict of interest.

References

1. Relief Web. Available online: https://reliefweb.int/disaster/eq-2018-000156-idn (accessed on 20 November 2018).
2. MSF. Available online: https://www.msf.org.uk/country/indonesia-tsunami-response (accessed on 20 November 2018).
3. CoastalWiki. Available online: http://www.coastalwiki.org/wiki/Detached_breakwaters#Reasons_for_selecting_a_submerged.2Flow_breakwater (accessed on 20 November 2018).
4. Synolakis, C.E. The runup of solitary waves. *J. Fluid Mech.* **2006**, *185*, 523–545. [CrossRef]
5. Briggs, M.J.; Synolakis, C.E.; Harkins, G.S.; Green, D.R. Laboratory experiments of tsunami runup on a circular island. *Pure Appl. Geophys.* **1995**, *144*, 569–593. [CrossRef]
6. Madsen, P.A.; Fuhrman, D.R.; Schäffer, H.A. On the solitary wave paradigm for tsunamis. *J. Geophys. Res.* **2008**, *113*. [CrossRef]
7. Qu, K.; Ren, X.Y.; Kraatz, S. Numerical investigation of tsunami-like wave hydrodynamic characteristics and its comparison with solitary wave. *Appl. Ocean Res.* **2017**, *63*, 36–48. [CrossRef]
8. Sté Grilli, P.T.; Losada, M.A. Characteristics of solitary wave breaking induced by breakwaters. *J. Waterway Port Coast. Ocean Eng.* **1994**, *120*, 74–92. [CrossRef]

9.   Lin, P.Z.; Chang, K.A.; Liu, L.F. Runup and rundown of solitary waves on sloping beaches. *J. Waterway Port Coast. Ocean Eng.* **1999**, *125*, 247–255. [CrossRef]

10.   Chang, K.A.; Hsu, T.J.; Liu, L.F. Vortex generation and evolution in water waves propagating over a submerged rectangular obstacle: Part II: Cnoidal waves. *Coast. Eng.* **2005**, *52*, 257–283. [CrossRef]

11.   Huang, C.J. On the interaction of a solitary wave and a submerged dike. *Coast. Eng.* **2001**, *43*, 265–286. [CrossRef]

12.   Liu, P.L.F.; Albanaa, K. Solitary wave runup and force on a vertical barrier. *J. Fluid Mech.* **2004**, *505*, 225–233. [CrossRef]

13.   Lin, P.Z. A numerical study of solitary wave interaction with rectangular obstacles. *Coast. Eng.* **2004**, *51*, 35–51. [CrossRef]

14.   Hsiao, S.C.; Lin, T.C. Tsunami-like solitary waves impinging and overtopping an impermeable seawall: Experiment and RANS modeling. *Coast. Eng.* **2010**, *57*, 1–18. [CrossRef]

15.   Wu, Y.T.; Hsiao, S.C.; Huang, Z.C. Propagation of solitary waves over a bottom-mounted barrier. *Coast. Eng.* **2012**, *62*, 31–47. [CrossRef]

16.   Monaghan, J.J. Simulating free surface flows with SPH. *J. Comput. Phys.* **1994**, *110*, 399–406. [CrossRef]

17.   Liu, G.R.; Liu, M.B. *Smoothed Particle Hydrodynamics: A Meshfree Particle Method*; World Scientific: Singapore city, Singapore, 2003.

18.   Koshizuka, S.; Nobe, A.; Oka, Y. Numerical analysis of breaking waves using the moving particle semi-implicit method. *Int. J. Numerical Methods Fluids* **1998**, *26*, 751–769. [CrossRef]

19.   Marrone, S.; Antuono, M.; Colagrossi, A.; Colicchio, G.; le Touzé, D.; Graziani, G. δ-SPH model for simulating violent impact flows. *Comput. Methods Appl. Mech. Eng.* **2011**, *13–16*, 1526–1542. [CrossRef]

20.   Lind, S.J.; Xu, R.; Stansby, P.K.; Rogers, B.D. Incompressible smoothed particle hydrodynamics for free-surface flows: A generalised diffusion-based algorithm for stability and validations for impulsive flows and propagating waves. *J. Comput. Phys.* **2012**, *231*, 1499–1523. [CrossRef]

21.   Liu, X.; Lin, P.; Shao, S. An ISPH simulation of coupled structure interaction with free surface flows. *J. Fluids Struct.* **2014**, *48*, 46–61. [CrossRef]

22.   Ren, B.; He, M.; Li, Y.; Dong, P. Application of smoothed particle hydrodynamics for modeling the wave-moored floating breakwater interaction. *Appl. Ocean Res.* **2017**, *67*, 277–290. [CrossRef]

23.   Amicarelli, A.; Albano, R.; Mirauda, D.; Agate, G.; Sole, A.; Guandalini, R. A Smoothed Particle Hydrodynamics model for 3D solid body transport in free surface flows. *Comput. Fluids* **2015**, *116*, 205–228. [CrossRef]

24.   Gotoh, H.; Khayyer, A. On the state-of-the-art of particle methods for coastal and ocean engineering. *Coast. Eng. J.* **2018**, *60*. [CrossRef]

25.   Monaghan, J.J.; Kos, A. Solitary waves on a cretan beach. *J. Waterway Port Coast. Ocean Eng.* **1999**, *125*, 145–155. [CrossRef]

26.   Shao, S.; Lo, E.Y.M. Incompressible SPH method for simulating newtonian and non-newtonian flows with a free surface. *Adv. Water Resour.* **2003**, *26*, 787–800. [CrossRef]

27.   Zhang, Y.L.; Tang, Z.Y.; Wan, D.C. Simulation of solitary wave interacting with flat plate by MPS method. *Chin. J. Hydrodyn.* **2016**, *31*, 395–401.

28.   Huang, Y.; Zhu, C. Numerical analysis of tsunami-structure interaction using a modified MPS method. *Nat. Hazards* **2015**, *75*, 2847–2862. [CrossRef]

29.   Khayyer, A.; Gotoh, H.; Shao, S. Enhanced predictions of wave impact pressure by improved incompressible SPH methods. *Appl. Ocean Res.* **2009**, *31*, 111–131. [CrossRef]

30.   Koh, C.G.; Gao, M.; Luo, C. A new particle method for simulation of incompressible free surface flow problems. *Int. J. Numerical Methods Eng.* **2012**, *89*, 1582–1604. [CrossRef]

31.   Luo, M.; Koh, C.G.; Gao, M.; Bai, W. A particle method for two-phase flows with large density difference. *Int. J. Numerical Methods Eng.* **2015**, *103*, 235–255. [CrossRef]

32.   Koh, C.G.; Luo, M.; Gao, M. Modelling of liquid sloshing with constrained floating baffle. *Comput. Struct.* **2013**, *122*, 270–279. [CrossRef]

33.   Luo, M.; Koh, C.G. Shared-Memory parallelization of consistent particle method for violent wave impact problems. *Appl. Ocean Res.* **2017**, *69*, 87–99. [CrossRef]

34.   Chorin, A.J. The numerical solution of the Navier-Stokes equations for an incompressible fluid. *Bull. Am. Math. Soc.* **1967**, *73*, 928–931. [CrossRef]

35.   Dilts, G.A. Moving least-squares particle hydrodynamics II: Conservation and boundaries. *Int. J. Numerical Methods Eng.* **2000**, *48*, 1503–1524. [CrossRef]
36.   Liu, X.; Xu, H.; Shao, S. An improved incompressible SPH model for simulation of wave–structure interaction. *Comput. Fluids* **2013**, *71*, 113–123. [CrossRef]
37.   Goring, D.G. Tsunamis—The Propagation of Long Waves onto a Shelf. Ph.D. Thesis, California Institute of Technology, Pasadena, CA, USA, 1978.
38.   Sugimoto, N.; Nakajima, N.; Kakutani, T. Edge-layer theory for shallow-water waves over a step–reflection and transmission of a soliton. *J. Phys. Soc. Jpn.* **1987**, *56*, 1717–1730. [CrossRef]
39.   Cooker, M.J.; Peregrine, D.H.; Vidal, C.; Dold, J.W. The interaction between a solitary wave and a submerged semicircular cylinder. *J. Fluid Mech.* **1990**, *215*, 1–22. [CrossRef]
40.   Young, D.M.; Testik, F.Y. Onshore scour characteristics around submerged vertical and semicircular breakwaters. *Coast. Eng.* **2009**, *56*, 868–875. [CrossRef]

Article

Numerical Analysis of the Impact Factors on the Flow Fields in a Large Shallow Lake

Haifei Liu [1], Zhexian Zhu [2,*], Jingling Liu [2,*] and Qiang Liu [2]

[1]   The Key Laboratory of Water and Sediment Sciences of Ministry of Education, Beijing Normal University, Beijing 100875, China; haifei.liu@bnu.edu.cn
[2]   School of Environment, Beijing Normal University, Beijing 100875, China; qiang.liu@bnu.edu.cn
*   Correspondence: 201621180007@mail.bnu.edu.cn (Z.Z.); jingling@bnu.edu.cn (J.L.);
    Tel.: +86-10-58800709 (Z.Z.); +86-10-58805092 (J.L.)

Received: 27 November 2018; Accepted: 11 January 2019; Published: 16 January 2019

Abstract: Wetland acts as an important part of climatic regulation, water purification, and biodiversity maintenance. As an integral part of wetlands, large shallow lakes play an essential role in protecting ecosystem diversity and providing water sources. Baihe Lake in the Momoge Wetland is one such example, so it is necessary to study the flow pattern characteristics of this lake under different conditions. A new model, based on the lattice Boltzmann method, was used to investigate the effects of different impact factors on flow fields, such as water discharge from surrounding farmland, rainfall, wind speed, and aquatic vegetation. Importantly, this study provides a hydrodynamic basis for local ecological protection and restoration work.

Keywords: wetland; lattice Boltzmann method; shallow lake; drag force

1. Introduction

There are many wetlands in China that play an important role in water resource conservation and ecological diversity maintenance. Wetlands have irreplaceable ecological significance for human survival [1]. Although China has emphasized wetland protection over the past few decades, there remains a sharp decline in the area of wetlands [2]. According to official statistics of the China Forestry Administration, China's wetland area fell by 8.83% in 2013, compared with that of 2003. The shrinking of wetland area is closely related to the decrease of water resources in wetlands, and the atrophy of lake area is directly impacted by the shrinking of wetland area [3]. Therefore, it is of vital importance to study the water flow in lakes to find out the feasible measures for wetland conservation [1,4].

A large number of microbes, animals, and plants live in lakes. Therefore, improving purification ability is essential to maintaining the wetland ecological environment [5]. In China, the area of lakes in wetlands has reduced greatly, and as a result, finding ways to slow down the shrinkage of lakes has become an increasingly important topic [6]. The reduction of lake area is bound to the flow of water, and there is no doubt that the characteristics of the water flow is of great importance [5–7]. In general, gravity, bed friction, rainfall, wind speed, and topography may influence the hydrodynamic conditions of lakes of different levels [6]. In addition, vegetation in lakes can also have an important impact on water flow with its drag effect [8,9].

Physical modelling and numerical simulation are the two main approaches to studying flow characteristics [10]. With the development of computer science, numerical modeling has become prevalent [11]. The MIKE series model, developed by the Danish Institute of Water Resources and Water Environment (DHI, Hørsholm, Denmark), is a general commercial package with a wide range of applications for water simulation [12]. The Environmental Fluid Dynamics Code (EFDC) model, created by the Virginia Institute of Marine Science at the College of William and Mary (Williamsburg,

VA, USA), has many functions for water quality modelling and its partially open-source code offers good flexibility [12–14]. However, these commercial models are only based on the traditional finite difference method or the finite volume method, and further development is inhibited given that these models do not have a flexible external force term.

The lattice Boltzmann method (LBM) is a relatively new numerical method for fluid flows [15]. The related theoretical basis of LBM was carried out in the 1980s [16]. Salmon (1999) [17] developed a lattice Boltzmann model for shallow water. Compared with traditional computational methods based on the direct approach of the flow equation, it proposes a new solution for the flow equation [18]. The method is characterized by simple calculation and easy handling of boundary conditions [19]. In recent years, it has become a promising approach in computational fluid dynamics [20]. The use of the LBM to simulate different kinds of flow (e.g., open channel flows, tidal flows, and dam-break flows) has become popular. Ottolenghi et al. (2018) [21] used lattice Boltzmann method to investigate the properties of graphene oxide for environmental applications. O'Brien et al. (2002) [22] developed a lattice Boltzmann scheme to study reactive transport in porous media. Zhou (2007) [23] developed a lattice Boltzmann model for groundwater flow. Tubbs and Tsai (2009) [24] developed the parallel computation for multi-layer shallow water flows. Liu et al. (2012) [25] worked out a large eddy simulation of turbulent shallow water flows using the LBM. Prestininzi (2016) [26] presented a 2D multi-layer shallow water lattice Boltzmann model able to predict the salt wedge intrusion in river estuaries. Furthermore, Yang et al. (2017) [8] developed a rigid vegetation model with 2D shallow water equations using the LBM.

In this paper, the LBM is used to study the hydrodynamic characteristics of Baihe Lake. This is also the first numerical simulation study of Baihe Lake in China, which is helpful for local ecological and hydrological restoration. Through the numerical study of different scenarios, the model provides a quantitative investigation of different impact factors and the results could provide a theoretical basis for local wetland restoration and hydraulic engineering construction.

2. Methodology

2.1. Study Area

The study area was Baihe Lake in the Momoge Wetland of Zhenlai County, Jilin Province, China. The Momoge Wetland (45°45′~46°10′ N, 122°27′~124°04′ E) is one of the most important wetlands in Northeastern China, as shown in Figure 1. It is the main migration path for white crane in China and the total area is 1440 km². In recent years, the area of the Momoge Wetland has shrunk significantly. The local government has developed a series of wetland restoration projects. Baihe Lake is one of the largest lakes in the Momoge Wetland. The area of the lake is 15 km² and the eastern region of the lake is next to the Nenjiang River. The lake is also the main fishing area for local fishermen and its upstream is surrounded by local farmland. Every year a large amount of irrigation water recedes to Baihe Lake as a replenishment. At the same time, the lake also plays a role in purifying the water quality. The lake's outlet is next to the Taoer River. Therefore, Baihe Lake plays an extremely important role in local services such as economic activities, water quality purification, hydrological connectivity, and ecological protection.

2.2. Governing Equations

Baihe Lake is a shallow lake, where the horizontal scale is much larger than the vertical scale. According to field measurements, the largest water depth of Baihe Lake is 2.82 m. Therefore, the two-dimensional hydrodynamic model can be used to study its water flow. The depth-averaged hydrodynamic equations are derived from the incompressible Navier–Stokes equations, which are called the shallow water equations [25].

$$\frac{\partial h}{\partial t} + \frac{\partial (hu_j)}{\partial x_j} = 0 \tag{1}$$

242

$$\frac{\partial(hu_i)}{\partial t} + \frac{\partial(hu_iu_j)}{\partial x_j} = -\frac{g}{2}\frac{\partial h^2}{\partial x_i} + v\frac{\partial^2(hu_i)}{\partial x_j\partial x_j} + F_i \qquad (2)$$

where h is the water depth; t is the time; the subscripts i and j are the space directions based on the Einstein summation convention; x_j and u_j are the distance and instantaneous velocity components in the j direction; g is the gravitational acceleration and equals 9.81 m^2/s; v is the kinematic viscosity; and F_i is the external force term.

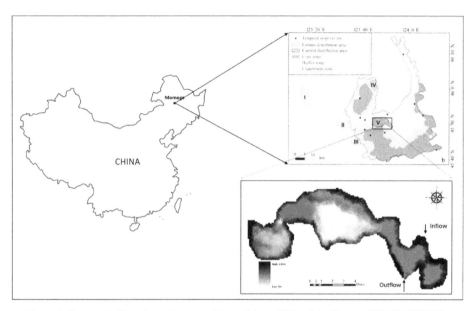

Figure 1. Geographical location and topographic conditions of Baihe Lake (Datum: 45°56′ N, 122°45′ E).

2.3. External Force Term

Shallow water can be significantly influenced by external forces. Generally, the external forces caused by gravity, wind speed, and riverbed friction should be considered [27]. In Equation (2), ignoring the Coriolis force, F_i is the force term and can be expressed as:

$$F_i = -gh\frac{\partial z_b}{\partial x_i} + \frac{\tau_{wi}}{\rho} - \frac{\tau_{bi}}{\rho} + S_{vi} \qquad (3)$$

ρ is the water density, which is equal to 1000 kg/m^3. Z_b is the bed elevation; τ_{bi} is the bed friction and can be expressed as:

$$\tau_{bi} = \rho c_b u_i \sqrt{u_ju_j} \qquad (4)$$

where $c_b = gn^2/h^{1/3}$; n is Manning's coefficient; and τ_{wi} is the wind shear stress that can be expressed as:

$$\tau_{wi} = \rho_a c_w u_{wi} \sqrt{u_{wj}u_{wj}} \qquad (5)$$

where ρ_a is the air density; c_w is the resistance coefficient; and u_{wi} and u_{wj} are the wind velocities in the i and j directions.

In addition, if there is aquatic vegetation in lakes, the drag effect of vegetation on the water body should be considered as an external force [8,28]. A large amount of vegetation is distributed in Baihe Lake, where reeds and bulrush are dominant [29]. This type of aquatic vegetation shown in Figure 2 with high toughness is generally higher than the water free surface and can be simplified as unsubmerged rigid vegetation [29,30]. A rigid vegetation model based on two-dimensional shallow water equations is presented in Reference [8]. The model treated the unsubmerged rigid vegetation as vertical cylinders and the drag force can be expressed as:

$$S_{vi} = -\frac{1}{2}\lambda C_d h u_{vi} \sqrt{u_{vj} u_{vj}}$$ (6)

where u_{vi} is the average velocity on the vegetation elements in the i direction; C_d is the drag force coefficient and is usually in the range of 1 and 1.5 [31]; and λ is the projected area (normal to the flow) of vegetation per unit volume of water and is calculated by:

$$\lambda = \frac{4\alpha_v c}{\pi D_v}$$ (7)

where α_v represents the shape factor; c is the density of the vegetation zones and represents the projected area of vegetation per unit bed area; D_v is the vegetation stems diameter; and u_{vi} is equal to the average velocity u_i.

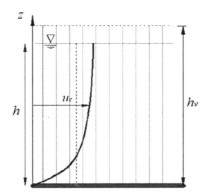

Figure 2. Flow over unsubmerged rigid vegetation.

2.4. Lattice Boltzmann Method (LBM)

The LBM is a modern numerical technique for computational fluid dynamics. The LBM for non-linear two-dimensional shallow water equations has been widely used [10]. It is a discrete computational method based on the lattice gas automata—a simplified, fictitious molecular model. The three main components in the LBM are lattice pattern, kinetic equation, and equilibrium distribution. The lattice Boltzmann equation can be expressed as:

$$f(x + e_\alpha \Delta t, t + \Delta t) = f_\alpha(x, t) - \frac{1}{\tau_t}[f_\alpha(x, t) - f_\alpha^{eq}(x, t)] + \Delta t F_\alpha$$ (8)

where f_α is the particle distribution function; Δx is the lattice size; Δt is time step; the external force F_α is calculated by:

$$F_\alpha = 3 w_\alpha \frac{1}{e^2} e_{\alpha i} F_i$$ (9)

where F_i is force term computed by Equation (3); $e = \Delta x/\Delta t$; w_α is the weight factor: $w_\alpha = 4/9$ for $\alpha = 0$; $w_\alpha = 1/9$ for $\alpha = 1, 3, 5, 7$; and $w_\alpha = 1/36$ for $\alpha = 2, 4, 6, 8$.

Lattice pattern in the LBM has two functions: indicating grid points and resolving particle motions. The former represents a similar role in the traditional numerical simulation methods. The latter shows a microscopic model for molecular dynamics. $e_{\alpha i}$ is the particle velocity in the i direction. The nine-velocity square lattice is shown in Figure 3. Each particle moves one lattice unit at its velocity along the eight links represented by numbers 1–8, while 0 represents a particle at rest with zero speed. The velocity vector of the particles is defined by:

$$e_\alpha = \begin{cases} (0,0) & \alpha=0 \\ e\left[\cos\frac{(\alpha-1)\pi}{4}, \sin\frac{(\alpha-1)\pi}{4}\right] & \alpha=1,\,3,\,5,\,7 \\ \sqrt{2}e\left[\cos\frac{(\alpha-1)\pi}{4}, \sin\frac{(\alpha-1)\pi}{4}\right] & \alpha=2,\,4,\,6,\,8 \end{cases} \tag{10}$$

A local equilibrium distribution function decides what flow equations are solved by means of the lattice Boltzmann equation. For 2D shallow water Equations (1) and (2), the local equilibrium distribution function f_α^{eq} is defined as:

$$f_\alpha^{eq} = \begin{cases} h - \frac{5gh^2}{6e^2} - \frac{2h}{3e^2}u_iu_i, & \alpha=0 \\ \frac{gh^2}{6e^2} + \frac{h}{3e^2}e_{\alpha i}u_i + \frac{h}{2e^4}e_{\alpha i}e_{\alpha j}u_iu_j - \frac{h}{6e^2}u_iu_i, & \alpha=1,\,3,\,5,\,7 \\ \frac{gh^2}{24e^2} + \frac{h}{12e^2}e_{\alpha i}u_i + \frac{h}{8e^4}e_{\alpha i}e_{\alpha j}u_iu_j - \frac{h}{24e^2}u_iu_i, & \alpha=2,\,4,\,6,\,8 \end{cases} \tag{11}$$

Then the remaining task is to determine the physical quantities. The macroscopic variables, water depth h and flow velocity u_i can be expressed as:

$$h = \sum_\alpha f_\alpha, u_i = \frac{1}{h}\sum_\alpha e_{\alpha i}f_\alpha \tag{12}$$

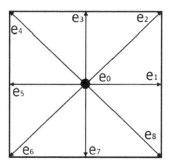

Figure 3. D2Q9 lattice pattern (D2 represent two-dimensional and Q9 represent nine velocity directions in each particle).

2.5. Rainfall

Rainfall has an effect on the surface water and runoff, causes soil erosion and floods, and contaminants transport. A numerical model should be applied to investigate how rainfall influences water flow [32]. In the LBM, the local equilibrium distribution function g_α^{eq} for shallow water equations with source term was developed. It can be expressed as follows:

$$g_\alpha^{eq} = \begin{cases} h - R\Delta t - \frac{5gh^2}{6e^2} - \frac{2h}{3e^2}u_iu_i & \alpha=0 \\ \frac{gh^2}{6e^2} + \frac{h}{3e^2}e_{\alpha i}u_i + \frac{h}{2e^2}e_{\alpha i}e_{\alpha j}u_iu_j - \frac{h}{6e^2}u_iu_j & \alpha=1,\,3,\,5,\,7 \\ \frac{gh^2}{24e^2} + \frac{h}{12e^2}e_{\alpha i}u_i + \frac{h}{8e^2}e_{\alpha i}e_{\alpha j}u_iu_j - \frac{h}{24e^2}u_iu_j & \alpha=2,\,4,\,6,\,8 \end{cases} \tag{13}$$

where R is the rainfall intensity and the influence of dynamic pressure caused by precipitation is neglected.

2.6. Boundary Conditions

A general treatment at the inlet is to set a constant velocity and a water depth, whereas a specific water depth is imposed at the outlet. In addition, a zero-gradient condition is used to obtain the velocity components u and v at the outlet. The standard bounce-back scheme, in which an incoming particle towards the boundary is bounced back into the fluid, is widely used. At the upper boundary:

$$f_6 = f_2, f_7 = f_3, f_8 = f_4 \tag{14}$$

and at the lower boundary:

$$f_2 = f_6, f_3 = f_7, f_4 = f_8 \tag{15}$$

The grids in the corner are involved with changes in boundary conditions, as shown in Figure 4. At the boundary corner point, there will be multiple directions without particle input due to the proximity of the model boundary. It cannot be calculated using the bounce-back scheme or macro variable boundary conditions. It is necessary to calculate the missing particle distribution according to the depth and velocity of the adjacent grids. The computational formula can be expressed as:

$$\begin{cases} f_1 = \frac{2h_{x+1,y+1}u_{x,y}}{3e} + f_5, \\ f_2 = f_6 + \frac{h_{x+1,y+1}u_{x,y}}{6e} + \frac{h_{x+1,y+1}v_{x,y}}{2e}, \\ f_3 = f_7, \\ f_4 = \frac{1}{2}(h_{x+1,y+1} - f_1 - f_2 - f_3 - f_5 - f_6 - f_7 - f_9), \\ f_8 = f_4. \end{cases} \tag{16}$$

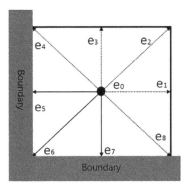

Figure 4. Corner grid.

3. Results

3.1. Initial Conditions

Based on data from 2017, the drainage of farmland converged towards the lake inlet and discharged into Baihe Lake at an average rate of 26.5 m³/s from 11 May to 25 June, and at an average rate of 14.4 m³/s from 10 October to 21 November. The average wind speed was equal to 1.78 m/s in the northeast (132°) from 11 May to 25 June, and average wind speed was equal to 2.36 m/s in the northeast (110°) from 10 October to 21 November.

Figure 5 shows the vegetation distribution obtained by a geographic information system and a field investigation. There are a large number of reeds in Baihe Lake that are higher than the water surface and have high rigidity. From inspection of Figure 5, it can be seen that the vegetation is mainly distributed in the northwest and southeast regions of the lake. Based on a field survey, it was found that the vegetation density varies significantly throughout the year. The density is highest from July to

August and lowest from January to March. In order to confirm the effect of drag force caused by the vegetation, two situations involving high- and low-vegetation density were tested in the present model. In the computation, the vegetation element shape factor and drag coefficient C_d are equal to 1.0 [28].

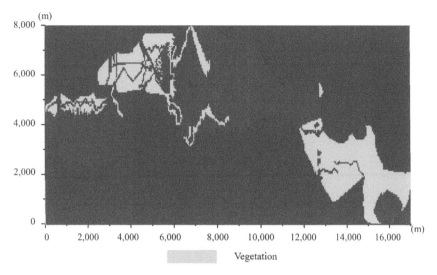

Figure 5. Vegetation distribution in Baihe Lake.

3.2. Numerical Tests

A 2D hydrodynamic model was established using the LBM. Baihe Lake is covered by 340×161 grids with each lattice being 50 m $\times$ 50 m. The inflow velocity from the northeast inlet is given by 0.5 m/s, which is controlled by sluice gate. The initial water depth in June is shown in Figure 6. Overall, the water depth in the western and eastern regions was relatively shallow compared to the middle region. The average water depth was 1.4 m, and the deepest area was 2.14 m.

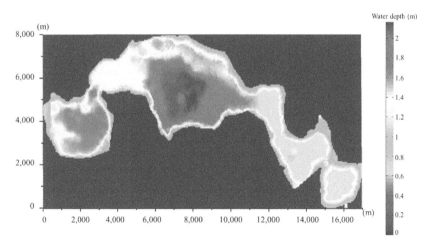

Figure 6. Initial water depth in Baihe Lake.

In order to discuss the hydrodynamic characteristics of the lake, the present study simulated the flow field with time from one day to five days. Figure 7 presents the flow field in Baihe Lake after five days for vegetation density c = 0.2. The western region of the lake is far away from the inlet, resulting in a very small flow velocity. The average velocity is about 0.0016 m/s, with an average water depth of 1.32 m. The middle region of the lake is the main area for local fishery. It has sparse vegetation and a larger water depth. The average velocity is about 0.36 cm/s, with an average water depth of 1.89 m. A large amount of vegetation is distributed in the southeast region of the lake. The velocity of the flow is small except for the areas near the outlet or the inlet where the velocities are about 26 cm/s and 0.063 m/s, respectively. The average velocity in the east area is about 0.32 cm/s.

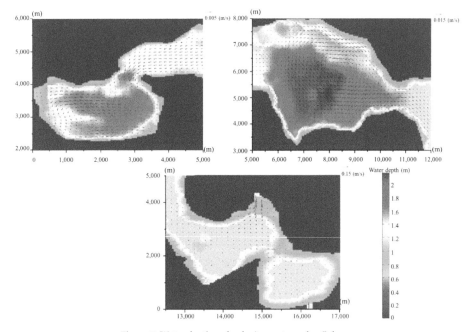

Figure 7. Water depth and velocity vectors after 5 days.

3.3. Sensitivity Analysis

3.3.1. Wind Speed

The wind speed in the Momoge Wetland varies all year round. As a result, it is necessary to investigate the flow field under different wind conditions. The wind speed in the northeast direction was varied by 10%, 20%, 30%, −10%, −20%, and −30%. The flow field became steady after the running time reached five days. The simulation results revealed that the greater the wind speed, the greater the velocity. Wind speed fluctuated by 30%, the outflow velocity varied by 18.77%, and the average velocity of the whole lake varied by 22.16% (see Table 1).

Table 1. Effect of wind speed on flow velocity.

Variation Range	Wind Speed (m/s)	Outflow Velocity (m/s)	Outflow Velocity Variation	Average Velocity (m/s)	Average Velocity Variation
−30%	1.25	0.0381	−5.93%	0.00149	−7.18%
−20%	1.42	0.0386	−4.69%	0.00152	−5.01%
−10%	1.60	0.0395	−2.47%	0.00155	−2.98%
0%	1.78	0.0405	0.00%	0.00160	0.00%
10%	1.96	0.0411	1.48%	0.00164	2.55%
20%	2.14	0.0433	6.91%	0.00172	7.81%
30%	2.31	0.0457	12.84%	0.00184	14.98%

3.3.2. Inflow Discharge

The inflow discharge is significantly affected by drainage from farmlands around the Momoge Wetland and varies across the seasons. The variation percentage of the inflow discharge was varied by 10%, 20%, 30%, −10%, −20%, and −30%. The velocity at the outlet and the flow field in Baihe Lake were the focus. The simulation results revealed a significant positive correlation between inflow and outflow velocity. Due to the strong drag effect of the reeds, a 30% fluctuation in inflow velocity could give rise to only up to 24.11% variation in outflow velocity. Average velocity in the whole lake varied from −21.99% to 27.65% (see Table 2).

Table 2. Effect of inflow velocity on flow velocity.

Variation Range	Inflow Discharge (m^3/s)	Outflow Velocity (m/s)	Outflow Velocity Variation	Average Velocity (m/s)	Average Velocity Variation
−30%	18.55	0.0338	−16.56%	0.00125	−21.99%
−20%	21.20	0.0358	−11.53%	0.00137	−14.58%
−10%	23.85	0.0370	−8.71%	0.00144	−10.24%
0%	26.50	0.0405	0.00%	0.00160	0.00%
10%	29.15	0.0436	7.66%	0.00178	11.42%
20%	31.80	0.0467	15.23%	0.00188	17.63%
30%	34.45	0.0503	24.11%	0.00204	27.65%

3.3.3. Vegetation Density

There is a large amount of aquatic vegetation distributed in Baihe Lake. Aquatic vegetation density becomes higher in summer but lower in autumn and winter. The simulation results revealed that there was a negative correlation between vegetation density and outflow velocity. The velocity with higher vegetation density was slower than that with lower density. In addition, a 30% density fluctuation would result in around −5% variation in outflow velocity, which indicates that the vegetation drag force can affect the water flow. Meanwhile, because of the uneven distribution of vegetation, more vegetation is distributed at the inlet and outlet. The average velocity in the whole lake varied from −3.72% to 2.97%, and was less than the variation at the outlet (see Table 3).

Table 3. Effect of vegetation density on flow velocity.

Variation Range	Vegetation Density	Outflow Velocity (m/s)	Outflow Velocity Variation	Average Velocity (m/s)	Average Velocity Variation
−30%	0.14	0.0424	4.71%	0.00165	2.97%
−20%	0.16	0.0417	3.03%	0.00164	2.57%
−10%	0.18	0.0411	1.49%	0.00162	1.09%
0%	0.20	0.0405	0.00%	0.00160	0.00%
10%	0.22	0.0399	−1.57%	0.00158	−1.23%
20%	0.24	0.0392	−3.11%	0.00155	−2.98%
30%	0.26	0.0385	−4.88%	0.00154	−3.72%

3.3.4. Rain Density

Due to the temperate continental climate, rainfall is less in autumn compared to summer. Therefore, the effect of rainfall on the flow field should be considered. The changes in rain density were 10%, 20%, 30%, −10%, −20%, and −30%. The flow field became steady when the simulation time reached five days. The simulation results revealed a positive correlation between rainfall density and outflow velocity. The outflow velocity varied from −14.23% to 21.11%, and the average velocity in the flow field varied from −14.17% to 20.44% (see Table 4).

Table 4. Effect of rainfall on flow velocity.

Variation Range	Rainfall (mm)	Outflow Velocity (m/s)	Outflow Velocity Variation	Average Velocity (m/s)	Average Velocity Variation
−30%	50.82	0.0347	−14.23%	0.00137	−14.17%
−20%	58.08	0.0370	−8.56%	0.00147	−8.34%
−10%	65.34	0.0385	−4.78%	0.00153	−4.66%
0%	72.60	0.0405	0.00%	0.00160	0.00%
10%	79.86	0.0433	6.97%	0.00171	6.85%
20%	87.12	0.0446	10.23%	0.00176	9.98%
30%	94.38	0.0490	21.11%	0.00193	20.44%

3.4. Scenario Simulation

Scenario simulations of different seasons were also run using data from July 2017 and September 2017.

In July 2017, the peak inflow discharge of the drainage from the surrounding farmland was 32 m^3/s. The monthly rainfall was 184 mm; the average wind speed was 2.13 m/s in the northeast direction (132°); and the vegetation density was 0.26. When the simulation was stable and the result was steady, the outflow rate was 0.54 m/s. The change in velocity and depth is shown in Figure 8a.

In September 2017, the inflow discharge was 12.6 m^3/s in the receding trough of the farmland drainage. The monthly rainfall was 38 mm; the average wind speed was 2.85 m/s in the northeast direction (110°); and the vegetation density was 0.12. The change in velocity and depth is shown in Figure 8b.

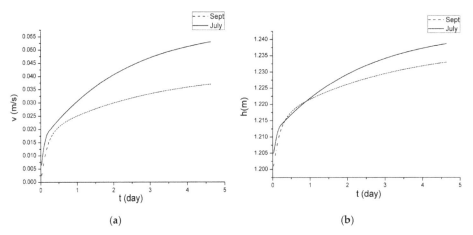

(a) (b)

Figure 8. Comparison of the simulated results with different initial conditions: (**a**) Case 1 (water velocity); (**b**) Case 2 (water depth).

4. Discussion

In this paper, the LBM method is applied to Baihe Lake to reveal the temporal and spatial characteristics of water body. Overall, the water body of Baihe Lake suffered poor exchange, where the lowest flow rate based on simulation reached just 0.016 m/s. Simulations of different impact factors (e.g., wind speed, rainfall, vegetation density) were conducted here. The simulation results revealed that changes in inflow velocity, primarily as the result of water drainage from surrounding farmland, have the strongest effect. A 30% variation in the impact factors resulted in a 24.11% variation in the outflow rate at most. In addition, when the impact of wind speed was relatively small, the variation of the outflow rate reached just 18.87%. Relatedly, given the high distribution of aquatic vegetation in some areas of Baihe Lake, and that vegetation density varies with the season, the impact of vegetation on the flow field should be considered separately. The present model includes the change in vegetation density and investigates its effect on the flow fields. The simulation result revealed that vegetation density was negatively correlated with outflow rate. However, the influence of aquatic vegetation on flow distribution was relatively small, being only 9.59% when the vegetation density fluctuated 30%. On the whole, the outflow of Baihe Lake strongly depends on the drainage from upstream farmland, followed by the rainfall, and then by local wind speed. Aquatic vegetation can also have an impact. The inflow rate is the only factor that is easy to control. As a result, ecological protection and hydraulic engineering construction can be established upstream to control the water level and flow rate.

The monitored rainfall, wind speed, flowrate, and vegetation density data in different months was input into the present model in order to simulate and obtain the flow characteristics of Baihe Lake in July and September. Although the influence of wind speed and vegetation was relatively small in the single factor analysis, they can still have an impact on the flow as these two factors have potential uncertainty due to the change in climate. The monitored data revealed large drainage from the surrounding farmland in July, together with high rainfall. Consequently, based on the simulation results, the impact of wind speed and vegetation density was relatively small. In contrast, a reduction in rainfall with a similar farmland drainage happens in September. Importantly, this increases the influence of wind speed and vegetation density. The results indicated that the mean flow rate is much bigger in July than September, which also verifies the significance of different impact factors. Relatedly, the simulation results also demonstrate that the water exchange of Baihe Lake during the flood season from June to August is much better than it is in the dry season from September to November. Therefore, when developing an ecological restoration program, both the farmland drainage and the season should be emphasized.

5. Conclusions

Based on the lattice Boltzmann method (LBM), the present study has established a two-dimensional shallow water model that includes vegetation drag force. This paper for the first time puts forward a method to study the variation of flow field by establishing a numerical model of the Baihe Lake, which plays a guiding role in the construction of the ecological engineering of the Baihe Lake. The hydrodynamic model in this paper can be widely used in shallow lakes. Meanwhile, the drag force of aquatic vegetation to Baihe Lake is specially considered in the model, leading to that the model is most suitable for lakes with aquatic vegetation.

The model was used to explore the hydrodynamic impact of different influencing factors in Baihe Lake. Through a sensitivity test, drainage from surrounding farmland showed a dominant influence on the flow field, followed by rainfall, wind speed, and vegetation density. Furthermore, these factors may produce a noticeable variation in local flow characteristics in different seasons. As a result, the impact of these factors should be considered when developing an ecological restoration program. We should pay attention to different schemes corresponding to different seasons and control hydraulic construction to influence farmland receding water and improve lake fluidity.

Author Contributions: Conceptualization, H.L.; Data curation, Z.Z.; Formal analysis, H.L.; Investigation, Z.Z. and Q.L.; Methodology, H.L. and J.L.; Project administration, J.L.; Resources, J.L. and Q.L.; Visualization, Z.Z.; Writing–original draft, Z.Z.; Writing–review & editing, H.L.

Funding: This research was funded by the National Key R&D Program (Grant No. 2016YFC0500402) and the National Natural Science Foundation of China (Grant No. 51779011).

Conflicts of Interest: The authors declare no conflict of interest.

References

1. Zeng, L.; Chen, G.Q. Flow distribution and environmental dispersivity in a tidal wetland channel of rectangular cross-section. *Commun. Nonlinear Sci.* **2009**, *17*, 4192–4209. [CrossRef]
2. Calhoun, A.J.; Mushet, D.M.; Bell, K.P.; Boix, D.; Fitzsimons, J.A.; Isselin-Nondedeu, F. Temporary wetlands: Challenges and solutions to conserving a 'disappearing' ecosystem. *Biol. Conserv.* **2017**, *211*, 3–11. [CrossRef]

3. Hui, F.; Xu, B.; Huang, H.; Yu, Q.; Gong, P. Modelling spatial-temporal change of Poyang Lake using multitemporal Landsat imagery. *Int. J. Remote. Sens.* **2008**, *29*, 5767–5784. [CrossRef]

4. Yang, R.; Cui, B. A wetland network design for water allocation based on environmental flow requirements. *Clean Soil Air Water* **2012**, *40*, 1047–1056. [CrossRef]

5. Stephan, U.; Gutknecht, D. Hydraulic resistance of submerged flexible vegetation. *J. Hydrol.* **2002**, *269*, 27–43. [CrossRef]

6. Kouwen, N. Friction factors for coniferous tress along rivers. *J. Hydraul. Eng.* **2000**, *126*, 732–740. [CrossRef]

7. Huai, W.X.; Zeng, Y.H.; Xu, Z.G. Three-layer model for vertical velocity distribution in open channel flow with submerged rigid vegetation. *Adv. Water. Resour.* **2009**, *32*, 487–492. [CrossRef]

8. Yang, Z.; Bai, F.; Huai, W.; An, R.; Wang, H. Modelling open-channel flow with rigid vegetation based on two-dimensional shallow water equations using the lattice Boltzmann method. *Ecol. Eng.* **2017**, *106*, 75–81. [CrossRef]

9. Yang, S.; Bai, Y.; Xu, H. Experimental analysis of river evolution with riparian vegetation. *Water* **2018**, *10*, 1500. [CrossRef]

10. Zhou, J.G. A lattice Boltzmann model for the shallow water equations. *Int. J. Mod. Phys. C* **2002**, *13*, 1135–1150. [CrossRef]

11. Liu, Q.; Qin, Y.; Zhang, Y.; Li, Z. A coupled 1D–2D hydrodynamic model for flood simulation in flood detention basin. *Nat. Hazards* **2015**, *75*, 1303–1325. [CrossRef]

12. Patro, S.; Chatterjee, C.; Mohanty, S.; Singh, R.; Raghuwanshi, N.S. Flood inundation modeling using MIKE FLOOD and remote sensing data. *J. Indian Soc. Remote Sens.* **2009**, *37*, 107–118. [CrossRef]

13. Arifin, R.R.; James, S.C.; de Alwis Pitts, D.A.; Hamlet, A.F.; Sharma, A.; Fernando, H.J. Simulating the thermal behavior in Lake Ontario using EFDC. *J. Great Lakes Res.* **2016**, *42*, 511–523. [CrossRef]

14. Bai, H.; Chen, Y.; Wang, D.; Zou, R.; Zhang, H.; Ye, R.; Ma, W.; Sun, Y. Developing an EFDC and numerical source-apportionment model for nitrogen and phosphorus contribution analysis in a Lake Basin. *Water* **2018**, *10*, 1315. [CrossRef]

15. Chen, S.; Doolen, G.D. Lattice Boltzmann method for fluid flows. *Ann. Rev. Fluid Mech.* **1998**, *30*, 329–364. [CrossRef]

16. Succi, S.; Santangelo, P.; Benzi, R. High-resolution lattice-gas simulation of two-dimensional turbulence. *Phys. Rev. Lett.* **1988**, *60*, 2738–2740. [CrossRef] [PubMed]

17. Salmon, R. The lattice Boltzmann method as a basis for ocean circulation modeling. *J. Mar. Res.* **1999**, *57*, 503–535. [CrossRef]

18. Dellar, P.J. Non-hydrodynamic modes and a priori construction of shallow water lattice Boltzmann equations. *Phys. Rev. E* **2002**, *65*, 036309. [CrossRef]

19. Liu, H.; Ding, Y.; Wang, H.; Zhang, J. Lattice Boltzmann method for the age concentration equation in shallow water. *J. Comput. Phys.* **2015**, *299*, 613–629. [CrossRef]

20. Liu, H.; Zhou, J.G. Inlet and outlet boundary conditions for the Lattice-Boltzmann modelling of shallow water flows. *Prog. Comput. Fluid. Dyn.* **2012**, *12*, 11–18. [CrossRef]

21. Ottolenghi, L.; Prestininzi, P.; Montessori, A.; Adduce, C.; La Rocca, M. Lattice Boltzmann simulations of gravity currents. *Eur. J. Mech. B-Fluid* **2018**, *67*, 125–136. [CrossRef]

22. O'Brien, G.S.; Bean, C.J.; Mcdermott, F. A comparison of published experimental data with a coupled lattice Boltzmann-analytic advection–diffusion method for reactive transport in porous media. *J. Hydrol.* **2002**, *268*, 143–157. [CrossRef]

23. Zhou, J.G. A rectangular lattice Boltzmann method for groundwater flows. *Mod. Phys. Lett. B* **2007**, *21*, 531–542. [CrossRef]

24. Tubbs, K.R.; Tsai, T.C. Multilayer shallow water flow using lattice Boltzmann method with high performance computing. *Adv. Water. Resour.* **2009**, *32*, 1767–1776. [CrossRef]

25. Liu, H.; Li, M.; Shu, A. Large eddy simulation of turbulent shallow water flows using multi-relaxation-time lattice Boltzmann model. *Int. J. Numer. Meth. Fluids* **2012**, *70*, 1573–1589. [CrossRef]

26. Prestininzi, P.; Montessori, A.; Rocca, M.; Sciortino, G. Simulation of arrested salt wedges with a multi-layer Shallow Water Lattice Boltzmann model. *Adv. Water Resour.* **2016**, *96*, 282–289. [CrossRef]

27. Chávarri, E.; Crave, A.; Bonnet, M.P.; Mejía, A.; Da Silva, J.S.; Guyot, J.L. Hydrodynamic modelling of the Amazon River: Factors of uncertainty. *J. S. Am. Earth. Sci.* **2013**, *44*, 94–103. [CrossRef]

28. Stone, B.M.; Shen, H.T. Hydraulic resistance of flow in channels with cylindrical roughness. *J. Hydraul. Eng.* **2002**, *128*, 500–506. [CrossRef]

29. Konings, A.G.; Katul, G.G.; Thompson, S.E. A phenomenological model for the flow resistance over submerged vegetation. *Water. Resour. Res.* **2012**, *48*, 2478. [CrossRef]

30. Poggi, D.; Krug, C.; Katul, G.G. Hydraulic resistance of submerged rigid vegetation derived from first-order closure models. *Water. Resour. Res.* **2009**, *45*, 2381–2386. [CrossRef]

31. Guan, M.; Liang, Q. A two-dimensional hydro-morphological model for river hydraulics and morphology with vegetation. *Environ. Modell. Softw.* **2017**, *88*, 10–21. [CrossRef]

32. Ding, Y.; Liu, H.; Peng, Y.; Xing, L. Lattice Boltzmann method for rain-induced overland flow. *J. Hydrol.* **2018**, *562*, 789–795. [CrossRef]

Article

Numerical Study of the Velocity Decay of Offset Jet in a Narrow and Deep Pool

Xin Li, Maolin Zhou *, Jianmin Zhang * and Weilin Xu

State Key Laboratory of Hydraulics and Mountain River Engineering, Sichuan University, Chengdu 610065, China; bixizhen@foxmail.com (X.L.); xuwl@scu.edu.cn (W.X.)
*   Correspondence: water636@126.com (M.Z.); zhangjianmin@scu.edu.cn (J.Z.)

Received: 29 November 2018; Accepted: 25 December 2018; Published: 31 December 2018

Abstract: The present study examines the configuration of an offset jet issuing into a narrow and deep pool. The standard k-ε model with volume-of-fluid (VOF) method was used to simulate the offset jet for three exit offset ratios (OR = 1, 2 and 3), three expansion ratios (ER = 3, 4 and 4.8), and different jet exits (circular and rectangular). The results clearly show significant effects of the circumference of jet exits (L_{exit}) in the early region of flow development, and a fitted formula is presented to estimate the length of the potential core zone (L_{PC}). Analysis of the flow field for OR = 1 showed that the decay of cross-sectional streamwise maximum mean velocity (U_m) in the transition zone could be fitted by power law with the decay rate n decreased from 1.768 to 1.197 as the ER increased, while the decay of U_m for OR = 2 or 3 was observed accurately estimated by linear fit. Analysis of the flow field of circular offset jet showed that U_m for OR = 2 decayed fastest due to the fact that the main flow could be spread evenly in floor-normal direction. For circular jets, the offset ratio and expansion ratio do not affect the spread of streamwise velocity in the early region of flow development. It was also observed that the absence of sudden expansion of offset jet is analogous to that of a plane offset jet, and the flow pattern is different.

Keywords: offset jet; potential core; decay rate; k-ε model

---

1. Introduction

Offset jets are common in drainage systems [1], slot fishways [2], and hydraulics engineering [3]. The offset jet is formed when a fluid jet discharges into an ambient medium above the floor and parallel to the axis of the jet exit but which is offset by a certain height. The submerged offset jets in the narrow and deep pool are more complex flow because it can easily be affected by side walls.

The offset jet flow was widely used as energy dissipation downstream of hydraulic structures—for example, the submerged hydraulic jump at an abrupt drop could be considered as an offset jet flow. From an engineering point-of-view, the length of the stilling basin should be as short as possible. The focus of this paper is to find the factors affect the efficiency of energy dissipation of the offset jet flow. Figure 1 shows a sketch of the submerged offset jet in a pool. The jet is mainly divided into three regions according to the relationship with the floor, viz., recirculation region, impingement region, and wall jet region. The Cartesian coordinate system is used with x, y, and z representing the streamwise, lateral direction, and floor-normal direction, respectively. The exit of the jet is at $x = 0$, and the symmetry plane is at $y = 0$, the pool bottom is at $z = 0$. The symbols L_x, L_y, L_z, S, d, h_t, U_j, and U_m represent the length, width, and height of jet pool, offset height, the diameter of jet exit, depth of tailwater, bulk velocity, and cross-sectional streamwise maximum mean velocity, respectively. Likewise, if the jet exit is rectangular, a_0, b_0 represents the height and width of the rectangle, respectively. The flow field in jet pool can be divided into three zones by the decay of U_m: (a) the potential core zone, where U_m was equal to the jet exit velocity, U_j; (b) transition zone, where the decay of U_m was rapidly in this zone; and (c) fully development zone, where the decay of U_m became very slow.

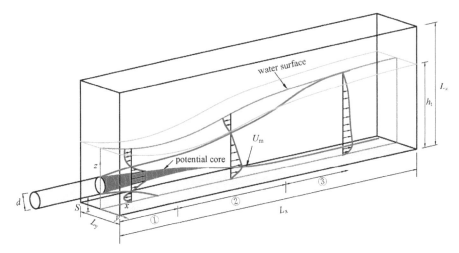

Figure 1. the sketch of the submerged offset jet in a chamber. Flow divided by a relationship with bottom floor: ① recirculation region, ② impingement region, and ③ wall jet region.

The recent studies on submerged offset jets were carried out by Subhasish Dey [4] to investigate the vertical profiles of time-averaged velocity components, and Reynolds stresses in jet flows for offset ratios OR = 2.8–5.1, submergence ratio SR = 2.4–5.6, expansion ratio ER = 1, and jet Reynold number R_0 = 28,475–80,730, where OR = S/a_0 (for circular jet exit $a_0 = d$), SR = h_t/S, ER = L_y/b_0, and $R_0 = (U_0 a_0)/v$, v is the coefficient of kinematic viscosity of water. They concluded important characteristic length of submerged offset jets, such as the length of the recirculation region and impingement region, are expressed as a function of the R_0, OR, and SR. In another attempt, Bhuiyan [5] analyzed the characteristics of submerged turbulent plane offset jets (OR = 0.5–3.6) in channels with rough beds and shallow tailwater depths. The results indicated that for an offset height larger than the jet thickness, the peak velocity, the flow momentum decay faster in the downstream direction in an offset jet than in a turbulent plane wall jet. Durand et al. [6] studied the effect of Reynolds number on 3D offset jets both experimentally and numerically for R_0 = 34,000, 53,000 and 86,000. The results indicated the floor-normal location of maximum mean velocity and jet spread to be independent of Reynolds number. The investigated by Nyantekyi-Kwakye [7] on a 3D rectangular offset jet, performed at three offset ratios (OR = 0, 2, and 4) and revealed that large-scale structures dominate the inner layer of the wall jet region. For 3D circular offset jets (with OR ranging from 0–3.5), Agelin-chaab and Tachie [8] using a planar particle image velocimetry (PIV) system conducted the experiments to study the velocity. They observed that OR influenced both the decay of U_m and growth of the shear layer within the developing region. Besides, there have been other experimental studies on 3D offset jets [9–11].

Given the brief overview, it can be concluded that the mechanics of offset jet has been studied extensively. However, the hydraulic properties of submerged offset jets in a narrow and deep pool have not been completely understood, although it was widely used as an efficient energy dissipator in hydraulic engineering located in narrow canyons.

This paper addresses the decay of velocity of offset jets in the narrow and deep pool with various offset ratios, expansion ratios and jet exit shapes, through numerical simulations using a 3D computational fluid dynamics (CFD) model with the k-ε turbulence model coupled with VOF method, which was confirmed perform well in jet flow [10,12–17].

2. Numerical Simulation

2.1. Mathematical Model

An RNG k-ε model has been compared to a standard k-ε model for computing transient jets by Abraham and Maji [18] and concluded that the RNG k-ε model results in predictions of greater mixing in the jets relative to the standard model. The study carried by Nasr and Lai [19] indicated that the standard k-ε turbulence model predicts better than RNG k-ε and Reynolds stress turbulence model. In this study, the standard k-ε model [20] was used as the turbulence closure

$$\frac{\partial(\rho k)}{\partial t} + \frac{\partial(\rho u_i k)}{\partial x_i} = \frac{\partial}{\partial x_i}\left[\left(\mu + \frac{\mu_t}{\sigma_k}\right)\frac{\partial k}{\partial x_i}\right] + G_k - \rho\varepsilon \tag{1}$$

$$\frac{\partial(\rho\varepsilon)}{\partial t} + \frac{\partial(\rho u_i \varepsilon)}{\partial x_i} = \frac{\partial}{\partial x_i}\left[\left(\mu + \frac{\mu_t}{\sigma_\varepsilon}\right)\frac{\partial\varepsilon}{\partial x_i}\right] + C_{1\varepsilon}\frac{\varepsilon}{k}G_k - C_{2\varepsilon}\rho\frac{\varepsilon^2}{k} \tag{2}$$

where ρ is the density of mixture; t is time; k is turbulent kinetic energy; μ is the dynamic viscosity of fluid; u_i is component of velocity in the x_i direction; ε is turbulent energy dissipation rate; μ_t is dynamic turbulent viscosity; σ_k, σ_ε is turbulent Prandtl number for k and ε, respectively; G_k is generation of turbulent kinetic energy due to mean velocity gradients; and μ_t and G_k can be determined as

$$\mu_t = \rho C_\mu \frac{k^2}{\varepsilon} \tag{3}$$

$$G_k = \mu_t\left(\frac{\partial u_i}{\partial x_j} + \frac{\partial u_j}{\partial x_i}\right)\frac{\partial u_i}{\partial x_j} \tag{4}$$

The values of the empirical constants in the turbulence model are $C_\mu = 0.09$, $\sigma_k = 1.0$, $\sigma_\varepsilon = 1.3$, $C_{1\varepsilon} = 1.44$, and $C_{2\varepsilon} = 1.92$.

For this study, the volume-of-fluid (VOF) method was used as the two-phase flow model to track the water surface in the domain. The VOF method is widely used to determine the position of the interface of two or more immiscible flows [21,22]. Air and water were the primary and secondary phases, respectively. In the calculation, all fluids share the turbulence model. For air-water two-phase flow, α_a and α_w are the volume fraction of air and water, respectively. For each control cell,

$$\alpha_w + \alpha_a = 1 \tag{5}$$

If the cell contains only air, the value of $\alpha_w = 0$; if the cell is full of water, the value of $\alpha_w = 1$; and if the interface cuts the cell, then $0 < \alpha_w < 1$. The volume fraction of water, α_w, is calculated from [23]

$$\frac{\partial\alpha_w}{\partial t} + V\cdot\nabla\alpha_w = 0 \tag{6}$$

where the V is the fluid velocity. The fluid properties, such as density ρ and molecular viscosity μ, are adjusted according to the volume fraction. It should be noted that other numerical methods can also be used to study the detailed flow properties in complex contexts [24–29], the choice of numerical methods depends on the research focus.

2.2. Simulation Setup

ANSYS ICEM 16.0 (ANSYS®, Canonsburg, PA, USA) was utilized to develop the numerical models for 13 types of offset jet. Hexahedral grids were used throughout the computational domain. The grid meshing is shown in Figure 2. A Cartesian coordinate system is used so that the origin is at the center of the intersection of the offset wall and floor. The computational domain consists of the pipe and jet pool. The jet is discharged from offset wall offset by a height S above the floor.

The total pipe length is 1 m, and the type of pipe can be divided into the circular pipe and rectangular pipe. The length, L_x, and height, L_z, of the pool are 8 m and 0.8 m, respectively. For the circular pipe, the offset ratio (OR) varied from 1 to 3, the expansion ratio (ER) varied from 3 to 4.8 as the width of the jet pool increased from 0.3 m to 0.48 m. For the rectangular pipe, the aspect ratio (AR) of the exit varied from 0.33 to 11.46. Moreover, the area of exit remains constant for all shape of exits. The parameters of all cases are given in Table 1, and the jet exits and the offset wall are shown in Figure 3.

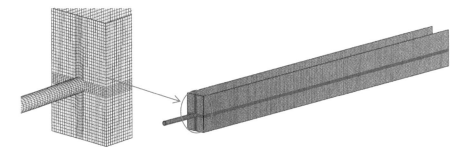

Figure 2. Layout of the calculation (C-O3-E5).

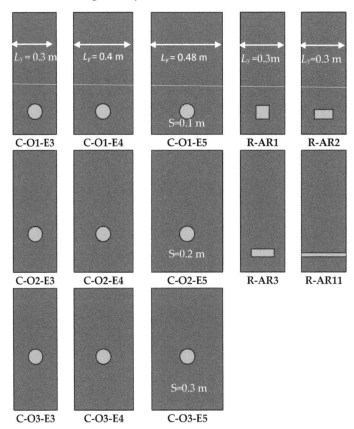

Figure 3. Sketch of jet exits.

Table 1. Parameters of computational cases.

Case Name	Jet Exit Shape	d (m)	a_0 (m)	b_0 (m)	S (m)	L_y (m)	Offset Ratio (OR)	Expansion Ratio (ER)	AR (b_0/a_0)
C-O1-E3	circular	0.1	-	-	0.1	0.3	1	3	-
C-O1-E4	circular	0.1	-	-	0.1	0.4	1	4	-
C-O1-E5	circular	0.1	-	-	0.1	0.48	1	4.8	-
C-O2-E3	circular	0.1	-	-	0.2	0.3	2	3	-
C-O2-E4	circular	0.1	-	-	0.2	0.4	2	4	-
C-O2-E5	circular	0.1	-	-	0.2	0.48	2	4.8	-
C-O3-E3	circular	0.1	-	-	0.3	0.3	3	3	-
C-O3-E4	circular	0.1	-	-	0.3	0.4	3	4	-
C-O3-E5	circular	0.1	-	-	0.3	0.48	3	4.8	-
R-AR1	rectangular	-	0.089	0.089	0.1	0.3	1.13	3.39	1
R-AR2	rectangular	-	0.125	0.063	0.1	0.3	1.60	2.39	2
R-AR3	rectangular	-	0.153	0.051	0.1	0.3	1.95	1.95	3
R-AR11	rectangular	-	0.1	0.026	0.1	0.3	3.82	1.00	11.46

2.3. Initial Conditions and Boundary Conditions

All the variables except at boundaries (such as u_i, P, k, and ε) are initialized at zero. The computed data are stored at every alternate time step for post-processing. All computations are conducted on an Inter core i7 3.60 GHz Windows machine.

The boundary conditions were set as follows.

- Inflow boundary: the inlet was treated as an inlet velocity boundary with the velocity was set as 5 m/s;
- Outflow boundary: pressure outlet boundary was selected at the outlet, the depth of tailwater was fixed at 0.55 m with the help of user-defined function (UDF);
- Free surface: pressure inlet was employed, and its value was the standard atmospheric pressure, the operating pressure and density were selected as 101,325 Pa and 1.225 kg/m^3, respectively.
- Wall boundary: for the parameters investigated in this study, the data near the wall was ineffective. No slip boundary condition is considered for velocity. To avoid the fine mesh required to resolve the viscous sub-layer near the boundary, so standard wall function method has been used.

2.4. Numerical Discretizations

ANSYS Fluent 16.0 (ANSYS®, Canonsburg, PA, USA) was utilized to perform the simulation. The governing equations are discretized using the implicit Finite Volume Method (FVM). The SIMPLE algorithm, using a relationship between velocity and pressure corrections to enforce mass conservation and to obtain the pressure field, was applied to couple the velocity and pressure. The least-squares cell-based method was used to calculate the gradient. PRESTO! was used to discretize the pressure and Geo-Reconstruct was used for the volume fraction. The second-order upwind scheme was used for the momentum and the first-order upwind for the turbulent kinetic energy and the dissipation rate with ANSYS Fluent's default under relaxation values for all parameters. The time step was $\Delta t = 0.0001\sim0.001$ s and the iteration number was always less than 2.

The computational results are considered to be converged when the residual becomes smaller than 0.001 for all equations. Here, the residual is defined as the square root of the summation of the squares of the difference between right and lefts sides of the discretized equations for a single control volume [30]. The results analyzed and presented in this study were taken from the simulation when quasi-steady state was reached.

2.5. Grids Sensitivity and Model Validation

Grid independence was examined to ensure the reliability of the numerical simulation results. The Case C-O3-E5 was chosen to test mesh sensitivity. Three sets of grids had 1,065,575 (grid 1), 438,221 (grid 2) and 181,036 (grid 3) cells, respectively. The grid convergence index (GCI) is widely used to estimate discretization uncertainty [31]. The GCI is given by

$$GCI = \frac{1.25|(\varnothing_3 - \varnothing_2)/\varnothing_3|}{(h_2/h_3)^P - 1} \tag{7}$$

$$P = \frac{1}{\ln(h_2/h_3)} \left| \ln|(\varnothing_1 - \varnothing_3)/(\varnothing_2 - \varnothing_3)| + \ln\left| \frac{(h_2/h_3)^P - \mathrm{sgn}[(\varnothing_2 - \varnothing_3)/(\varnothing_2 - \varnothing_3)]}{(h_1/h_2)^P - \mathrm{sgn}[(\varnothing_1 - \varnothing_3)/(\varnothing_2 - \varnothing_3)]} \right| \right| \tag{8}$$

where $\varnothing_i$ is the solution on the ith grid and h_i is the average grid size on the ith grid and $h_1 > h_2 > h_3$.

Figure 4 presents an axial velocity profile along the jet exit axis. In this figure, 80% out of 40 points exhibited oscillatory convergence. The maximum uncertainties in velocity were approximately 12.6%, which corresponds to a maximum uncertainty in velocity of about ±0.25 m/s. Considering the simulation accuracy and the computation efficiency, the final grid number in this study was taken as 438,221.

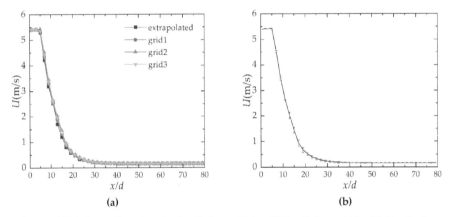

Figure 4. (**a**)Velocity profiles on jet exit axis with three grid sizes; (**b**) Results from grid 2, with discretization error bars computed using GCI.

To examine simulation accuracy, the physical model experiment of Case C-O3-E5 was performed at the State Key Laboratory of Hydraulics and Mountain River Engineering, Sichuan University, Chengdu. The experimental model is shown in Figure 5. It can be seen from Figure 6 that the calculated results of the height of water surface and velocity distribution were fairly consistent with that of laboratory tests. The results are indicating that the numerical simulation produced reliable and acceptable results.

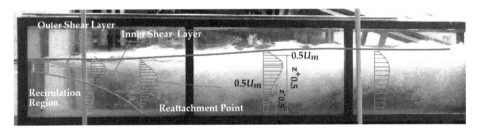

Figure 5. Experiment model.

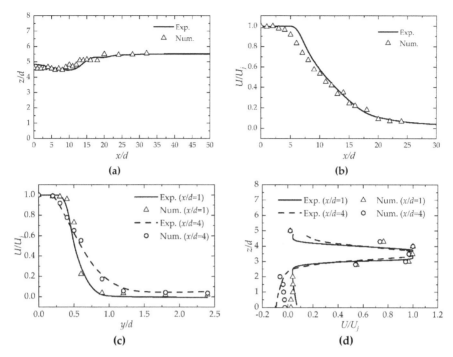

Figure 6. Model validation: (**a**) Heights of the water surface in streamwise direction; (**b**) Velocity distribution in streamwise direction (central line of exit); (**c**) Velocity distribution in lateral direction; (**d**) Velocity distribution in floor-normal direction.

3. Results and Analysis

3.1. Velocity Attenuation

The attenuation of the jet flow in the narrow and deep pool was characterized using the decay of the cross-sectional maximum streamwise mean velocity (U_m). Previous studies indicated that U_m is constant within the potential core region, followed by a rapid decay with streamwise distance in transition zone [32,33]. The entrained ambient fluid gets momentum since the jet enters the pool, the section of jet flow continues to expand, and the velocity is decreasing as the results of the mixing of jet and ambient fluid. Figure 7a shows the distribution of U_m normalized by the jet exit velocity, U_j. Normalized values of U_m were observed decayed sharply within the region $x/d > L_{PC}$ and the trend of decay of U_m varies greatly in the transition zone for different circular offset jet. It can be seen from Figure 7b that the length of the potential core zone (L_{PC}) varied from 5.0 to 5.5 with changes in ER, and a higher ER indicated a higher L_{PC}.

As can be observed in Figure 7c–d, the L_{PC} varies greatly with AR of the rectangular jet exit due to the circumference of exit (L_{exit}) were changed. When L_{exit} increased, the L_{PC} decreased sharply, a fitting formula was used to estimate the relationship of L_{exit} and L_{PC} as indicated in Figure 7d, the empirical equation for L_{PC} was employed as $L_{PC}/d = 2100e^{-2(L_{exit}/d)} + 0.5$.

The decay of the U_m over distance x (scaled by d) for Case R-AR11 is shown in Figure 7c. The most striking feature of Case R-AR11 is that there is no sudden expansion (ER = 1). Thus, the Case R-AR11 is analogous to that of a plane offset jet. The decay of U_m in the recirculation region is quite sharp, the U_m drops to a local minimum value due to an increase in pressure resulting from the jet impingement after the reattachment of the jet. The decay of U_m is rather gradual in the wall jet region. A similar observation was reported by Gu [16] and Dey [4].

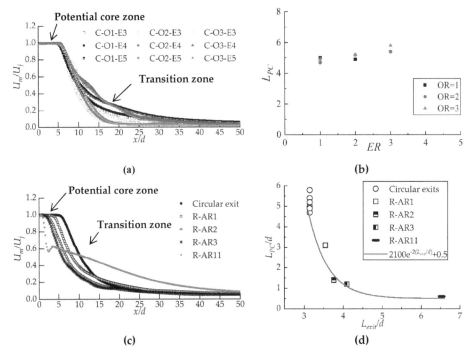

Figure 7. (a) U_m decay for circular offset jets; (b) the relationship of L_{PC}, ER, and OR for circular offset jets; (c) U_m decay for rectangular offset jets (AR ≤ 3); (d) the relationship of L_{PC} and L_{exit}.

The U_m/U_j decay profiles can be grouped based on their characteristics. The power law of the form $U_m/U_j = C(x/d)^{-n}$ is typically used to describe the velocity decay in the transition zone [6,32,34,35], where C and n are a proportionality symbol and decay rate, respectively. The decay rate indicated the extent of entrainment and boundary effects on the jet. Figure 8 shows the variation of U_m/U_j for all cases with the normalized longitudinal coordinate x/d within the transition zone. Rajaratman reported $n = 0.5$ for a plane free jet by considering simplified conservation laws and entrainment hypotheses [36]. The wall jet was observed to be accurately estimated by using the power law. However, the distribution of U_m/U_j for the offset jets are not accurately described by the power law, which should be replaced by a linear fit [32].

For circular offset jet, using the power law fit, the decay rate for the lower offset ratio (OR = 1) circular jet was observed to be accurately estimated with the decay rate n varies from 1.197 to 1.768 as shown in Figure 8a. As can be seen in Figure 8b–c, the distributions of U_m/U_j, for the higher offset ratio (OR = 2 or 3) circular jet are not accurately described by the power law. Therefore, a linear fit was used to estimate the decay rates for higher offset ratio jet, the slope κ the linear equation was considered as the decay rate. Decay rate κ values of 0.110, 0.096 and 0.090 with R^2 above 0.94 were obtained for ER = 3, 4 and 4.8, respectively, when OR = 2. Decay rate values of 0.081, 0.048 and 0.049 with R^2 above 0.968 were obtained for ER = 3, 4 and 4.8, respectively, when OR = 3. Above analysis and as shown in Figure 8a–c indicated that when the offset ratio is moderate (OR = 2) produces a faster decay than that of higher or lower offset ratio (OR = 1 or 3). The reason is that two circulating vortices were developed below and above jet, respectively. The circulating vortices develop reverse flows against the inflow-jet direction. Negative momentum of the reverse flows reduced the inflow momentum slowing the jet rapidly down.

The decay rates for the various circular jet cases are reported in Table 2. The decay rate value is the largest as the expansion ratio (ER) is 3. There is a stronger jet interaction with the ambient fluid due to the more confined environment.

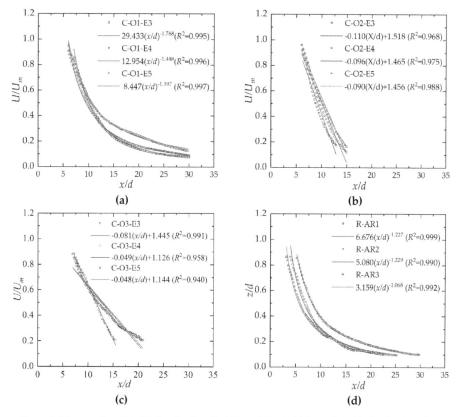

Figure 8. (a) power law fit to U_m for circular offset jets as OR = 1; (b) linear fit to U_m for circular offset jets as OR = 2; (c) linear fit to U_m for circular offset jets as OR = 3; (d) power law or linear fit to U_m for rectangular offset jets.

Table 2. The decay rate of the circular jet.

Decay Rate	n (OR = 1)	κ (OR = 2)	κ (OR = 3)
Fit Equation	Power Law	Linear Equation	
ER = 3	1.768	0.110	0.081
ER = 4	1.448	0.096	0.049
ER = 4.8	1.197	0.090	0.048

For rectangular offset jet, the U_m/U_j for $1 \leq AR \leq 3$ is expressed well by power law form with the decay rate n varies from 1.068 to 1.228 (Table 3), and the U_m decay in transition zone does not show substantial differences (Figure 8d). Therefore, it appears reasonable to assume the effect of the aspect ratio of the rectangular jet on the mainstream attenuation is mainly to influence the length of the potential core zone.

Table 3. The decay rate of the rectangular jet.

Decay Rate	n (OR = 1)
Fit Equation	Power Law
AR = 1	1.227
AR = 2	1.229
AR = 3	1.068

The algorithm for calculating the length of the potential core (L_{PC}) and decay profiles in the transition zone could be used to roughly calculate the length of the energy dissipation region (L_{ED}). For example, a square offset jet is issued to a pool. The side length of the square is 2.67 m (equivalent diameter is 3 m), the offset height is 3 m, the width of the pool is 12 m, and the height of tail water is 16.5 m. If the U/U_m = 0.2 as the end of the length of the energy dissipation region, the distance U/U_m decreases from 1 to 0.2 is called L_{tr}. The L_{PC} can be described by:

$$L_{ED} = L_{PC} + L_{tr} \tag{9}$$

The L_{PC} = 6.60 m calculated by $L_{PC}/d = 2100e^{-2(L_{exit}/d)} + 0.5$. The L_{tr} = 39.53 m calculated by $0.2 = -0.096(x/d) + 1.465$. As the result, the length of energy dissipation region is 43.13 m.

3.2. Vertical Velocity Spread

Figure 9 shows the flow field of the offset jet with different offset ratios. There is a large range of vortices formed near the jet flow, the location and size of the vortex on the streamwise section have notable differences. When OR = 1, in the range of x/d = 0–25, a large vortex was formed above the jet flow; When OR = 3, in the range of x/d = 0–25, a large vortex was formed below the jet flow. However, there is a smaller vortex above and below the jet flow in the range of x/d = 0–15.

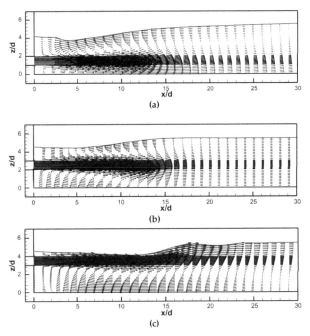

Figure 9. Velocity vectors of offset jet with different offset ratio: (**a**) OR = 1 (C-O1-E4); (**b**) OR = 2 (C-O2-E4); (**c**) OR = 3 (C-O3-E4).

Profiles of the streamwise mean velocities (U) extracted at x/d = 1, 4, 9, 13 and 17 are shown in Figure 10. It may be noted that to cover a range of ER = 4, Cases C-O1-E4, C-O2-E4, and C-O3-E4 were selected; otherwise, using the data of all Cases would make the plots clumsy. The velocity and length scale used was U_m and d, respectively. Here, z_c is the distance from the center of exit to the floor, that is, $z - z_c = 0$, where is the axis of the jet exits. Vertical profiles of U are shown in Figure 10a for ER is 4. The effect of offset ratio (OR) on the vertical velocity spread was not evident within the region $x/d < 9$. It can be noted that the velocity distribution in the floor-normal direction conforms with the Gaussian distribution within the region of $5 \le x/d \le 9$. However, as the jet travels away from the jet exits ($x/d > 9$), the data of jets at different offset height are no longer overlapped. The location of the maximum of U/U_m for OR = 2 is almost at the axis of the jet exits. The location of the maximum of U/U_m is above and below the axis of the jet exits for OR = 3 and OR = 1, respectively. The results reveal offset height changes travel direction of the bulk of the jet.

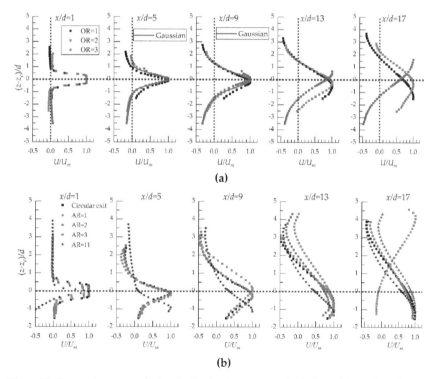

Figure 10. Streamwise mean velocity distribution in symmetry (x–z) plane: (**a**) circular offset jets (ER = 4); (**b**) rectangular offset jets.

Figure 10b shows the development of the U of the offset jet with rectangular exits. The results demonstrate that the bulk of the jet turns travel direction and then attaches to the floor within the region of $x/d < 13$. Further, the figure shows the spread rate of U is larger as the aspect ratio increases, and the flow direction for AR = 3 was changed at $x/d = 17$.

The U distribution in floor-normal direction in Case R-AR11 is different from those in other cases, the difference can be attributed to the jet exit of Case R-AR11 without sudden expand (ER = 1), there is an obviously lower pressure zone below the jet at the jet exit, then the main jet bends to the floor in short distance ($x/d < 2.5$). Thus, the profile is close to that of a wall jet.

Figure 11 shows contours of U/U_m in the symmetry plane for the offset jets, and the lower and upper lines represent the loci of $z_{0.5}^-$ ($U = 0.5\ U_m$, the inner separation line), and $z_{0.5}^+$ ($U = 0.5\ U_m$, the outer separation line), respectively. The jet between $z_{0.5}^-$ and $z_{0.5}^+$ was considered as the main flow in the present paper. The $L_{Z0.5}^-$ is the distance from the exit to where the $z_{0.5}^-$ reaches the floor, and $L_{Z0.5}^+$ is the distance from the exit to where the $z_{0.5}^+$ reaches the water surface. For circular exits, the main flow for OR = 1 bends to the floor at x/d = 12 ($L_{Z0.5}^-$ < 12 and $L_{Z0.5}^+$ > 46), then the flow spreads from floor to water surface (Figure 11a). The main flow for OR = 2 spreads evenly in the floor-normal direction at x/d > 20, the values of $L_{Z0.5}^-$ and $L_{Z0.5}^+$ are roughly equivalent (Figure 11b). Furthermore, the main flow for OR = 3 bends to the water surface, with the $L_{Z0.5}^+$ < 20 and $L_{Z0.5}^-$ > 42, which means the flow in the pool is spreading from water surface to floor (Figure 11c). For rectangular exits (Figure 11d), the main flow for AR = 1 is identical to that of a circular jet. The main flow for AR = 2 spreads from floor to water surface. The main flow for AR = 3 bends sharply to water surface as the jet enters the pool, the reason is that jet is too thin to effectively maintain a stable flow pattern. As the expansion ratio increases to 11.48, the spread range of the main flow in floor-normal direction is small, and the high velocity region is close to the floor.

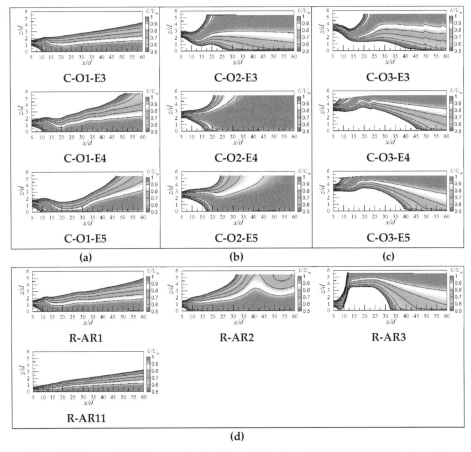

Figure 11. Contour of U/U_m in symmetry (x–z) plane: (**a**) circular offset jets (ER = 3); (**b**) circular offset jets (ER = 4); (**c**) circular offset jets (ER = 4.8); (**d**) rectangular offset jets.

3.3. Lateral Velocity Spread

As observed in Figure 12, there is a negligible effect of the expansion ratio on the lateral velocity spread within the region x/d < 9. This observation reveals that the profiles of U were independent of the expansion ratios, the Gaussian distribution is overlapped on the present data plots within the region $5 \leq x/d \leq 9$. The results indicated that the offset height did not greatly alter the flow within early development regions of the offset jet. However, at $x/d = 13$ and $x/d = 17$, U/U_m are lower for the expansion ratios, ER, are lower within the region $y/d > 0.5$. Figure 10 indicates that a more confining enclosure enhances the diffusion of the velocity than a less confining condition. The velocity profiles in the x-y plane of the jet with rectangular exits are not obtained. The flow patterns are not stable as the expansion ratio was changed [37].

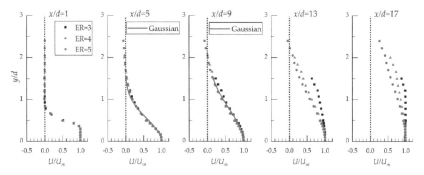

Figure 12. Profiles of U/U_m in (x–y) plane of circular jet.

4. Conclusions

The present investigation sought to elucidate the velocity development of a submerged 3D offset jet flow occurred in the narrow and deep pool using the standard k-ε model with VOF method. The flow field was studied for different offset ratio, expansion ratio, and jet exits. The numerical results showed that the solid boundaries of the narrow pool produce circulation in the surrounding fluid opposite the jet flow, which produces a rapid jet decay. The following conclusions can be drawn:

1. The development of the jet within the potential core zone was observed to be dependent on the circumference of jet exits. It was noted that the length of the potential core zone decreases with the circumference of exit.
2. The results showed that the development of the U_m in the transition zone could be divided into two modes. That is, (a) the decay of U_m could be estimated from power law fit with a decay rate n of 1.089–1.451 as the offset ratio is lower (OR = 1). (b) A linear fit was used to estimate the decay rates for higher offset ratio jet (OR = 2 or 3). It was observed that U_m decay is the fastest as the offset ratio (OR) is 2. Further, the results indicate that a more confining enclosure produces a more rapid jet decay than a less confining condition.
3. The spread rate of the circular jet is not affected by the expansion ratios and offset ratio in the early region of jet decay. The main flow for OR = 2 was traveled straightly, and the main flow was toward the floor and the water surface as the OR = 1 and OR = 3, respectively.
4. The absence of the sudden expansion coupled with the high aspect ratio of the exit contributed to form a unique flow pattern compared with other cases.

Author Contributions: Conceptualization and Methodology, X.L. and M.Z.; Software, X.L.; Formal analysis, X.L. and M.Z.; Investigation; X.L. and M.Z.; Writing—original draft, X.L.; Writing—Review & Editing, J.Z. and M.Z.; Supervision, J.Z. and W.X.; Project administration, J.Z. and W.X.; Funding acquisition, J.Z.

Funding: This research was funded by the National Key Research and Development Program of China (No. 2016YFC0401707), National Science Fund for Distinguished Young Scholars (No. 51625901), and National Natural Science Foundation of China (No. 51579165).

Conflicts of Interest: The authors declare no conflict of interest.

Abbreviations

x	Stream wise direction
y	Lateral direction
z	Floor direction
L_x	Length of the jet pool in the streamwise direction (m)
L_y	The width of the jet pool in the lateral direction (m)
L_z	The height of the jet pool in the floor-normal direction (m)
d	The diameter of circular jet exit (m) or equivalent diameter of rectangular exit (m)
a_0	The height of the rectangular exit (m)
b_0	Width of rectangular exit (m)
h_t	The height of tailwater (m)
S	Offset height of jet exit (m)
U	Mean velocity in streamwise (m/s)
U_m	Cross-Sectional streamwise maximum mean velocity (m/s)
U_j	Velocity in jet exit (m/s)
L_{PC}	Length of potential core
L_{exit}	The circumference of jet exit
z_c	Distance from center of exit to floor, $z_c = S + 0.5d$ or $z_c = S + 0.5a_0$
$z_{0.5}^+$	The loci of where $U = 0.5\ U_m$ above the jet
$z_{0.5}^-$	The loci of where $U = 0.5\ U_m$ below the jet
$L_{Z0.5}^-$	Distance from the exit to where the $z_{0.5}^-$ reaches floor
$L_{Z0.5}^+$	Distance from the exit to where the $z_{0.5}^+$ reaches water surface
n	Decay rate used in power law
κ	Decay rate used in the linear fit
R_0	Reynolds number
OR	Offset ratio of jet exit (S/d or S/a_0)
ER	The expansion ratio of jet exit (L_y/d or L_y/b_0)
AR	Aspect ratio of jet exit (b_0/a_0)

References

1. SoaresFrazão, S. *Wastewater Hydraulics*; Springer: Berlin/Heidelberg, Germany, 2010; pp. 842–843.
2. Liu, M.; Rajaratnam, N.; Zhu David, Z. Mean Flow and Turbulence Structure in Vertical Slot Fishways. *J. Hydraul. Eng.* **2006**, *132*, 765–777. [CrossRef]
3. Chen, J.-G.; Zhang, J.-M.; Xu, W.-L.; Li, S.; He, X.-L. Particle image velocimetry measurements of vortex structures in stilling basin of multi-horizontal submerged jets. *J. Hydrodyn.* **2013**, *25*, 556–563. [CrossRef]
4. Dey, S.; Ravi Kishore, G.; Castro-Orgaz, O.; Ali, S.Z. Hydrodynamics of submerged turbulent plane offset jets. *Phys. Fluids* **2017**, *29*, 065112. [CrossRef]
5. Bhuiyan, F.; Habibzadeh, A.; Rajaratnam, N.; Zhu David, Z. Reattached Turbulent Submerged Offset Jets on Rough Beds with Shallow Tailwater. *J. Hydraul. Eng.* **2011**, *137*, 1636–1648. [CrossRef]
6. Durand, Z.M.J.; Clark, S.P.; Tachie, M.F.; Malenchak, J.; Muluye, G. Experimental Study of Reynolds Number Effects on Three-Dimensional Offset Jets. In Proceedings of the 12th International Conference on Nanochannels, Microchannels, and Minichannels, Chicago, IL, USA, 3–7 August 2014. [CrossRef]
7. Nyantekyi-Kwakye, B.; Tachie, M.F.; Clark, S.P.; Malenchak, J.; Muluye, G.Y. Experimental study of the flow structures of 3D turbulent offset jets. *J. Hydraul. Res.* **2015**, *53*, 773–786. [CrossRef]
8. Agelin-Chaab, M.; Tachie, M.F. Characteristics and structure of turbulent 3D offset jets. *Int. J. Heat Fluid Flow* **2011**, *32*, 608–620. [CrossRef]
9. Li, J.-N.; Zhang, J.-M.; Peng, Y. Characterization of the mean velocity of a circular jet in a bounded basin. *J. Zhejiang Univ. Sci. A* **2017**, *18*, 807–818. [CrossRef]
10. Kumar Rathore, S.; Kumar Das, M. Effect of Freestream Motion on Heat Transfer Characteristics of Turbulent Offset Jet. *J. Therm. Sci. Eng. Appl.* **2015**, *8*, 011021. [CrossRef]

11. Camino, G.A.; Zhu, D.Z.; Rajaratnam, N. Jet diffusion inside a confined chamber. *J. Hydraul. Res.* **2012**, *50*, 121–128. [CrossRef]

12. Zhang, Z.; Guo, Y.; Zeng, J.; Zheng, J.; Wu, X. Numerical Simulation of Vertical Buoyant Wall Jet Discharged into a Linearly Stratified Environment. *J. Hydraul. Eng.* **2018**, *144*, 06018009. [CrossRef]

13. Assoudi, A.; Habli, S.; Mahjoub Saïd, N.; Bournot, H.; Le Palec, G. Three-dimensional study of turbulent flow characteristics of an offset plane jet with variable density. *Heat Mass Transf.* **2016**, *52*, 2327–2343. [CrossRef]

14. Mondal, T.; Guha, A.; Das, M.K. Computational study of periodically unsteady interaction between a wall jet and an offset jet for various velocity ratios. *Comput. Fluids* **2015**, *123*, 146–161. [CrossRef]

15. Vishnuvardhanarao, E.; Das, M.K. Computation of Mean Flow and Thermal Characteristics of Incompressible Turbulent Offset Jet Flows. *Numer. Heat Transf. A* **2007**, *53*, 843–869. [CrossRef]

16. Gu, R. Modeling Two-Dimensional Turbulent Offset Jets. *J. Hydraul. Eng.* **1996**, *122*, 617–624. [CrossRef]

17. Hnaien, N.; Marzouk, S.; Ben Aissia, H.; Jay, J. CFD investigation on the offset ratio effect on thermal characteristics of a combined wall and offset jets flow. *Heat Mass Transf.* **2017**, *53*, 2531–2549. [CrossRef]

18. Abraham, J.; Magi, V. *Computations of Transient Jets: RNG k-e Model Versus Standard k-e Model*; SAE: Warrendale, PA, USA, 1997.

19. Nasr, A.; Lai, J.C.S. A turbulent plane offset jet with small offset ratio. *Exp. Fluids* **1998**, *24*, 47–57. [CrossRef]

20. Launder, B.E.; Sharma, B.I. Application of the energy-dissipation model of turbulence to the calculation of flow near a spinning disc. *Lett. Heat Mass Transf.* **1974**, *1*, 131–137. [CrossRef]

21. Bai, Z.; Wang, Y.; Zhang, J. Pressure distributions of stepped spillways with different horizontal face angles. In *Proceedings of the Institution of Civil Engineers-Water Management*; Thomas Telford Ltd.: London, UK; pp. 1–12.

22. Li, S.; Zhang, J. Numerical Investigation on the Hydraulic Properties of the Skimming Flow over Pooled Stepped Spillway. *Water* **2018**, *10*, 1478. [CrossRef]

23. Hirt, C.W.; Nichols, B.D. Volume of Fluid (VOF) Method for the Dynamics of Free Boundaries. *J. Comput. Phys.* **1981**, *39*, 201–225. [CrossRef]

24. Peng, Y.; Mao, Y.F.; Wang, B.; Xie, B. Study on C–S and P–R EOS in pseudo-potential lattice Boltzmann model for two-phase flows. *Int. J. Mod. Phys. C* **2017**, *28*, 1750120. [CrossRef]

25. Peng, Y.; Wang, B.; Mao, Y. Study on Force Schemes in Pseudopotential Lattice Boltzmann Model for Two-Phase Flows. *J. Math. Probl. Eng.* **2018**, *2018*, 9. [CrossRef]

26. Peng, Y.; Zhang, J.; Meng, J. Second-order force scheme for lattice Boltzmann model of shallow water flows. *J. Hydraul. Res.* **2017**, *55*, 592–597. [CrossRef]

27. Peng, Y.; Zhang, J.M.; Zhou, J.G. Lattice Boltzmann Model Using Two Relaxation Times for Shallow-Water Equations. *J. Hydraul. Eng.* **2016**, *142*, 06015017. [CrossRef]

28. Peng, Y.; Zhou, J.G.; Burrows, R. Modeling Free-Surface Flow in Rectangular Shallow Basins by Using Lattice Boltzmann Method. *J. Hydraul. Eng.* **2011**, *137*, 1680–1685. [CrossRef]

29. Peng, Y.; Zhou, J.G.; Burrows, R. Modelling solute transport in shallow water with the lattice Boltzmann method. *Comput. Fluids* **2011**, *50*, 181–188. [CrossRef]

30. Van Doormaal, J.P.; Raithby, G.D. Enhancements of the Simple Method for Predicting Incompressible Fluid Flows. *Numer. Heat Transf.* **1984**, *7*, 147–163. [CrossRef]

31. Celik, I.; Ghia, U.; Roache, P.; Freitas, C.J. Procedure for estimation and reporting of uncertainty due to discretization in CFD applications. *J. Fluids Eng.* **2008**, *130*, 078001. [CrossRef]

32. Nyantekyi-Kwakye, B.; Clark, S.P.; Tachie, M.F.; Malenchak, J.; Muluye, G. Flow characteristics within the recirculation region of three-dimensional turbulent offset jet. *J. Hydraul. Res.* **2015**, *53*, 230–242. [CrossRef]

33. Li, X.; Wang, Y.; Zhang, J. Numerical Simulation of an Offset Jet in Bounded Pool with Deflection Wall. *Math. Probl. Eng.* **2017**, *2017*, 11. [CrossRef]

34. Nyantekyi-Kwakye, B.; Tachie, M.F.; Clark, S.P.; Malenchak, J.; Muluye, G.Y. Acoustic Doppler velocimeter measurements of a submerged three-dimensional offset jet flow over rough surfaces. *J. Hydraul. Res.* **2017**, *55*, 40–49. [CrossRef]

35. Padmanabham, G.; Lakshmana Gowda, B.H. Mean and Turbulence Characteristics of a Class of Three-Dimensional Wall Jets—Part 1: Mean Flow Characteristics. *J. Fluids Eng.* **1991**, *113*, 620–628. [CrossRef]

36. Rajaratnam, N. *Turbulents Jets*; Elsevier: New York, NY, USA, 1976.

37. Ohtsu, I.; Yasuda, Y.; Ishikawa, M. Submerged Hydraulic Jumps below Abrupt Expansions. *J. Hydraul. Eng.* **1999**, *125*, 492–499. [CrossRef]

Article

Numerical Investigation on the Hydraulic Properties of the Skimming Flow over Pooled Stepped Spillway

Shicheng Li and Jianmin Zhang *

State Key Laboratory of Hydraulics and Mountain River Engineering, Sichuan University,
Chengdu 610065, China; tyutscu@163.com
* Correspondence: zhangjianmin@scu.edu.cn

Received: 21 September 2018; Accepted: 15 October 2018; Published: 19 October 2018

Abstract: Pooled stepped spillway is known for high aeration efficiency and energy dissipation, but the understanding for the effects of pool weir configuration on the flow properties and energy loss is relatively limited, so RNG $k - \varepsilon$ turbulence model with VOF method was employed to simulate the hydraulic characteristics of the stepped spillways with four types of pool weirs. The calculated results suggested the flow in the stepped spillway with staggered configuration of' two-sided pooled and central pooled steps (TP-CP) was highly three dimensional and created more flow instabilities and vortex structures, leading to 1.5 times higher energy dissipation rate than the fully pooled configuration (FP-FP). In FP-FP configuration, the stepped spillway with fully pooled and two-sided pooled steps (FP-TP) and the spillway with fully pooled and central pooled steps (FP-CP), the pressure on the horizontal step surfaces presented U-shaped variation, and TP-CP showed the greatest pressure fluctuation. For FP-TP and FP-CP, the vortex development in the transverse direction presented the opposite phenomenon, and the maximum vortex intensity in TP-CP occurred at Z/W = 0.25, while FP-FP illustrated no significant change in the transverse direction. The overlaying flow velocity distribution in the spanwise direction demonstrated no obvious difference among FP-FP, FP-TP, and FP-CP, while the velocity in TP-CP increased from the axial plane to the sidewalls, but the maximum velocity for all cases were approximately the same.

Keywords: numerical simulation; pooled stepped spillway; pool weir; flow property

---

1. Introduction

Stepped spillway, an important water-transferring hydraulic infrastructure, has long been used in water related projects owing to its high efficiency of energy dissipation and air entrainment, making it possible to release massive and high speed flood flows while reducing and preventing dangerous scour and cavitation. Additionally, economic benefits will be enjoyed due to the reduction in the construction of the downstream energy dissipation basin [1,2]. Due to the advantages of stepped spillway, a growing number of this particular hydraulic structure is being applied to engineering projects.

For the past decades, stepped spillway has been studied globally by the means of laboratory experiments and numerical simulations [3–7], providing significant insights into the complicated air-water flow properties, momentum exchange process, and energy loss mechanism. For example, model tests performed by Felder and Chanson [8] to investigate the influence of uniform and non-uniform step height on the hydraulic performance illustrated that there was no significant change between the two configurations regarding the residual head. Channel slope also plays a vital role in the air-water flow properties, such as velocity profile and turbulence intensity reported by Felder and Chanson [9]. The time-averaged pressure on the step faces was experimentally measured by Zhang et al. [10], drawing the conclusions that the pressure distributions on the horizontal step surfaces exhibited S-shaped variations with the maximum and minimum values occurring at the downstream end and upstream start, respectively. Physical models were established by Zhang and Chanson [11] to examine the effect of step edge and cavity geometry on the air entrainment and energy loss performance in stepped chutes. Apart from experimental studies, thanks to the advances in the computational performance of computers, numerical methods have also been successfully used to study the detailed flow properties in complex contexts [12–15]. Baylar et al. [16] numerically examined the energy loss and aeration efficiency of the stepped chutes and found that both of them in nappe flow regime were greater than that of the skimming flow regime. Numerical simulations by researchers [17–19] demonstrated its high performance in investigating the detailed flow field and pressure distribution over stepped spillway.

In recent years, an innovative design has been made to stepped spillway to improve the energy dissipation performance. Unlike the conventional stepped spillway whose horizontal step faces are flat, this novel kind of stepped spillway, pooled stepped spillway, is equipped with pool weirs at the step edges. A few of the researches have been carried out to investigate the air-water structures in this unique type of stepped spillway. For instance, the differences concerning the air entrainment processes of flat, pooled, and a combination of flat and pooled stepped spillway have been clarified by André [20] and Kökpinar [21]. Thorwarth [22] reported self-induced instabilities over pooled stepped spillways with channel slopes of 8.9° and 14.6°. The detailed air-water flow properties over several kinds of stepped pooled spillway were also physically studied by Felder [7] and Felder and Chanson [9].

While the existing researches have gained understanding into this particular type of stepped spillway, little has been known about the impact of pool weir configurations on the flow properties. Therefore, in order to expand the limited insights into the effect of pool weir's geometrical properties on the flow characteristics, numerical investigations of pooled stepped spillways with different pool weir configurations were carried out.

2. Numerical Simulation

In the present study, ANSYS 16.0 (ANSYS®, Canonsburg, PA, USA) was utilized to investigate the flow characteristics over stepped spillway with different pool weir configurations. The numerical model consists of 4 parts: the upstream water tank, the broad-crested weir with an upstream rounded corner (radius = 0.08 m), the stepped spillway and the downstream channel. The length, L_{crest}, and width, W, of the broad-crested weir are 1.01 m and 0.52 m, respectively. The spillway has 10 steps with vertical step height, h, and horizontal step length, l, of 0.1 m and 0.2 m, respectively. The channel slope is 26.6° and constant in each configuration. Moreover, the height, d, and the thickness, l_w, of pool weir are, respectively, 0.09 m and 0.015 m. The schematics of the spillways are shown in Figure 1.

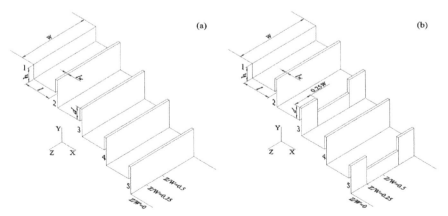

Figure 1. *Cont.*

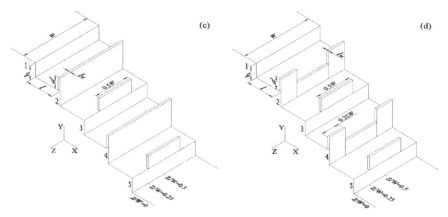

Figure 1. Sketch of pooled stepped spillway configurations. (**a**) Fully pooled steps (FP-FP); (**b**) fully pooled and two-sided pooled steps (FP-TP); (**c**) fully pooled and central pooled steps (FP-CP); and (**d**) two-sided pooled and central pooled steps (TP-CP).

2.1. Volume of Fluid Method

Developed by Hirt and Nichols [23], the volume of fluid (VOF) model is believed to be a highly effective method to determine the position of the interface of two or more immiscible flows [19,24–26]. For the purpose of tracking the air-water interface, the continuity equation for the volume fraction of water is adopted in the following format:

$$\frac{\partial \alpha_w}{\partial t} + u_i \frac{\partial \alpha_w}{\partial x_i} = 0 \tag{1}$$

where α_w is the volume fraction of water, u_i and x_i are the velocity components and coordinates ($i = 1, 2, 3$), respectively. Hence, by solving the continuity equation, the free surface of water can be determined. In addition, it is clear that the sum of their volume fractions, α_w and α_a, respectively, is unity in every computational cell, which can be described as:

$$\alpha_w + \alpha_a = 1 \tag{2}$$

Obviously, the maximum and minimum value for the volume fraction of water, α_a, is 1 and 0, indicating the given cell is full of water or air, respectively, between which there must be a mixture of them containing the interface.

As for the density, ρ, and molecular viscosity, μ, they can be given as

$$\rho = \alpha_w \rho_w + (1 - \alpha_w) \rho_a \tag{3}$$

$$\mu = \alpha_w \mu_w + (1 - \alpha_w) \mu_a \tag{4}$$

where ρ_w and ρ_a represents the density of water and air, respectively, while μ_w and μ_a are the molecular viscosity of water and air, respectively. Therefore the values of ρ and μ can be known by iterating the Equations (3) and (4).

2.2. Turbulence Model

Proposed by Yakhot and Orszag [27], the Renormalization Group (RNG) $k - \varepsilon$ closure model has been employed to simulate flow over stepped spillways in many studies [28–30], yielding reliable results compared with laboratory experiments. The continuity equation, momentum equation, and transport equations of turbulent kinetic energy, and the turbulent dissipation rate, used in the model, are shown below.

Continuity equation:

$$\frac{\partial \rho}{\partial t} + \frac{\partial \rho u_i}{\partial x_i} = 0 \tag{5}$$

Momentum equation:

$$\frac{\partial \rho u_i}{\partial t} + \frac{\partial}{\partial x_j}(\rho u_i u_j) = -\frac{\partial p}{\partial x_i} + \frac{\partial}{\partial x_j}\left[(\mu + \mu_t)\left(\frac{\partial u_i}{\partial x_j} + \frac{\partial u_j}{\partial x_i}\right)\right] \tag{6}$$

k equation:

$$\frac{\partial(\rho k)}{\partial t} + \frac{\partial(\rho_i k)}{\partial x_i} = \frac{\partial}{\partial x_i}\left[\left(\mu + \frac{\mu_t}{\sigma_k}\right)\frac{\partial k}{\partial x_i}\right] + G_k - \rho\varepsilon \tag{7}$$

ε equation:

$$\frac{\partial(\rho\varepsilon)}{\partial t} + \frac{\partial(\rho u_i \varepsilon)}{\partial x_i} = \frac{\partial}{\partial x_i}\left[\left(\mu + \frac{\mu_t}{\sigma_\varepsilon}\right)\frac{\partial \varepsilon}{\partial x_i}\right] + C_{1\varepsilon}\rho\frac{\varepsilon}{k}G_k - C_{2\varepsilon}\rho\frac{\varepsilon^2}{k} \tag{8}$$

where ρ and μ respectively represent the average density of the volume fraction and the molecular viscosity. p is the modified pressure, and μ_t is the turbulent viscosity, which can be calculated from the following equations consisting of k and ε. $\mu_t = C_\mu \frac{k^2}{\varepsilon}$, $C_{1\varepsilon} = 1.42 - \frac{\eta\left(1 - \frac{\eta}{\eta_0}\right)}{1 + \beta\eta^3}$, $\eta = \frac{Sk}{\varepsilon}$, $S = \sqrt{2S_{ij}S_{ij}}$. The constants in the above expressions are given in the Table 1.

Table 1. Constants of governing equations.

η_0	438
β	0.012
C_μ	0.085
$C_{2\mu}$	1.68
σ_k	0.7179
σ_ε	0.7179

2.3. Boundary Conditions

Combined with VOF method, RNG $k - \varepsilon$ turbulence model was applied to numerically study the flow characteristics over stepped spillway with different pool weirs and the entire simulation domain was meshed with structured grids, as shown in Figure 2. The boundary conditions for the mathematical models were set as follows:

(a) Inflow boundary: Velocity of the flow calculated from the discharge, which is 0.113 m^3/s in all cases, was given as the inlet boundary;

(b) Outflow boundary: Pressure outlet boundary was selected at the outlet and there was no normal gradient for all variables;

(c) Free surface: Pressure inlet was employed and its value was the standard atmospheric pressure; and

(d) Wall boundary: No-slip velocity boundary and wall function were chosen for the wall surfaces and near-wall regions, respectively.

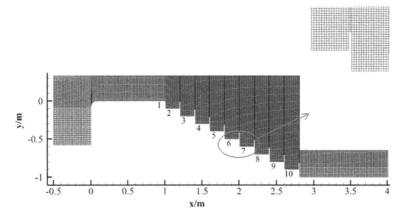

Figure 2. Meshing pattern in the computational domain.

2.4. Grid Testing and Model Verification

Before applying the numerical model to further analysis, it is necessary to validate its reliability. Therefore, the accuracy of the calculation results were tested by utilizing grid convergence index (GCI) [31] with grid numbers of approximately 0.31 million, 0.55 million and 0.76 million. Figure 3 presents the impact of the grid sizes on the uncertainty of the computational velocity distribution at the axial plane. The discretization uncertainties at most locations were quite small. The maximum GCI values for the velocity profiles over step 6 and step 7 were 7.7% and 5.6%, corresponding to ±0.19 m/s and ±0.15 m/s, respectively. Considering the simulation efficiency and accuracy, the grid number set as 0.55 million is reasonable.

In addition, experimental data obtained from Felder et al. [32] were used to verify the numerical results by comparing the velocity distribution over the pool at steps 8, 9, and 10, as shown in Figure 4. The schematic of the physical model used for validation was the same as the FP-FP configuration in the present article, except the pool weir height. In the research of Felder et al. [32], the pool weir height was 0.031 m. It can be seen from Figure 4 that, at most locations, the calculated results of velocity distribution were fairly consistent with that of laboratory tests. The maximums of the relative error for steps 8, 9 and 10 were 14.8%, 7.1% and 6.5%, respectively, indicating that the numerical simulations produced reliable and acceptable results.

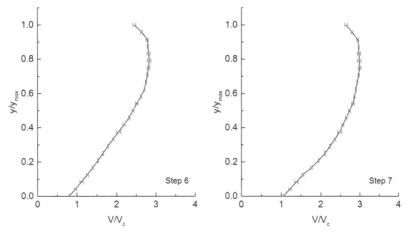

Figure 3. Fine-grid solution with discretization error bars computed using the GCI index.

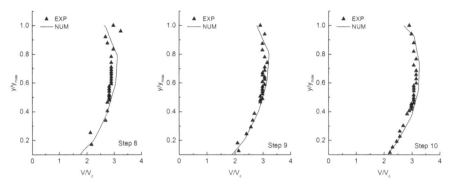

Figure 4. Comparison of numerical and experimental data. Notes: V$_c$: critical flow velocity; y represents the distance normal to the pseudo-bottom formed by the pool edges with $y = 0$ at the pool edges.

3. Results and Analysis

3.1. Flow Pattern

After ensuring the accuracy of the calculation, the effects of pool weir geometrical parameters on the hydraulic behaviors were numerically examined. Figure 5 illustrates the flow pattern at the axial plane in the calculated cases. It is noticeable that the free water surface of FP-FP configuration was parallel to the pseudo-bottom, and showed the least instability among the simulated configurations. FP-TP and FP-CP showed the similar free surface profile, but slight flow turbulences caused by the partially pooled steps can be observed. An obvious feature of the flow in TP-CP was the highly three-dimensional flow motion in the spanwise direction, so TP-CP highlighted strong fluctuations in free water surface.

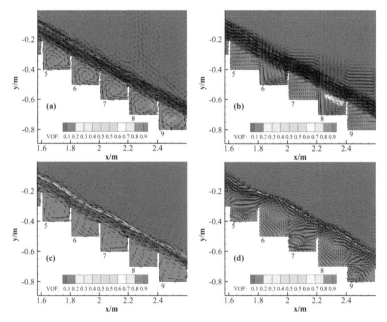

Figure 5. Flow pattern at the axial plane in different configurations. (**a**) FP-FP; (**b**) FP-TP; (**c**) FP-CP; and (**d**) TP-CP.

Figure 6 highlights the streamlines on the representative steps in each configuration. As can be seen, like the traditional flat stepped spillway, the flow surface over the pool weir was approximately parallel to the pseudo-bottom formed by the pool edges, although TP-CP presented some instabilities caused by the staggered configurations of central pooled and two-sided pooled steps. For FP-FP configuration, vortexes with the similar size and position were formed in every pool and there was no significant change in the transverse direction, which was different from other configurations, that is, the flow pattern in FP-TP, FP-CP, and TP-CP was three-dimensional. In FP-TP, the scale of the vortex on the two-sided pooled steps increased from the axial plane to the sidewalls, while the scale of the vortex on the fully pooled steps decreased from the axial plane to the sidewalls. FP-CP presented the opposite phenomenon compared with FP-TP. In TP-CP configuration, the maximum vortex intensity occurred at $Z/W = 0.25$ and no eddy can be found on the central pooled steps at the sidewalls and on the two-sided pooled steps at the axial plane.

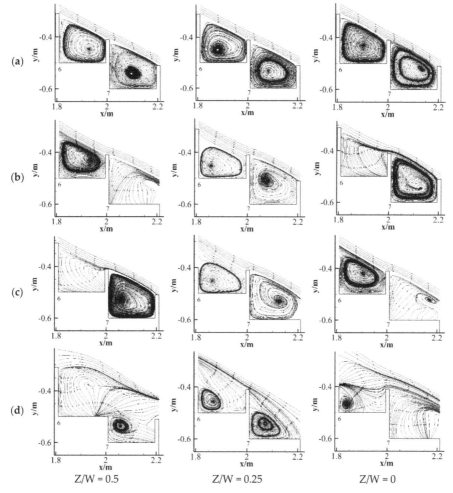

Figure 6. Instantaneous streamlines in the transeverse direction. (**a**) FP-FP; (**b**) FP-TP; (**c**) FP-CP; and (**d**) TP-CP.

For the identification of turbulence structures, Q criterion [33] is introduced to examine the effects of pool weir geometry on the vortex flow. The Q criterion is defined as follows:

$$Q = \frac{1}{2}\left(\|\Omega_{ij}\|^2 - \|S_{ij}\|^2\right) \tag{9}$$

where

$$\Omega_{ij} = \frac{1}{2}\left(\frac{\partial u_i}{\partial x_j} - \frac{\partial u_j}{\partial x_i}\right) \tag{10}$$

represents the rate of rotation tensor corresponding to the pure rotational motion, and

$$S_{ij} = \frac{1}{2}\left(\frac{\partial u_i}{\partial x_j} + \frac{\partial u_j}{\partial x_i}\right) \tag{11}$$

represents the rate of strain tensor corresponding to the irrotational motion.

The parameter Q is an effective approach to reveal the main motion pattern of fluid element. It can been seen from Equation (9) that the rotation motion will dominate the region where the value of Q is positive, while the deformation motion will dominate when Q is negative. However, it should be noted that the vortex structures cannot be accurately represented in the region with $Q = 0$, due to experimental error and viscous diffusion.

Figure 7 demonstrates the iso-surface of $Q = 100$. In FP-FP configuration, the number of lumps of the Q iso-surface distributing on each step was approximately the same and there was no apparent change in the transverse direction. The distribution of lumps in FP-TP and FP-CP exhibited the similar pattern. The major difference was that small lumps can be found near the sidewalls on the two-sided pooled steps in FP-TP, while small lumps were located near the axial plane in FP-CP, meaning turbulent structures existed at the two locations. It is noticeable that distribution of lumps in TP-CP configuration was much denser than that of other cases, indicating greater vortex and more turbulent structures.

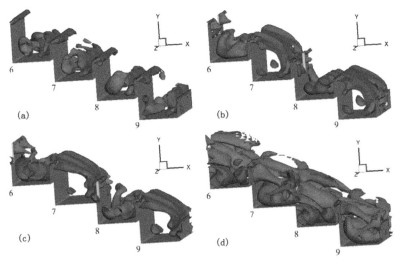

Figure 7. Distributions of lumps of the iso-surface of $Q = 100$. (**a**) FP-FP; (**b**) FP-TP; (**c**) FP-CP; and (**d**) TP-CP.

3.2. Velocity Distribution

Figure 8 illustrates the velocity vectors at the axial plane. It implies that, in all configurations, the skimming flow transferring from the upstream to the downstream over the steps was partly trapped in the recirculating zones under the pseudo-bottom. However, it is evident that stable vortices can only be found on some steps at the axial plane in FP-TP, FP-CP, and TP-CP, which is consistent with the analysis from flow pattern. Furthermore, the velocity magnitude of the overlaying flow was significantly greater than that of the vortex.

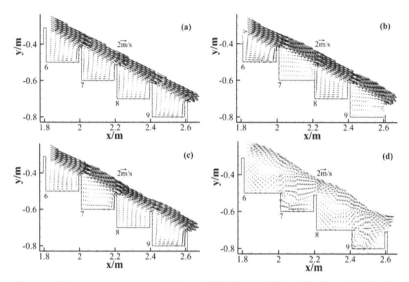

Figure 8. Velocity vectors at the axial plane. (**a**) FP-FP; (**b**) FP-TP; (**c**) FP-CP; and (**d**) TP-CP.

Figure 9 displays the velocity distributions above the center of the horizontal step surfaces. In FP-FP, the velocity showed the same distribution in the transverse direction; the velocity distributions on each step presented the similar pattern, and no significant change between different steps can be observed for $y/y_{max} < 0.8$, but in the flow surface region, the velocity above the downstream step surfaces was greater than that of the upstream ones; the minimum and maximum velocity above each step occurred at the same position: $y/y_{max} = 0.3$ and $y/y_{max} = 0.9$, respectively. In FP-TP configuration, at the axial plane, the velocity distributions above the fully pooled and two-sided pooled steps were obviously different in the bottom region, with the former greater than the latter; However, at $Z/W = 0.25$, there was no major difference in the velocity distribution above different steps except the free surface region. In FP-CP, at $Z/W = 0.5$, the velocity at $0.5 < y/y_{max} < 0.8$ for different steps indicated the same distribution; different from $Z/W = 0.5$, the velocity over different steps at $Z/W = 0.25$ exhibited similar distributions. In TP-CP, the velocity distributions presented a totally different pattern from other configurations; the velocity demonstrated the different distribution on two-sided pooled steps and central pooled steps, and velocity fluctuated greatly in the spanwise direction, proving its highly three dimensional characteristics.

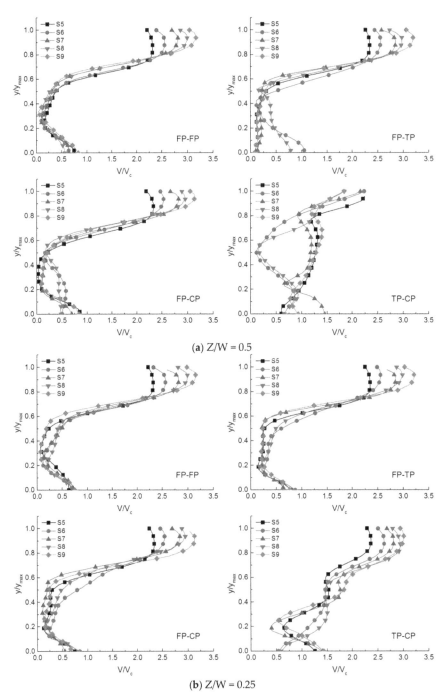

Figure 9. Velocity distributions above the step surfaces in each configuration. (**a**) Z/W = 0.5; and (**b**) Z/W = 0.25.

In order to better compare the velocity field in the transverse direction, velocity contour maps at different cross-sections are shown in Figure 10. It can be noticed that, in all cases, the velocity distribution along the transverse direction was symmetrical and the overlaying flow over the pool weir showed the maximum velocity magnitude, while the minimum velocity value occurred near the bottom, but both the minimum and maximum indicated no apparent change in different configurations; In FP-FP, FP-TP and FP-CP, the velocity of the mainstream illustrated no significant difference along the spanwise direction, but the velocity in TP-CP fluctuated greatly with its value increasing from the axial plane to the sidewalls.

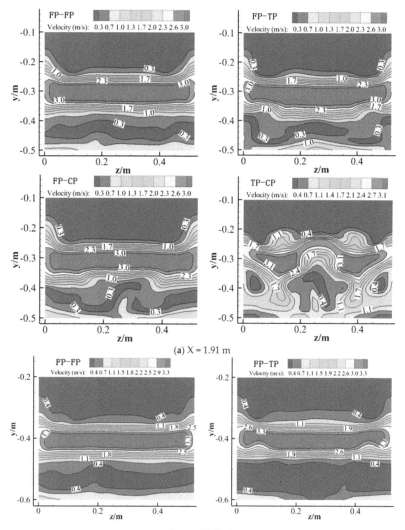

(a) X = 1.91 m

Figure 10. *Cont.*

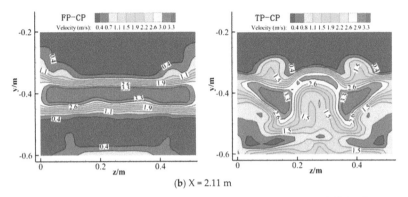

(b) X = 2.11 m

Figure 10. Velocity countour maps at the cross-sections. (**a**) X = 1.91 m; and (**b**) X = 2.11 m.

3.3. Pressure Distribution

Figure 11 highlights the pressure distributions on the horizontal step surfaces of two representative steps (steps 6 and 7). In FP-FP configuration, the minimum and maximum pressure on different horizontal step surfaces showed no significant change, as well as the locations where the minimum and maximum pressure occurred. The pressure distribution patterns on the fully pooled steps in FP-TP and on the central pooled steps in FP-CP were similar with the extreme pressure occurring at the axial plane near the downstream end of the step face and the minimum pressure occurring in the middle the step surface; the pressure distributions on the two-sided pooled steps in FP-TP and on the fully pooled steps in FP-CP demonstrated the similar patterns with the extreme pressure occurring at the sidewalls near the downstream end of the step face. As for TP-CP configuration, the extreme pressures on the two-sided pooled and central pooled steps occurred near the sidewalls and axial plane, respectively.

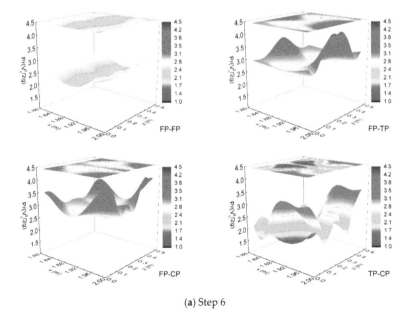

(a) Step 6

Figure 11. *Cont.*

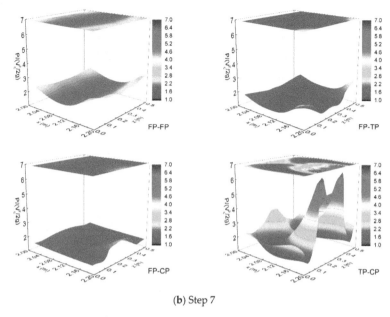

(**b**) Step 7

Figure 11. Pressure distributions on the horizontal step surfaces. (**a**) Step 6; (**b**) Step 7.

Figure 12 illustrates the pressure distributions on the vertical step surfaces of two representative steps (steps 6 and 7). As can been seen, in FP-FP, FP-TP and FP-CP, the pressure distributions on the vertical step surfaces highlighted the same pattern: the value of pressure decreased with the increasing in water depth, but the pressure for TP-CP presented some variations in the vertical direction. The maximum pressures on odd-number step surfaces in different configurations were in the following order: FP-CP > TP-CP > FP-TP > FP-FP, while for the even-number steps, it was TP-CP > FP-FP > FP-TP > FP-CP.

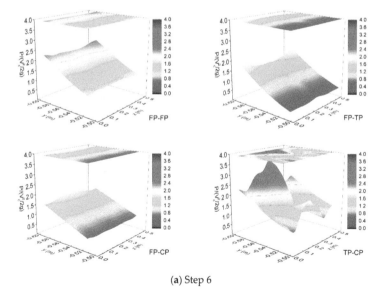

(**a**) Step 6

Figure 12. *Cont.*

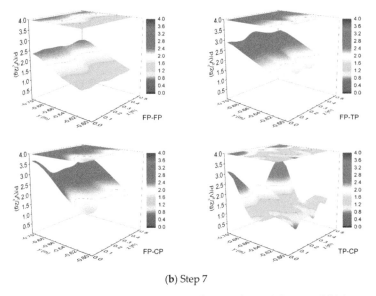

(**b**) Step 7

Figure 12. Pressure distributions on the vertical step surfaces. (**a**) Step 6; and (**b**) Step 7.

Figure 13 presents the pressure distributions on the horizontal step surfaces in the transverse direction. In FP-FP, FP-TP and TP-CP, the pressure on the upstream step surfaces was greater than that of the downstream ones, but it was quite chaotic and showed no obvious trend in TP-CP. The pressure distributions on horizontal step surfaces of FP-FP, FP-TP, and TP-CP showed U-shaped variation with the maximum pressures occurring at the downstream end of the step surface. In addition, the greatest fluctuation in pressure in the spanwise and streamwise direction can be found in TP-CP, indicating its flow complexities and highly three-dimensional characteristics.

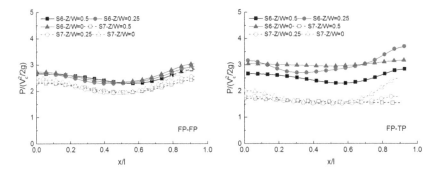

Figure 13. *Cont.*

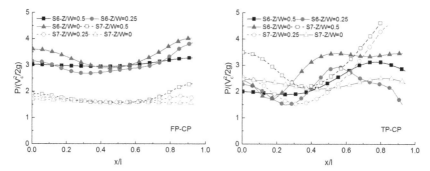

Figure 13. Pressure distributions on horizontal step surfaces in the transverse direction.

Figure 14 presents the pressure distributions on the vertical step surfaces in the transverse direction. In FP-FP, FP-TP, and FP-CP, the pressure distributions on the vertical step surfaces displayed no significant change in the transverse direction, but great pressure fluctuation can be seen in TP-CP. In FP-TP and FP-CP, the pressure on the vertical step surfaces of the downstream steps was moderately greater than that of the upstream ones, while FP-FP showed no evident difference.

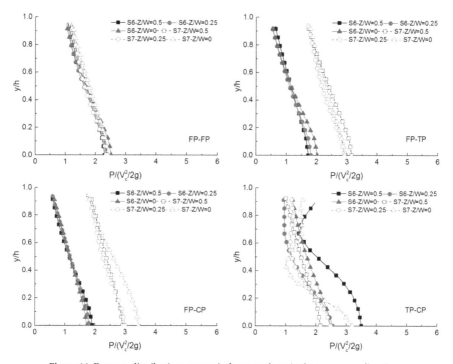

Figure 14. Pressure distributions on vertical step surfaces in the transverse direction.

3.4. Residual Head and Energy Dissipation

For the conservation of the downstream hydraulic structures, it is critical to quantify the residual head at the downstream of the spillway and the relative energy dissipation rate over the stepped spillway. The residual head, H_{res}, at the last pool weir can be estimated through the following equation:

$$H_{res} = d_e \times \cos\theta + \frac{U_w^2}{2 \times g} + d \tag{12}$$

where d_e is the equivalent clear water depth, θ represents the channel slope, U_w and d indicate the flow velocity and the pool weir height, respectively.

Together with the previous studies in Table 2, the dimensionless residual head, H_{res}/d_c, is shown in Figure 15. The flow discharge of the studies in Table 2 was the same and the flow regime was skimming flow. The residual energy in FP-FP configuration between the present work and the study of Morovati et al. [29] showed little difference, indicating the accuracy of the simulation. For configuration 1~3, the residual head showed no significant change for a given pool height, but the residual data increased with the decrease in pool height. In addition, the residual energy for the spillway with the combination of two-sided pooled steps and central pooled steps (TP-CP) decreased significantly compared with the spillway with only two-sided pooled steps or central pooled steps.

Table 2. Summary of the pooled stepped spillways in previous and present studies.

Reference	Step Geometry
Morovati et al. [29]	Fully pooled steps: $h = 0.1$ m, $l = 0.2$ m, $d = 0.09$ m In-line configuration (pooled and flat steps in-line): $h = 0.1$ m, $l = 0.2$ m, $d = 0.09$ m Staggered configuration (pooled and flat steps staggered): $h = 0.1$ m, $l = 0.2$ m, $d = 0.09$ m Two-sided pooled steps: $h = 0.1$ m, $l = 0.2$ m, $d = 0.09$ m Central-pooled steps: $h = 0.1$ m, $l = 0.2$ m, $d = 0.09$ m
Felder et al. [32]	Fully pooled steps: $h = 0.1$ m, $l = 0.2$ m, $d = 0.031$ m In-line configuration (pooled and flat steps in-line): $h = 0.1$ m, $l = 0.2$ m, $d = 0.031$ m Staggered configuration (pooled and flat steps staggered): $h = 0.1$ m, $l = 0.2$ m, $d = 0.031$ m
Present study	FP-FP: $h = 0.1$ m, $l = 0.2$ m, $d = 0.09$ m FP-TP: $h = 0.1$ m, $l = 0.2$ m, $d = 0.09$ m FP-CP: $h = 0.1$ m, $l = 0.2$ m, $d = 0.09$ m TP-CP: $h = 0.1$ m, $l = 0.2$ m, $d = 0.09$ m

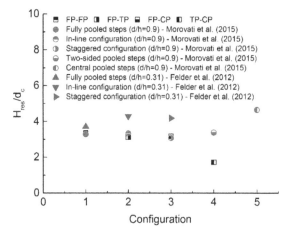

Figure 15. Residual energy at the downstream end of the spillway.

The energy dissipation rate, $\Delta H / H_{max}$, is an important indicator to evaluate the relative energy loss from the start to the end of the stepped spillway. The energy dissipation rate is given as:

$$\frac{\Delta H}{H_{max}} = \frac{H_{max} - H_{res}}{H_{max}} = \frac{H_{dam} + 1.5d_c - H_{res}}{H_{dam} + 1.5d_c} \tag{13}$$

where H_{dam} represents the dam height. The rate of energy dissipation is presented in Figure 16. It can be seen that TP-CP showed the best energy dissipation performance compared with other configurations, and the energy dissipation rate for TP-CP was almost 1.5 times larger than that of FP-FP, but no obvious change can be observed in FP-FP, FP-TP and FP-CP. That is because there were more turbulent structures in TP-CP caused by the concave-convex pool weirs, leading to greater momentum change between the overlaying flow and the vortex flow, thus creating more energy loss.

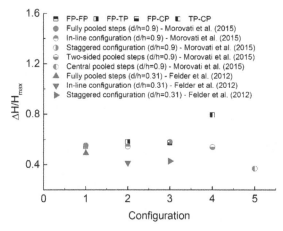

Figure 16. Energy dissipation rate.

4. Conclusions

In the present study, four types of pooled stepped spillway with different pool weirs were numerically investigated by using RNG $k - \varepsilon$ turbulence model with VOF method. The numerical results agreed well with the experimental data from Felder et al. [32] and the following conclusions can be drawn:

1. The flow pattern in FP-FP presented no significant change in the transverse direction and was similar on different steps. But changing half of the fully pooled steps into partially pooled steps (two-sided pooled or central pooled) can result in three dimensional flow motion. In FP-TP, the maximum vortex intensity on the fully pooled steps and partially pooled steps occurred at the axial plane and sidewalls, respectively, while the opposite phenomenon can be seen in FP-CP. When replacing all the fully pooled steps by staggered configuration of two-sided pooled and central pooled steps (TP-CP), the flow was characterized by highly three dimensional motion, and the vortex development in the transverse direction showed a unique pattern with the maximum intensity of vortex occurring at $Z/W = 0.25$. Besides, TP-CP created more flow instabilities and turbulent structures.

2. The velocity distributions over the steps in FP-FP highlighted the similar pattern with the minimum and maximum values occurring at $y/y_{max} = 0.3$ and 0.9, respectively. In FP-TP, FP-CP and TP-CP, the velocity distributions on the odd-number steps was different from that of the even-number steps, but the maximum velocity in all configurations indicated no difference. In the transverse direction, the velocity distribution in FP-FP showed the smallest variation, while TP-CP presented the greatest change.

3. The pressure on the horizontal step surfaces of FP-FP, FP-TP and FP-CP showed U-shaped variations with the maximum pressures occurring at the downstream end of the step surface, and TP-CP highlighted the greatest pressure fluctuation in both streamwise and transverse direction. The pressure distribution patterns on the fully pooled steps in FP-TP and on the central pooled steps in FP-CP were similar; the pressure distributions on the two-sided pooled steps in FP-TP and on the fully pooled steps in FP-CP demonstrated the similar patterns. For the vertical steps, the maximum pressures on odd-number step surfaces were in the following order: FP-CP > TP-CP > FP-TP > FP-FP, while for the even-number steps, it was TP-CP > FP-FP > FP-TP > FP-CP. There was no negative pressure on the step surfaces.

4. From the largest to the smallest, the energy dissipation rates were in the following order: TP-CP > FP-CP > FP-TP > FP-FP. Specifically, the energy loss for FP-TP and FP-CP was quite close, but slightly higher than FP-FP, and the energy dissipation rate for TP-CP was 1.5 times larger than FP-FP. Thus, changing half of the fully pooled steps into central pooled or two-sided pooled steps presented no obvious effect on the energy dissipation ratio, while shifting all the fully pooled steps into the combination of central pooled and two-sided pooled steps can significantly improve the energy dissipation performance.

Author Contributions: Conceptualization, S.L.; Methodology, S.L.; Software, S.L.; Formal Analysis, S.L.; Investigation, S.L.; Writing-Original Draft Preparation, S.L.; Writing-Review & Editing, J.Z.; Supervision, J.Z.; Project administration, J.Z.; Funding acquisition, J.Z.

Funding: This research was funded by the National Key Research and Development Program of China (No. 2016YFC0401707), National Natural Science Foundation of China (No. 51579165), and National Science Fund for Distinguished Young Scholars (No. 51625901).

Conflicts of Interest: The authors declare no conflict of interest.

References

1. Boes, R.M.; Hager, W.H. Hydraulic design of stepped spillways. *J. Hydraul. Eng.* **2002**, *129*, 671–679. [CrossRef]
2. Chanson, H. Hydraulic Design of stepped spillways and downstream energy dissipators. *Dam Eng.* **2001**, *11*, 205–242.
3. Chanson, H.; Luke, T. Hydraulics of stepped chutes: The transition flow. *J. Hydraul. Res.* **2004**, *42*, 43–54. [CrossRef]
4. Chanson, H.; Toombes, L. Experimental investigations of air entrainment in transition and skimming flows down a stepped chute. *Can. J. Civ. Eng.* **2002**, *29*, 145–156. [CrossRef]
5. Felder, S.; Chanson, H. Turbulence, dynamic similarity and scale effects in high-velocity free-surface flows above a stepped chute. *Exp. Fluids* **2009**, *47*, 1–18. [CrossRef]
6. Wang, J.; Fu, L.; Xu, H.; Jin, Y. Numerical study on flow over stepped spillway using Lagrangian method. In Proceedings of the International Conference on Energy Engineering and Environmental Protection, Sanya, China, 20–22 November 2017.
7. Felder, S. Air-Water Flow Properties on Stepped Spillways for Embankment Dams: Aeration, Energy Dissipation and Turbulence on Uniform, Non-Uniform and Pooled Stepped Chutes. Ph.D. Thesis, University of Queensland, Brisbane, Australia, 2013.
8. Felder, S.; Chanson, H. Energy dissipation down a stepped spillway with non-uniform step heights. *J. Hydraul. Eng.* **2011**, *137*, 1543–1548. [CrossRef]
9. Felder, S.; Chanson, H. Aeration, flow instabilities, and residual energy on pooled stepped spillways of embankment dams. *J. Irrig. Drain. Eng.* **2013**, *139*, 880–887. [CrossRef]
10. Zhang, J.M.; Chen, J.G.; Wang, Y.R. Experimental study on time-averaged pressures in stepped spillway. *J. Hydraul. Res.* **2012**, *50*, 236–240. [CrossRef]
11. Zhang, G.; Chanson, H. Effects of step and cavity shapes on aeration and energy dissipation performances of stepped chutes. *J. Hydraul. Eng.* **2018**, *144*, 04018060. [CrossRef]
12. Peng, Y.; Zhou, J.G.; Burrows, R. Modelling the free surface flow in rectangular shallow basins by lattice Boltzmann method. *J. Hydraul. Eng.* **2011**, *137*, 1680–1685. [CrossRef]

13. Peng, Y.; Zhou, J.G.; Burrows, R. Modelling solute transport in shallow water with the lattice Boltzmann method. *Comput. Fluids* **2011**, *50*, 181–188. [CrossRef]

14. Peng, Y.; Mao, Y.F.; Wang, B.; Xie, B. Study on C-S and P-R EOS in pseudo-potential lattice Boltzmann model for two-phase flows. *Int. J. Mod. Phys. C* **2017**, *28*, 1750120. [CrossRef]

15. Peng, Y.; Wang, B.; Mao, Y.F. Study on force schemes in pseudopotential lattice Boltzmann model for two-phase flows. *Math. Probl. Eng.* **2018**, 6496379. [CrossRef]

16. Baylar, A.; Unsal, M.; Ozkan, F. The effect of flow patterns and energy dissipation over stepped chutes on aeration efficiency. *KSCE J. Civ. Eng.* **2011**, *15*, 1329–1334. [CrossRef]

17. Dong; Zhi-yong; Joseph; Hun-wei. Numerical simulation of skimming flow over mild stepped channel. *J. Hydrodyn.* **2006**, *18*, 367–371. [CrossRef]

18. Zhenwei, M.U.; Zhang, Z.; Zhao, T. Numerical simulation of 3-D flow field of spillway based on VOF method. *Procedia Eng.* **2012**, *28*, 808–812. [CrossRef]

19. Tabbara, M.; Chatila, J.; Awwad, R. Computational simulation of flow over stepped spillways. *Comput. Struct.* **2005**, *83*, 2215–2224. [CrossRef]

20. André, S. High Velocity Aerated Flows on Stepped Chutes with Macro-Roughness Elements. Ph.D. Thesis, EPFL, Lausanne, Switzerland, 2004.

21. Kökpinar, M.A. Flow over a stepped chute with and without macro-roughness elements. *Can. J. Civ. Eng.* **2004**, *31*, 880–891. [CrossRef]

22. Thorwarth, J. Hydraulics of Pooled Stepped Spillways—Self-Induced Unsteady Flow and Energy Dissipation. Ph.D. Thesis, University of Aachen, Aachen, Germany, 2008.

23. Hirt, C.W.; Nichols, B.D. Volume of fluid (VOF) method for the dynamics of free boundaries. *J. Comput. Phys.* **1981**, *39*, 201–225. [CrossRef]

24. Li, X.; Wang, Y.R.; Zhang, J.M. Numerical simulation of an offset jet in bounded pool with deflection wall. *Math. Probl. Eng.* **2017**, *2017*, 1–11. [CrossRef]

25. Zhang, J.M.; Chen, J.G.; Xu, W.L.; Wang, Y.R. Three-dimensional numerical simulation of aerated flows downstream sudden fall aerator expansion-in a tunnel. *J. Hydrodyn.* **2011**, *23*, 71–80. [CrossRef]

26. Chen, Q.; Dai, G.; Liu, H. Volume of Fluid Model for Turbulence Numerical Simulation of Stepped Spillway Overflow. *J. Hydraul. Eng.* **2002**, *128*, 68–688. [CrossRef]

27. Yakhot, V.; Orszag, S.A. Renormalization group analysis of turbulence. I. Basic theory. *J. Sci. Comput.* **1986**, *1*, 3–51. [CrossRef]

28. Bai, Z.L.; Zhang, J.M. Comparison of different turbulence models for numerical simulation of pressure distribution in V-shaped stepped spillway. *Math. Probl. Eng.* **2017**, *2017*, 1–9. [CrossRef]

29. Morovati, K.; Eghbalzadeh, A.; Javan, M. Numerical investigation of the configuration of the pools on the flow pattern passing over pooled stepped spillway in skimming flow regime. *Acta Mech.* **2015**, *227*, 1–14. [CrossRef]

30. Carvalho, R.F.; Rui, M. Stepped spillway with hydraulic jumps: Application of a numerical model to a scale model of a conceptual prototype. *J. Hydraul. Eng.* **2014**, *135*, 615–619. [CrossRef]

31. Celik, I.B.; Ghia, U.; Roache, P.J.; Freitas, C.J. Procedure of estimation and reporting of uncertainty due to discretization in CFD applications. *J. Fluids Eng.* **2008**, *130*, 078001–078004. [CrossRef]

32. Felder, S.; Guenther, P.; Hubert, C. *Air-Water Flow Properties and Energy Dissipation on Stepped Spillways: A Physical Study of Several Pooled Stepped Configurations*; The University of Queensland: Brisbane, Australia, 2012; ISBN 9781742720555.

33. Hunt, J.C.R.; Wray, A.A.; Moin, P. Eddies, streams and convergence zones in turbulent flows. In *Proceedings of the Summer Program 1988*; Stanford University: Stanford, CA, USA; 1988.

Article

Assessing the Analytical Solution of One-Dimensional Gravity Wave Model Equations Using Dam-Break Experimental Measurements

Wenjun Liu, Bo Wang *, Yunliang Chen, Chao Wu and Xin Liu

State Key Laboratory of Hydraulics and Mountain River Engineering, Sichuan University, Chengdu 610065, China; Liuwenjun_SCU@163.com (W.L.); liangyunchen@163.com (Y.C.); wuchao_w@163.com (C.W.); Liuxin_SCU@163.com (X.L.)
* Correspondence: wangbo@scu.edu.cn; Tel.: +86-028-8540-1144

Received: 19 August 2018; Accepted: 13 September 2018; Published: 15 September 2018

Abstract: The one-dimensional gravity wave model (GWM) is the result of ignoring the convection term in the Saint-Venant Equations (SVEs), and has the characteristics of fast numerical calculation and low stability requirements. To study its performances and limitations in 1D dam-break flood, this paper verifies the model using a dam-break experiment. The experiment was carried out in a large-scale flume with depth ratios (initial downstream water depth divided by upstream water depth) divided into 0 and 0.1~0.4. The data were collected by image processing technology, and the hydraulic parameters, such as water depth, flow discharge, and wave velocity, were selected for comparison. The experimental results show that the 1D GWM performs an area with constant hydraulic parameters, which is quite different from the experimental results in the dry downstream case. For a depth ratio of 0.1, the second weak discontinuity point, which is connected to the steady zone in the 1D GWM, moves upstream, which is contrary to the experimental situation. For depth ratios of 0.2~0.4, the moving velocity of the second weak discontinuity point is faster than the experimental value, while the velocity of the shock wave is slower. However, as the water depth ratio increases, the hydraulic parameters calculated by 1D GWM in the steady zone gradually approach the experimental value.

Keywords: gravity wave model; dam-break flood; image processing technology; experimental study

---

1. Introduction

The dam-breaking flood problem has been of wide concern in academic and engineering circles due to its importance and complexity; once the dam fails, it will cause huge numbers of casualties and property losses [1–3]. Ritter derived a theoretical solution of the dam-breaking flood problem in a rectangular prismatic channel that has a flat bottom and no resistance with a dry bed downstream [4]. Based on Ritter's solution, Stoker derived the theoretical solution of instantaneous dam failure in the rectangular prismatic channel for stationary water downstream [5]. Considering that actual rivers often have a certain slope, many scholars have tried to add this factor into the analytical model. Hunt derived an approximate solution for the dam failure of a rectangular prismatic channel in the case of a declining bottom [6]. In addition, actual river channels are mostly irregular, and their cross-sections cannot be simplified into rectangles. Wang et al. derived an analytical solution for the collapse of a prismatic channel with an arbitrary cross-sectional shape under a tilted channel [7]. Due to the occurrence of undular bore waves in nature, some physical models have been proposed based on a nonhydrostatic pressure assumption in dam break flows [8–10]. Since the current analytical model is still quite simplified compared to actual situations, it is necessary to conduct much experimental research. Stansby et al. systematically studied the initial shape of 1D dam failure. The analysis shows that at the initial stage of dam break, a mushroom-like jet occurs downstream [11]. Aleixo et al. used particle tracking velocimetry (PTV) to measure the cross-sectional velocity of the 1D dam-breaking flow and obtained a more accurate distribution of the cross-sectional flow velocity [12]. LaRocque et al. used an ultrasonic velocity profiler (UVP) to study the vertical velocity distribution in a rectangular flume and compared the different turbulence models to simulate the dam flow, and the large eddy simulation (LES) performed better than other models [13]. Many dam break experiments have also been carried out in the case of a moving bed to further approach the actual situation [14–16]. Recently, image processing technology has been used to experimentally study dam-break flood movements [17–21]. Because the equipment is not immersed, the measurement does not interfere with the flow field, and the whole measuring process is easy to program; furthermore, the cost is relatively low compared with that of high-precision equipment such as particle image velocimetry (PIV) or PTV, and its application prospects are quite extensive [22].

Certain problems, such as a long calculation time and unstable or divergent calculations, often exist in solving shallow water equations (SWEs). To solve these problems, on the one hand, many scholars use some excellent numerical schemes or commercial software to simulate them [23–25], and the lattice boltzmann model (LBM) performs very well for dealing with SWEs under many different boundary conditions [26–34]; smoothed particle hydrodynamics (SPH) can also give sufficient simulation accuracy in urban flood and 2D dam break flood problems [35–37], but on the other hand, when some items in the equation are negligible compared to others in some actual physical phenomena, many researchers have also tried to simplify the equations [38–43]. Among the simplified models, zero-inertia (or diffusive wave) models seem to need more computing time for flood simulation, and their reliability is significantly affected by complex topographic information; they give poor predictions of events around buildings, especially in urban districts [38,40,41,43]. Local inertial approximation (or gravity wave model) has good agreement with full-dynamic models when $Fr < 0.5$, leads to milder water depth gradients in steady flow, and has slower flood propagation speeds in unsteady flow. However, it requires less computing time, and is a faster alternative for gradually varied subcritical flows where domain-average friction typically exceeds $n = 0.03$ [39,40,42]. Most previous methods for studying these simplified models include numerical analysis; Martins et al. derived a simplified analytical solution for the 1D gravity wave model (GWM) and compared it with the analytical solution of the Saint-Venant Equation (SVE) in dam break flow [44]. After using the MacCormack and Roe scheme to simulate the simplified equations [45], the author found that the 1D GWM was similar to the analytical solution of the SVE when there was a subcritical flow downstream and the shock wave velocity was relatively small, whereas the difference was large when there was no water downstream. The numerical simulation does not need special initial conditions to remain steady and has good agreement with the derived simplified analytical solution.

However, the work of Martins did not analyze the relationship between the 1D GWM and the actual dam-breaking flow in detail, and the only two cases (dry and wet bed downstream) selected for comparison with SVE seemed to be insufficient. The ability of the 1D GWM to capture the 1D dam break flow characteristics and whether it can be used as a simplified alternative model outside the SVE need to be verified by experiment. This study carried out a large number of dam-break experiments in a large-scale water flume and used a non-immersion image processing technology to collect experimental data. Depth ratios (initial downstream water depth divided by upstream water depth) of 0 and 0.1~0.4 were considered for an upstream head of 0.6 m. Hydraulic parameters, such as water depth, flow discharge, average flow velocity, Froude number, and wave velocity, were measured or calculated. Through detailed comparison of the experimental data and the analytical solution of the 1D GWM, the ability of the 1D GWM to describe the characteristics of 1D dam-break flood movement was analyzed, and a valuable reference for using this model to calculate 1D dam-break flows in the future was created.

2. Materials and Methods

2.1. Flume and Other Instruments

The flume was made of tempered glass supported by a steel frame. The upstream was 8.37 m long, and the downstream 9.63 m long. The width was 1 m, and the height was 1.09 m. The flume had a rectangular cross section and a flat bottom. The gate material was a 15-mm-thick fiberglass board, which was lifted instantly by the electromagnetic brake asynchronous motor. This method was used to simulate the instantaneous break of the dam. The lifting time was 0.27 s when the upstream water depth was 0.6 m, which satisfies the following transient failure condition [46]: $t < (2h_u/g)^{1/2} = 0.3499$ s (h_u is the upstream water depth, and g is the gravity acceleration, 9.8 m/s^2). Eight CCD (charge coupled device) cameras were arranged in parallel at the side of the flume at a distance of 1.4 m to capture the evolution of the water flow. To ensure the quality of the video and make it more convenient to use the image processing technology, the background was covered with a smooth white car sticker, and the water was dyed in the experiment. To avoid noise from the reflection of the flume glass during image processing, a light-colored curtain was arranged on both sides of the flume to eliminate the shadow caused by natural light. In the experiment, 8 CCD cameras were simultaneously turned on by a wireless remote control. A wide-angle lens with a resolution of 1440 × 1920 was used for shooting, and the shooting speed was 48 frames per second. The device schematic is shown in Figure 1.

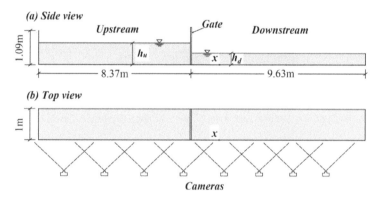

Figure 1. Schematic of the experimental device: (**a**) side view; (**b**) top view

2.2. Data Acquisition Technology and Experimental Repeatability Verification

The process of experimental data acquisition is divided into two main steps: the first step is to obtain the external and internal parameters of each camera by using the calibration plate and the calibration toolbox in MATLAB (MATLAB2016a, Mathworks company, Natick, MA, USA) and then to use the data and the corresponding calibration function to re-project and obtain the undistorted image. The second step is to analyze the image based on the undistorted image. In addition, the calibrated picture is then cropped and merged using the ruler above the flume. After binarizing the image, the water depth at each location can be obtained by boundary recognition functions. The calibration plate is shown in Figure 2a. The size of the square lattice in the plate is 100 mm × 100 mm. The number of grids is 11 on one side and 10 on the other. A margin of 50 mm is left around the calibration plate for the calibration toolbox to recognize the corner points so that the internal and external parameters of each camera can be accurately obtained to better correct the distortion. A set of 25 pictures in different orientations and angles is used to calibrate each camera, and the pictures before and after calibration are shown in Figure 2b,c. The actual length represented by one pixel is approximately 1.8 mm, and the depth of the part blocked by the flume steel column is obtained by interpolation. The picture after clipping and binarization is shown in Figure 3.

Figure 2. Picture calibration: (**a**) calibration board; (**b**) original image; (**c**) calibrated image.

Figure 3. Picture binarization. (**a**) original image; (**b**) binarized image.

The depth ratio is defined as $\alpha = h_d/h_u$, where h_u and h_d are the upstream and downstream initial water depths, respectively. The experimental conditions are shown in Table 1:

Table 1. Experimental conditions.

h_u (m)	h_d (m)				
	$\alpha = 0$	$\alpha = 0.1$	$\alpha = 0.2$	$\alpha = 0.3$	$\alpha = 0.4$
0.6	0	0.06	0.12	0.18	0.24

To ensure the stability and accuracy of the experimental data collection, the repeatability of the experiment is verified here by repeating the conditions of $h_u = 0.6$ m and $h_d = 0$ m and comparing the depth data at positions of $x = 0.3$ m (near the dam) and $x = 4.5$ m (far from the dam). The results are shown in Figure 4, where *Run 1* and *Run 2* represent the first and second experiments for the conditions of $h_u = 0.6$ m and $h_d = 0$ m, respectively. It can be seen that the overlap of the data between the two groups is quite good, and R^2 is close to 1, which proves that the experiment has good repeatability.

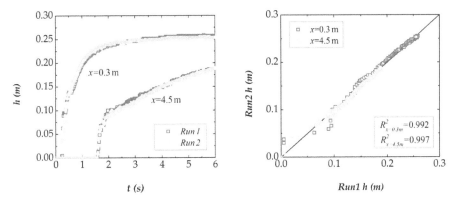

Figure 4. Experimental repeatability verification.

3. Results

3.1. Dry Bed Downstream Condition

3.1.1. Water Depth, Flow Discharge, Average Velocity and *Fr*

Here, the hydraulic parameters involved are dimensionless in the following formulas:

$$X = \frac{x}{t\sqrt{gh_u}} \tag{1}$$

$$T = \frac{t}{\sqrt{h_u/g}} \tag{2}$$

$$H = \frac{h}{h_u} \tag{3}$$

$$U = \frac{u}{\sqrt{gh_u}} \tag{4}$$

where h, u are water depth and flow velocity; X, T, H, and U are the dimensionless position, time, water depth, and flow velocity, respectively,.

The method for calculating flow discharge in the experiment is similar to those of Bento et al. [47] and Cestero et al. [48]; that is, calculate the water body area of the façade by the water depth data at each position for a certain time, because it is a one-dimensional case, then multiply the width of the flume to obtain the volume of the cross section at a certain moment. The following formula is used to calculate the flow discharge:

$$Q_{(t)} = \frac{V_{(t+\Delta t)} - V_{(t)}}{\Delta t} \tag{5}$$

where $Q_{(t)}$ is the flow discharge for time t, $V_{(t)}$ is the water volume between a cross section and the end of the upstream reservoir at time t, $V_{(t+\Delta t)}$ is the water volume between a cross section and the end of the upstream reservoir at time $t + \Delta t$, and Δt is the time difference of $1/48$ s.

For convenience of later description, the analytical solution of 1D GWM is simply explanation here. Martins divided the structure of dam break into four areas, as shown in Figure 5, where P_{1u} and P_{12} represent the first and second weak discontinuity point, respectively; P_{2d} represents the strong discontinuity point; ζ represents the downstream wave-front evolution velocity; h_1 and h_2 represent the water depth in the rarefaction wave zone and the steady zone, respectively; C_u and C_2 represent the velocity of the first and second weak discontinuity point, respectively.

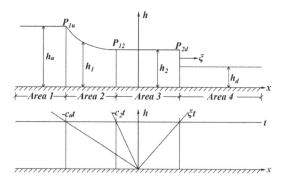

Figure 5. Dam break flood evolution model.

Using the method of characteristics (MOC) and performing some formula operations, the analytical formulas for the 1D GWM in four areas are shown in Table 2, where C_1 and C_d are celerity-velocity in Area2 and Area4, respectively; U_1 and U_2 are average flow velocity in Area2 and Area3, respectively.

Table 2. Analytical formulas for 1D GWM in four areas.

Variable	Area 1	Area 2	Area 3	Area 4
h	h_u	$h_1 = x^2/gt^2$	$h_2 = c_2^2/g$	h_d
c	$c_u = \sqrt{gh_u}$	$c_1 = -x/t$	$c_2^3 = c_u^3 + 3/2(c_d^2 - c_2^2)\sqrt{(c_d^2 + c_2^2)/2}$	$c_d = \sqrt{gh_d}$
u	0	$u_1 = 2(c_1^3 t^3 + x^3)/3x^2 t$	$u_2 = 2/3(c_u^3/c_2^2 - c_2)$	0
ζ	-	-	$\zeta = \sqrt{(c_d^2 + c_2^2)/2}$	-
P	$-c_u t$	$-c_2 t$	ζt	-

Under the dry bed downstream condition, the hydraulic parameters calculated by the 1D GWM are very different from those of the experiment, as shown in Figure 6, where Exp. and Fr represent the experimental results and Froude number. According to the analytical solution of Martins, the water depth of area 3 still exists when there is no water in the downstream, which is quite different from the actual situation. The 1D GWM cannot accurately describe the evolution of 1D dam-break flow when there is no water downstream.

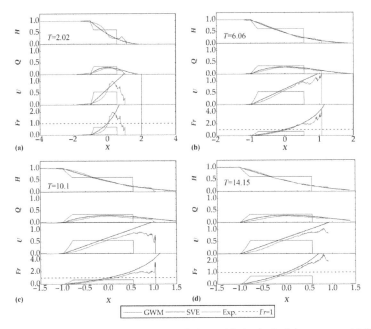

Figure 6. Water depth, flow discharge, average velocity and *Fr* for dry bed downstream: (**a**) *T* = 2.02; (**b**) *T* = 6.06; (**c**) *T* = 10.1; (**d**) *T* = 14.15.

3.1.2. Upstream Negative Wave and Downstream Wave-Front Velocities for Dry Bed Downstream

The calculation results of the upstream negative wave and the downstream wave-front velocities are shown in Figure 7, where U_n represents the dimensionless upstream negative wave evolution velocity and ζ_n represents the dimensionless downstream wave-front evolution velocity. Here, the relative root mean square error (*RRMSE*) is used to compare the error between the calculated value and the experimental value. The formula is as follows:

$$RRMSE = \sqrt{\frac{1}{n}\sum_{i=1}^{n}\left(\frac{y_i - \mu_i}{\mu_i}\right)^2} \times 100\% \tag{6}$$

where *RRMSE* represents the relative root mean square error, n is the total number of all measured data points, y_i is the model calculated value, and μ_i is the measured value.

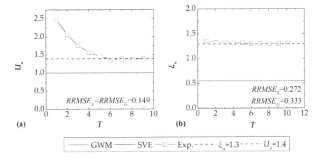

Figure 7. Velocities of (**a**) the upstream negative wave and (**b**) the downstream wave-front.

The result is shown in Figure 7, where RRMSE$_S$ and RRMSE$_G$ represent the relative root mean square error for SVE and 1D GWM, respectively. In the dry bed downstream case, the upstream negative wave velocities calculated by the 1D GWM and SVE are the same. The negative wave velocity gradually decreases and then stabilizes at approximately 1.4 in the experiment, Lauber and Hager explained that it is caused by excessive curvature of the streamline [46]. While the evolution velocity of the downstream wave-front basically remains stable, the dimensionless velocity is approximately 1.3 due to the resistance, which is smaller than the value of 2.0 from the SVE and much larger than the value of 0.55 calculated by the 1D GWM. The upstream negative wave and downstream wave-front speeds obtained by Leal et al. are all average values [49], and the characteristics of their propagation speed are not obtained over time. However, from the results of the stable speed, we can see that they are basically consistent with the results of Leal et al. [49].

3.2. Wet Bed Downstream Condition

3.2.1. Water Depth

Different degrees of water jump will occur downstream for the initial dam break ($T = 2.02$). It can be seen from Figure 8a that neither the SVE nor the 1D GWM describes this phenomenon well. Stansby et al. also observed that a mushroom-like jet occurs downstream of the initial dam break in the case of a wet bed downstream, and found that the movement form is quite complicated, but after the wave is broken, the motion of the water flow has good agreement with the analytical solution of the shallow water equations [11]. It can be seen from Figure 8 that the conclusion is still valid when the water depth ratio is between 0.1 and 0.4.

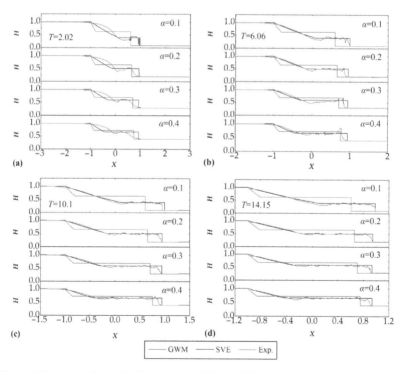

Figure 8. Water depth for wet bed downstream: (**a**) $T = 2.02$; (**b**) $T = 6.06$; (**c**) $T = 10.1$; (**d**) $T = 14.15$.

When $\alpha = 0.1$, the 1D GWM shows that the second weak discontinuity point moves upstream, but the experimental results are just the opposite. The evolution of water flow is much closer to that of the SVE. As Martins [44] stated, the 1D GWM shows only one subcritical flow state. When $\alpha = 0.2\sim0.4$, the second weak discontinuity point moves upstream in the experiment, but the range of the rarefaction wave zone calculated by the 1D GWM is shorter than that of the experiment. The speed at which the second weak discontinuity point moves upstream is much faster than the experimental value; the speed is 498%, 180%, and 94.7% faster than the experimental value for $\alpha = 0.2$, 0.3, and 0.4, respectively. In addition, the water depth of the steady zone calculated by the 1D GWM is higher than the experimental value, but it can be seen that as the water depth ratio increases, the water depth gradually approaches the experimental value.

3.2.2. Flow Discharge

In the case of wet bed downstream, the downstream water surface is distorted in the initial stage of dam break. The leaping water body will spin and aerate, and there will also be holes in the water body when it evolves downstream. Thus, the shape of the water surface line is complicated and not easy to capture. The flow discharge obtained by the above method cannot accurately describe the actual change, so the calculation of flow discharge in the case of wet bed downstream starts from $T = 6.06$. The calculation result is shown in Figure 9. It can be seen that due to the shortening of the rarefaction wave zone calculated by the 1D GWM, there is a significant deviation from the experimental value throughout the evolution period. The flow discharge in the steady zone is larger than the experimental value, but it can be seen that with the water depth ratio increased, the flow discharge value calculated by the 1D GWM in the steady zone also gradually approaches the experimental value.

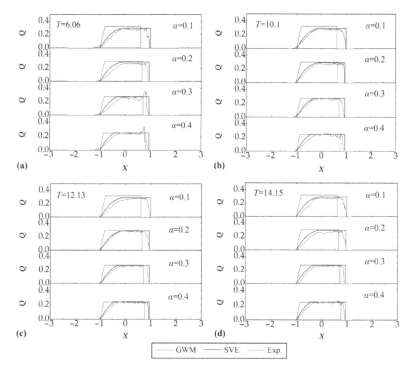

Figure 9. Flow discharge for wet bed downstream: (**a**) $T = 6.06$; (**b**) $T = 10.1$; (**c**) $T = 12.13$; (**d**) $T = 14.15$.

3.2.3. Average Velocity and *Fr*

The calculation results of the average velocity are shown in Figure 10. Similar to the flow discharge comparison, the average velocity calculated by the 1D GWM in the rarefaction wave zone is very different from the experimental value. The average velocity is smaller than the experimental value in the steady zone, especially when the water depth ratio is small, but it can be seen that as the water depth ratio increases, the average velocity gradually approaches the experimental value in the steady zone.

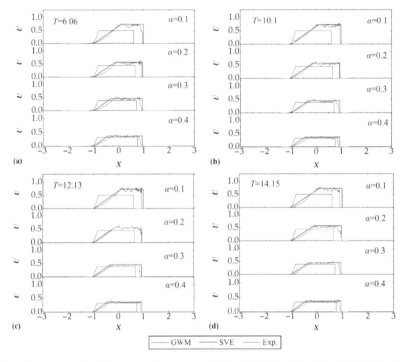

Figure 10. Average velocity for wet bed downstream: (a) $T = 6.06$; (b) $T = 10.1$; (c) $T = 12.13$; (d) $T = 14.15$.

The calculation results for *Fr* are shown in Figure 11. As described above, when $\alpha = 0.1$, the *Fr* value calculated by the 1D GWM is less than 1 in the entire flow region, and the *Fr* of the actual flow is greater than 1 in the steady zone. At this time, the second weak discontinuity point will not propagate upstream, and the description of the 1D GWM is flawed. When $\alpha = 0.2 \sim 0.4$, *Fr* is less than 1 in the whole flow region, and the flow is in a subcritical flow state. The result of the 1D GWM calculation still differs greatly in the rarefaction wave zone, and the *Fr* calculated in the steady zone is smaller than the experimental value. However, it also has the characteristic that its value gradually approaches the experimental value as the water depth ratio increases.

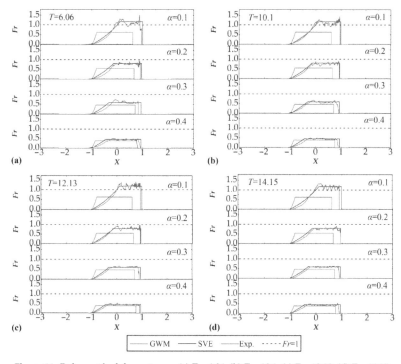

Figure 11. *Fr* for wet bed downstream: (**a**) *T* = 6.06; (**b**) *T* = 10.1; (**c**) *T* = 12.13; (**d**) *T* = 14.15.

3.2.4. Relative Error of Hydraulic Parameters in Steady Zone

In the previous sections, it was found that although the 1D GWM has large defects in the rarefaction wave zone, the results are much closer to the experimental value in the steady zone when the water depth ratio is relatively large. Here, the error is calculated by the following formula:

$$\delta = |y - \mu| / \mu \tag{7}$$

where y is the value calculated by the 1D GWM or SVE and μ is the experimental measurement value.

The calculation result is shown in Figure 12, where δ_H, δ_U, δ_{Fr}, and δ_Q represent the errors in the water depth, average velocity, *Fr* and flow discharge in the steady zone, respectively. It can be seen from Figure 12 that with increasing water depth ratio, the error between these hydraulic parameters and the experimental values tend to gradually decrease. When $\alpha = 0.4$, the errors for water depth, average velocity, *Fr* and flow discharge are 7.91%, 6.32%, 8.36%, and 3.94%, respectively.

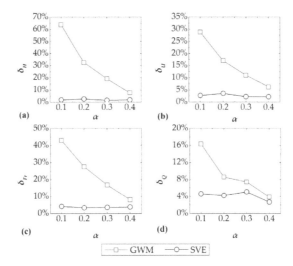

Figure 12. Error in the steady zone for different water depth ratios: (**a**) water depth; (**b**) average velocity; (**c**) *Fr*; (**d**) flow discharge.

3.2.5. Upstream Negative Wave and Downstream Wave-Front Velocities for Wet Bed Downstream

The calculation results of the velocity for upstream negative wave and downstream wave-front velocities for α = 0.1~0.4 are shown in Figure 13. It can be seen that the negative wave velocities calculated by the SVE and 1D GWM are the same for the wet bed downstream case, but the negative wave velocity is larger than those of the SVE or 1D GWM at the initial time point in the experiment. After the flow has evolved for a period of time, the velocity gradually decreases, and the error gradually decreases as the water depth ratio increases. The steady velocity is relatively stable in the experiment, and it is consistent with the SVE, the shock wave velocity is small for the 1D GWM. However, it should be noted that the error of shock wave velocity between the 1D GWM and the experimental value tends to gradually decrease as the water depth ratio increases.

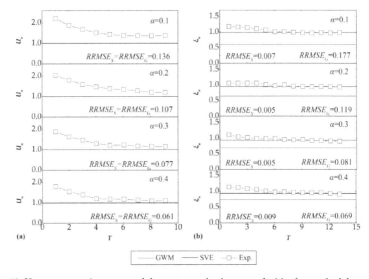

Figure 13. Upstream negative wave and downstream shock wave velocities for wet bed downstream.

4. Discussion

For the study of GWM, past researchers have mostly used different numerical calculation methods to solve the simplified governing equations directly [39,40,42]. Their work proves that the GWM requires less computing time and does not need to pay much effort to maintaining computational stability. It has certain prediction accuracy when the surface friction is not small ($n > 0.03$) in subcritical flows. In view of the above advantages of GWM, Martins tried to apply it to the 1D dam break flow and derived the analytical solution of 1D GWM under simplified conditions. However, the performances and limitations of analytical solution need to be further confirmed by experimental research [50,51]. This paper analyzes the analytical solution of 1D GWM by means of experimental research. Unlike the numerical model, which can arbitrarily control the boundary conditions and initial state to analyze the characteristics of the governing equations, both analytical and experimental research is performed under ideal boundary conditions here. 1D dam-breaking flow is a kind of unsteady flow in open channel, which has special evolution characteristics and need targeted analysis. In the experiment, we have observed that 1D GMW has a great difference with experimental conditions in the case of dry bed downstream, the calculated rarefaction wave zone is significantly shorter than the experimental one when the depth ratio is 0.1~0.4, and the wave-front evolution speed is slower. When the depth ratio is 0.4, the value calculated in the steady zone is close to the experimental value. It can be seen that the application of the 1D GWM to 1D dam break flow should be careful and significantly consider the influence of depth ratio.

Of course, unlike the previous numerical model studies, the experiment cannot capture the characteristics of dam break flood for very far from the dam (De Almeida and Bates 5000 m, Bates et al. 5000 m, Neal et al. 10,000 m). In this article, we compared the experimental result with the analytical solution of 1D GWM in a basic way. Future work could consider changing the boundary conditions for 1D GWM, such as the analytical solution of 1D GWM, considering the river bed slope or the shape of the cross-section (triangular, parabolic, etc.), and the corresponding experimental research could be carried out simultaneously. Because the frictional resistance has a great impact on GWM [39,40], considering different friction coefficients for the river bed material in the experiment is also a research direction.

5. Conclusions

This paper compares the data obtained by image processing technology with one-dimensional GWM in dam-break flood. Because the current test results are obtained under very simplified conditions, and for more complicated cases, further experiments or numerical simulation studies are needed in future, but at present we can draw the following conclusions:

(1) The calculated result of the 1D GWM is very different from the experimental value for dry bed downstream. There is still a region with the same hydraulic parameters value downstream that does not appear in the experiment. The 1D GWM gives a poor description for the dam-break flow movement in the dry bed downstream case.

(2) When $\alpha = 0.1$, the second weak discontinuity point calculated by the 1D GWM is continuously developed upstream, which is contrary to the experimental results. When $\alpha = 0.2$~0.4, the rarefaction wave zone of the 1D GWM is significantly shorter than that of the experiment, and the shock wave evolution distance is smaller than the experimental value.

(3) When $\alpha = 0.1$~0.4, the relative errors in the water depth, average velocity, Fr and flow discharge in the steady zone gradually decrease as the water depth ratio increases. When $\alpha = 0.4$, the errors are 7.91%, 6.32%, 8.36%, and 3.94%, respectively. The upstream negative wave speeds for the 1D GWM and SVE are the same, but both are smaller than the experimental value. The downstream shock speed for the 1D GWM is smaller than the experimental value. However, as the water depth ratio increases, the errors in the upstream negative wave and downstream shock wave velocities are also reduced. When $\alpha = 0.4$, the *RRMSE* values are 0.061 and 0.069, respectively.

Author Contributions: Conceptualization, B.W. and C.W.; Methodology, W.L., B.W. and X.L.; Software, W.L. and X.L.; Validation, B.W. and Y.C.; Formal Analysis, B.W.; Investigation, W.L. and B.W.; Resources, Y.C.; Data Curation, W.L.; Writing-Original Draft Preparation, W.L. and B.W.; Writing-Review & Editing, W.L. and B.W.; Visualization, W.L. and X.L.; Supervision, B.W.; Project Administration, B.W.; Funding Acquisition, B.W.

Funding: This research was funded by the National Key Research and Development Program of China (Grant No: 2016YFC0401707), the financial support of National Natural Science Foundation of China (No: 51879179) and the Sichuan Provincial Youth Science and Technology Innovation Research Team Special Funding Project (No.2016TD0020).

Conflicts of Interest: The authors declare no conflict of interest.

References

1. Haltas, I.; Tayfur, G.; Elci, S. Two-dimensional numerical modeling of flood wave propagation in an urban area due to Ürkmez dam-break, İzmir, Turkey. *Nat. Hazards* **2016**, *81*, 2103–2119. [CrossRef]

2. Kim, B.; Sanders, B.F. Dam-Break Flood Model Uncertainty Assessment: Case Study of Extreme Flooding with Multiple Dam Failures in Gangneung, South Korea. *J. Hydraul. Eng.* **2016**, *142*, 05016002. [CrossRef]

3. Bosa, S. A Numerical Model of the Wave that Overtopped the Vajont Dam in 1963. *Water Resour. Manag.* **2013**, *27*, 1763–1779. [CrossRef]

4. Ritter, A. The propagation of water waves. *Ver Deutsh Ing. Z.* **1892**, *36*, 947–954.

5. Stoker, J.J. *Water Waves*; Interscience Publishers, Inc.: New York, NY, USA, 1957.

6. Hunt, B. Asymptotic solution for dam-break on sloping channel. *J. Hydraul. Eng.* **1983**, *109*, 1698–1706. [CrossRef]

7. Wang, B.; Chen, Y.; Wu, C.; Peng, Y.; Ma, X.; Song, J.J. Analytical solution of dam-break flood wave propagation in a dry sloped channel with an irregular-shaped cross-section. *J. Hydro-Environ. Res.* **2016**, *14*, 93–104. [CrossRef]

8. Frazao, S.S.; Zech, Y. Undular bores and secondary waves -Experiments and hybrid finite-volume modelling. *J. Hydraul. Res.* **2003**, *40*, 33–43. [CrossRef]

9. Kim, D.H.; Lynett, P.J. Dispersive and Nonhydrostatic Pressure Effects at the Front of Surge. *J. Hydraul. Eng.* **2011**, *137*, 754–765. [CrossRef]

10. Cantero-Chinchilla, F.N.; Castro-Orgaz, O.; Dey, S.; Ayuso, J.L. Nonhydrostatic dam break flows. I: Physical equations and numerical schemes. *J. Hydraul. Eng.* **2016**, *142*, 04016068. [CrossRef]

11. Stansby, P.K.; Tcd, B.; Chegini, A. The initial stages of dam-break flow. *J. Fluid Mech.* **1998**, *374*, 407–424. [CrossRef]

12. Aleixo, R.; Soares-Frazão, S.; Zech, Y. Velocity-field measurements in a dam-break flow using a PTV Voronoï imaging technique. *Exp. Fluids* **2011**, *50*, 1633–1649. [CrossRef]

13. Larocque, L.A.; Imran, J.; Chaudhry, M.H. Experimental and numerical investigations of two-dimensional dam-break flows. *J. Hydraul. Eng.* **2013**, *139*, 569–579. [CrossRef]

14. Qian, H.; Cao, Z.; Liu, H.; Pender, G. New experimental dataset for partial dam-break floods over mobile beds. *J. Hydraul. Res.* **2017**, *56*, 1–12. [CrossRef]

15. Soares-Frazão, S.; Canelas, R.; Cao, Z.; Cea, L.; Chaudhry, H.M.; Die Moran, A.; El Kadi, K.; Ferreira, R.; Cadórniga, I.F.; Gonzalez-Ramirez, N.; et al. Dam-break flows over mobile beds: Experiments and benchmark tests for numerical models. *J. Hydraul. Res.* **2012**, *50*, 364–375. [CrossRef]

16. Goutiere, L.; Soares-Frazão, S.; Zech, Y. Dam-break flow on mobile bed in abruptly widening channel: Experimental data. *J. Hydraul. Res.* **2011**, *49*, 367–371. [CrossRef]

17. Kocaman, S.; Ozmen-Cagatay, H. The effect of lateral channel contraction on dam-break flows: Laboratory experiment. *J. Hydrol.* **2012**, *432–433*, 145–153. [CrossRef]

18. Kocaman, S.; Ozmen-Cagatay, H. Investigation of dam-break induced shock waves impact on a vertical wall. *J. Hydrol.* **2015**, *525*, 1–12. [CrossRef]

19. Ozmen-Cagatay, H.; Kocaman, S.; Guzel, H. Investigation of dam-break flood waves in a dry channel with a hump. *J. Hydro-Environ. Res.* **2014**, *8*, 304–315. [CrossRef]

20. Soares-Frazão, S. Experiments of dam-break wave over a triangular bottom sill. *J. Hydraul. Res.* **2007**, *45*, 19–26. [CrossRef]

21. Aureli, F.; Maranzoni, A.; Mignosa, P.; Ziveri, C. An image processing technique for measuring free surface of dam-break flows. *Exp. Fluids* **2011**, *50*, 665–675. [CrossRef]

22. Eaket, J.; Hicks, F.E. Use of stereoscopy for dam break flow measurement. *J. Hydraul. Eng.* **2005**, *131*, 24–29. [CrossRef]

23. Wu, G.; Yang, Z.; Zhang, K.; Dong, P.; Lin, Y.T. A non-equilibrium sediment transport model for dam break flow over moveable bed based on non-uniform rectangular mesh. *Water* **2018**, *10*, 616. [CrossRef]

24. Lu, C. Simulations of shallow water equations with finite difference lax-wendroff weighted essentially non-oscillatory Schemes. *J. Sci. Comput.* **2011**, *47*, 281–302. [CrossRef]

25. Macchione, F.; Costabile, P.; Costanzo, C.; Lorenzo, G.D.; Razdar, B. Dam breach modelling: Influence on downstream water levels and a proposal of a physically based module for flood propagation software. *J. Hydroinform.* **2016**, *18*. [CrossRef]

26. Peng, Y.; Zhou, J.G.; Burrows, R. Modelling the free surface flow in rectangular shallow basins by lattice boltzmann method. *J. Hydraul. Eng.* **2011**, *137*, 1680–1685. [CrossRef]

27. Peng, Y.; Zhou, J.G.; Zhang, J.M.; Liu, H.F. Lattice boltzmann modelling of shallow water flows over discontinuous beds. *Int. J. Numer. Method Fluids* **2014**, *75*, 608–619. [CrossRef]

28. Peng, Y.; Zhang, J.M.; Meng, J.P. Second order force scheme for lattice boltzmann model of shallow water flows. *J. Hydraul. Res.* **2017**, *55*, 592–597. [CrossRef]

29. Peng, Y.; Meng, J.P.; Zhang, J.M. Multispeed lattice boltzmann model with stream-collision scheme for transcritical shallow water flows. *Math. Probl. Eng.* **2017**, *2017*. [CrossRef]

30. Meng, J.P.; Gu, X.J.; Emerson, D.R.; Peng, Y.; Zhang, J.M. Discrete boltzmann model of shallow water equations with polynomial equilibria. *Int. J. Mod. Phys.* **2018**. [CrossRef]

31. Peng, Y.; Zhou, J.G.; Burrows, R. Modelling solute transport in shallow water with the lattice Boltzmann method. *Comput. Fluids* **2011**, *50*, 181–188. [CrossRef]

32. Peng, Y.; Zhou, J.G.; Zhang, J.M.; Burrows, R. Modeling moving boundary in shallow water by LBM. *Int. J. Mod. Phys. C* **2013**, *24*, 1–17. [CrossRef]

33. Peng, Y.; Zhou, J.G.; Zhang, J.M. Mixed numerical method for bed evolution. *Proc. Inst. Civil Eng. Water Manag.* **2015**, *168*, 3–15. [CrossRef]

34. Peng, Y.; Zhang, J.M.; Zhou, J.G. Lattice Boltzmann Model Using Two-Relaxation-Time for Shallow Water Equations. *J. Hydraul. Eng.* **2016**, *142*, 06015017. [CrossRef]

35. Ata, R.; Soulaïmani, A. A stabilized SPH method for inviscid shallow water flows. *Int. J. Numer. Methods Fluids* **2005**, *47*, 139–159. [CrossRef]

36. Amicarelli, A.; Albano, R.; Mirauda, D.; Agate, G.; Sole, A.; Guandalini, R. A smoothed particle hydrodynamics model for 3d solid body transport in free surface flows. *Comput. Fluids* **2015**, *116*, 205–228. [CrossRef]

37. Albano, R.; Sole, A.; Mirauda, D.; Adamowski, J. Modelling large floating bodies in urban area flash-floods via a smoothed particle hydrodynamics model. *J. Hydrol.* **2016**, *541*, 344–358. [CrossRef]

38. Moussa, R.; Bocquillon, C. On the use of the diffusive wave for modelling extreme flood events with overbank flow in the floodplain. *J. Hydrol.* **2009**, *374*, 116–135. [CrossRef]

39. Bates, P.D.; Horritt, M.S.; Fewtrell, T.J. A simple inertial formulation of the shallow water equations for efficient two-dimensional flood inundation modelling. *J. Hydrol.* **2010**, *387*, 33–45. [CrossRef]

40. Neal, J.; Villanueva, I.; Wright, N.; Fewtrell, T.; Bates, P. How much physical complexity is needed to model flood inundation? *Hydrol. Process.* **2012**, *26*, 2264–2282. [CrossRef]

41. Costabile, P.; Costanzo, C.; Macchione, F. Performances and limitations of the diffusive approximation of the 2-d shallow water equations for flood simulation in urban and rural areas. *Appl. Numer. Math.* **2016**. [CrossRef]

42. Almeida, G.A.M.D.; Bates, P. Applicability of the local inertial approximation of the shallow water equations to flood modeling. *Water Resour. Res.* **2013**, *49*, 4833–4844. [CrossRef]

43. Aricò, C.; Nasello, C. Comparative analyses between the zero-inertia and fully dynamic models of the shallow water equations for unsteady overland flow propagation. *Water* **2018**, *10*, 44. [CrossRef]

44. Martins, R.; Leandro, J.; Djordjević, S. Analytical solution of the classical dam-break problem for the gravity wave–model equations. *J. Hydraul. Eng.* **2016**, *142*, 06016003. [CrossRef]

45. Martins, R.; Leandro, J.; Djordjević, S. Analytical and numerical solutions of the local inertial equations. *Int. J. Non-Linear Mech.* **2016**, *81*, 222–229. [CrossRef]

46. Lauber, G.; Hager, W.H. Experiments to dambreak wave: Horizontal channel. *J. Hydraul. Res.* **1998**, *36*, 291–307. [CrossRef]

47. Bento, A.M.; Amaral, S.; Viseu, T.; Cardoso, R.; Rui, M.L.F. Direct estimate of the breach hydrograph of an overtopped earth dam. *J. Hydraul. Eng.* **2017**, *143*. [CrossRef]
48. Cestero, J.A.F.; Imran, J.; Chaudhry, M.H. Experimental investigation of the effects of soil properties on levee breach by overtopping. *J. Hydraul. Eng.* **2014**, *141*, 04014085. [CrossRef]
49. Leal, J.G.A.B.; Ferreira, R.L.; Cardoso, A.H. Dam-break waves on movable bed. In Proceedings of the Fluvial Hydraulics: River Flow 2002, Louvain-la-Neuve, Louvain-la-Neuve, Belgium, 4–6 September 2002; Bousmar, D., Zech, Y., Eds.; Balkema: Rotterdam, The Netherlands; pp. 553–563.
50. Bukreev, V.I.; Degtyarev, V.V.; Chebotnikov, A.V. Experimental verification of methods for calculating partial dam-break waves. *J. Appl. Mech. Tech. Phys.* **2008**, *49*, 754–761. [CrossRef]
51. Bukreev, V.I.; Gusev, A.V. Gravity waves due to discontinuity decay over an open-channel bottom drop. *J. Appl. Mech. Tech. Phys.* **2004**, *44*, 506–515. [CrossRef]

Article

An Analysis of the Factors Affecting Hyporheic Exchange based on Numerical Modeling

Jie Ren [1,*], Xiuping Wang [1], Yinjun Zhou [2,*], Bo Chen [1] and Lili Men [1]

[1] State Key Laboratory of Eco−hydraulics in Northwest Arid Region of China, Xi'an University of Technology, Xi'an 710048, China; wxpyl1992@163.com (X.W.); chenbo7289@163.com (B.C.); li15029069869@163.com (L.M.)
[2] Changjiang River Scientific Research Institute, Wuhan 430010, China
* Correspondence: renjie@xaut.edu.cn (J.R.); zhouyinjun1114@126.com (Y.Z.)

Received: 1 March 2019; Accepted: 28 March 2019; Published: 31 March 2019

Abstract: The hyporheic zone is a transition zone for the exchange of matter and energy between surface water and subsurface water. The study of trends and sensitivities of bed hyporheic exchanges to the various influencing factors is of great significance. The surface−groundwater flow process was simulated using a multiphysics computational fluid dynamics (CFD) method and compared to previous flume experiments. Based on that, the single-factor effects of flow velocity (u), water depth (H), dune wave height (h), and bed substrate permeability (κ) on hyporheic exchange in the bed hyporheic zone were investigated. The sensitivity analysis of various factors (H, u, dune wavelength (L), h, bed substrate porosity (θ), κ, and the diffusion coefficient of solute molecules (D_m)) in the surface−subsurface water coupling model was done using orthogonal tests. The results indicated that u, h, and κ were positively related, whereas H was negatively related to hyporheic exchange. H and u showed large effects, whereas κ, D_m, and θ had moderate effects, and L and h showed small effects on hyporheic exchange. This study provides valuable references for the protection and recovery of river ecology.

Keywords: hyporheic exchange; surface−groundwater flow process; influencing factors; orthogonal tests; sensitivity analysis

---

1. Introduction

The hyporheic zone plays an important role as an interface between subsurface and surface water [1,2], and has important influences on exchanges of water, nutrients and heat [3–5], pollutant migration [6,7], and quality of surface and subsurface water [8–10]. In regards to water quality, hyporheic exchange controls the temperature pattern [11,12], induces diffusion of solutes on the bed [13], increases the residence time of solutes [14], accelerates circulation of nutrients [15,16], and increases opportunities for biological and geochemical processing by decelerating migrations of dissolved and suspended substances [8,17]. Hence, hyporheic exchange has a significant effect on rivers and subsurface water systems [18,19].

Hyporheic exchanges exhibit complicated temporal and spatial variations due to influences of water fluctuations [20], in-stream structures and channel flow rates [21–23], bedform morphology [24–26], sediment penetrability [22,27–29], and rainfall patterns [20]. Recently, hyporheic exchanges have been the focus of intensive research [14,30–36]. For instance, hyporheic exchanges were measured in streams using ion or dye tracers [37]. The dynamics of the coupled system for unidirectional flow in the water column and a triangular interface on dunes were investigated [38]. Based on that, sensitivity analysis was performed via multiple computational fluid dynamics (CFD) simulations and interactions between turbulent water-column flows, current topography-driven flows in underlying permeable sediments, and ambient subsurface water discharge from deep subsurface waters were investigated [39]. Sawyer et al. [40] proposed equations for bed pressure profiles and hyporheic exchange rates in the vicinity of a channel-spanning log that can be used to evaluate the impacts of the removal or introduction of large woody debris on hyporheic mixing. Schmadel et al. [41] developed a framework relating diel hydrologic fluctuations to hyporheic exchange in simple bedform morphologies and simulated subsurface water flows under time-varying boundary conditions using an aquifer bounded by a straight stream and hill slope. The lattice Boltzmann method has been used successfully to study some complex flows [42–45]. The lattice Boltzmann method was used by Peng et al. [43] to investigate solute transport in shallow water flows. This suggests that the present model has great potential to predict morphological change in shallow water flows.

Laboratory flume tests have also been employed to investigate hyporheic exchanges [46–49]. Tonina and Buffington [50] reported a series of recirculating flume tests to investigate hyporheic exchanges in pool-riffle channels spanning a broad range of discharge and bedform morphologies. Wu and Hunkeler [51] investigated flow processes in a conduit—sediment system using both a model resembling a siphon and a numerical model. Packman et al. [6] studied solute exchange in flat and dune-shaped beds using laboratory flume tests, where dye injections indicated that a combination of convective pore water flow and turbulent diffusion near the stream–subsurface interface is responsible for solute exchange on flat beds. Sawyer et al. [52] used laboratory flume tests and numerical simulations to quantify hyporheic fluids and heat exchanges induced by current interactions with channel-spanning logs. Fox et al. [53] used a novel laboratory flume system to investigate the effects of losing and gaining flow conditions on hyporheic exchange fluxes in a sandy streambed. Lu et al. [54] constructed a two-dimensional (2D) sand tank to study the influence of a clay lens with low permeability on the hyporheic zone under different surface flow conditions. Despite the field of hyporheic zone research being fairly mature, few studies have investigated the effects of individual factors on hyporheic exchange.

In this study, sensitivity analysis was applied to factors in hyporheic exchange via orthogonal tests and hyporheic exchange was studied by flume tests [55]. The flume used had a length, width, and cross-sectional length of 2 m, 0.3 m, and 1.5 m, respectively. Seven identical triangular ripples with a wave height of 0.02 m, wavelength of 0.2 m, and distance from trough to crest of 15 cm were designed for the flume. A surface—subsurface flow coupling model was established and compared to previous flume experiments [55]. The effects of various factors, including water depth (H), flow velocity (u), dune wave height (h), bed substrate porosity (θ), bed substrate permeability (k), and coefficient of solute molecules (D_m) on hyporheic exchanges were investigated using orthogonal tests and sensitivity analysis. Additionally, numerical simulations of the injection of dye to the dune were developed for the surface—subsurface hyporheic exchange. The evaluation parameter selected was the time to equilibrium for solute concentrations at observation points.

2. Methodology

2.1. Governing Equations for Fluid Flow

The turbulent flow in the test section of the flume was simulated following the multiphysics computational fluid dynamic (CFD) approach of Cardenas and Wilson [38]. Turbulent flow is simulated by solving the Reynolds-averaged Navier—Stokes (RANS) equations with the $k-\omega$ turbulence closure model by Wilcox [56]. The pore water flow is simulated by solving a steady state groundwater flow model using COMSOL Multiphysics. These two sets of equations are coupled through the pressure distribution at the sediment—water interface [31]. Janseen et al. [55] used exactly the same combination of software. For an incompressible fluid, the steady state RANS equations are defined as:

$$\frac{\partial U_i}{\partial x_i} = 0 \tag{1}$$

$$\rho U_j \frac{\partial U_i}{\partial x_j} = -\frac{\partial P}{\partial x_i} + \frac{\partial}{\partial x_j}(2\mu S_{ij} - \rho \overline{u'_j u'_i}) \tag{2}$$

where ρ refers to the fluid density, μ refers to the dynamic viscosity (assumed standard for water), $U_{i\,or\,j}$ $(i, j = 1, 2$ where $i \neq j)$ refers to the time-averaged velocity, u_i' $(i = 1, 2)$ refer to the fluctuations in the instantaneous velocity components in $x_{i\,or\,j}$ $(i, j = 1, 2,$ where $i \neq j)$ directions, and P refers to the time-averaged pressure. The strain rate tensor (S_{ij}) is defined as:

$$S_{ij} = \frac{1}{2}\left(\frac{\partial U_i}{\partial x_j} + \frac{\partial U_j}{\partial x_i}\right) \tag{3}$$

The Reynolds stresses, τ_{ij}, are related to the mean strain rates by:

$$\tau_{ij} = -\overline{u_j' u_i'} = v_t(2S_{ij}) - \frac{2}{3}\delta_{ij}k \tag{4}$$

where v_t refers to the kinematic eddy viscosity, δ_{ij} refers to the Kronecker delta, and k refers to the turbulent kinetic energy. The $k-\omega$ turbulence closure scheme [56] was employed due to its advantages in simulating separated flows with adverse pressure gradients, including flow over dunes where a pronounced eddy is present [57,58]. The eddy viscosity in this closure scheme is:

$$v_t = \frac{k}{\omega} \tag{5}$$

where the specific dissipation, ω, is defined as the ratio of the turbulence dissipation rate ε to k:

$$\omega = \frac{\varepsilon}{\beta^* k} \tag{6}$$

where β^* refers to the closure coefficient.

The steady-state migration equations for k and ω are:

$$\rho \frac{\partial(U_j k)}{\partial x_j} = \rho \tau_{ij} \frac{\partial U_i}{\partial x_j} - \beta^* \rho \omega k + \frac{\partial}{\partial x_j}\left[(\mu + \mu_t \sigma_k)\frac{\partial k}{\partial x_j}\right] \tag{7}$$

$$\rho \frac{\partial(U_j \omega)}{\partial x_j} = \alpha \frac{\rho \omega}{k} \tau_{ij} \frac{\partial U_i}{\partial x_j} - \beta \rho \omega^2 + \frac{\partial}{\partial x_j}\left[(\mu + \mu_t \sigma_\omega)\frac{\partial \omega}{\partial x_j}\right] \tag{8}$$

The standard closure coefficients for the $k-\omega$ scheme are from Wilcox [56]: $\alpha = 5/9$, $\beta = 3/40$, $\beta^* = 9/100$, $\sigma_k = \sigma_\omega = 0.5$.

The 2D pore water flow in sand was modeled by solving the steady-state subsurface water flow equation:

$$\nabla\left(-\frac{\kappa}{\mu}\nabla \times P\right) = 0 \tag{9}$$

where κ refer to isotropic permeability. The parenthetical term is the Darcy flux or Darcy velocity (Q). The solution for simulating turbulent flow for pressure along the interface of sediment and water is described as a Dirichlet boundary at the top of the porous domain, whereas simulations of other porous flows corresponding to the flume have no flow boundaries. The flow field divided by porosity was used as the input for particle tracking, whereas dispersion is neglected as this study focuses on the convective flow paths. The pore water flows were simulated using the finite element code COMSOL Multiphysics.

The solute migration model was established based on convective diffusion equations [59]:

$$\frac{\partial C}{\partial t} = D_m \frac{\partial^2 C}{\partial x_i^2} - u_i \frac{\partial C}{\partial x_i} + \frac{\partial}{\partial x_i}\left(D_{ij}\frac{\partial C}{\partial x_i}\right) \tag{10}$$

where C is solute concentration, t is time, D_m is the molecular diffusion coefficient in porous media, and D_{ij} is the mechanical dispersion coefficient ensor. Index $i, j = 1, 2$.

The equation to obtain D_{ij} is [60]:

$$D_{ij} = \alpha_T U \delta_{ij} + (\alpha_L - \alpha_T) \frac{u_i u_j}{U} \tag{11}$$

where α_T and α_L are transverse and longitudinal dispersivities, U is the pore velocity magnitude, and δ_{ij} is the Kronecker delta function.

The Millington−Quirk model of the effective diffusion coefficient is:

$$\tau = \theta^{-\frac{1}{3}} \tag{12}$$

where θ refers to the porosity.

2.2. Calculation Model

The numerical calculation model proposed in the present study is based on the flume tests reported by Janssen et al. [55]. This model has a length of 1.5 m, including 0.05 m buffer segments at inlets and outlets and seven identical triangular dunes (crest curvature radius = 0.02 m, bed substrate height = 0.09 m, water depth H = 0.1 m). Figure 1 shows the 2D flume model in which arrows describe the flow directions.

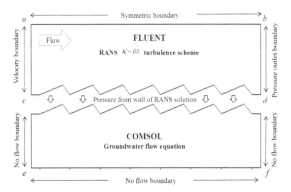

Figure 1. Two-dimensional flume model.

The present study assumed fluids to be steady and incompressible, bed substrates to be homogeneous, and isotropic and dune configurations to be free of any shifts. Fluent is a CFD software used for simulating and analyzing fluid flow and heat exchange problems in complex geometric regions. Using Fluent software to study the eddy current structure, flow separation, and water−sand interface pressure distribution results of surface water modeling under riverbed morphological disturbance has high precision. As shown in Figure 1, the overlying water was simulated by the CFD−Fluent; COMSOL Multiphysics is a software based on the finite element method. It is one of the softwares used for multiphysics coupling simulation by solving partial differential equations. It has good post-processing function. It is more accurate to use COMSOL software to describe the hyporheic exchange in a groundwater model. Therefore, we used the commercial finite element software COMSOL Multiphysics to solve the groundwater flow equations in the porous sediment. Pressure distributions at the interface of water and sand as determined by the CFD−Fluent were based on the coupling of surface and subsurface water.

The 6th dune was located in the middle of the flume, and the flow interference was lower. In addition, Janssen et al. [55] had measured the vertical velocity of the section above the water−sand interface of the 5th, 6th, and 7th sections. Based on this, the 6th dune was selected as the subject. With the height of the dune trough defined as "0", four feature observation points (A, B, C, and D) at −2 cm were selected, as shown in Figure 2. More specifically, observation points A and D were situated at the trough, observation point B was situated 8 cm to the right of observation point A (1/2 of the upstream face), and observation point C was situated at the crest. The dye injection into bed substrate were simulated as circular areas with observation points as the centers and radii of 0.5 cm to monitor the hyporheic exchange routes. In the present study, $CaCl_2$ solution was used as the dye and the time to equilibrium for concentrations at observation points was chosen as the evaluation parameter; the concentration variation was reflected by the dimensionless C/C_0 (C_0 = initial concentration).

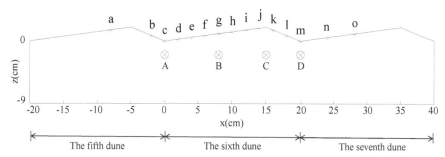

Figure 2. Layout of monitoring points.

Table 1 summarizes the parameters used in the proposed calculation model [55].

Table 1. Parameters used in the calculation model.

Fluid Density ρ (kg m^{-3})	Kinematic Eddy Viscosity v_t (m^2 s^{-1})	Porosity θ (%)	Permeability κ (m^2)	Molecular Diffusion Coefficient D_m (m^2 s^{-1})
998.8	1.10×10^{-6}	40	1.50×10^{-11}	5.00×10^{-11}

2.3. Grid Division and Boundary Conditions

The overlying water grid was generated using the Gambit. The boundary layer had a thickness of 1 cm, meshes near wall boundaries had sizes of 0.2 mm, the growth rate was 1.06, and a total of 25 layers were generated. The boundary layer used had a thickness of 1.1 cm, and 440,000 meshes were generated for the overlying water. The bed substrate was divided into 10,000 meshes with the mass of 0.95. The constant velocity inlet boundary is referred to as *ac*, *bd* refers to the constant pressure outlet boundary, and *ab* refers to the free fluid surface as a symmetric boundary. In shallow water applications the top of the *ab* domain is actually a free surface, but in our simulations the water depth was large enough to replace the free surface with the symmetry condition [55]. The variable *cd* refers to the interface of water and sand as a sliding-free wall boundary. Although recent studies have emphasized the effects of wall permeability on the mass and momentum exchange at the interface of water and sand, it is a valid approximation for interfacial flow with lower porosity and permeability porous media such as considered in this study [31]. Flux free boundaries are referred to as *ce*, *ef*, and *df*.

2.4. Model Evaluation

Root mean square error (RMSE), coefficient of determination (R^2), and the relative error (Re) were used to quantitatively evaluate the accuracy of the calculation model:

$$RMSE = \sqrt{\sum_{i=1}^{n} (O_i - S_i)^2 / n} \tag{13}$$

$$R^2 = 1 - \sum_{i=1}^{n}(O_i - S_i)^2 / \sum_{i=1}^{n}(O_i - \overline{O})^2 \tag{14}$$

$$Re = \sqrt{\sum_{i=1}^{n}(O_i - S_i)^2 / \sum_{i=1}^{n}O_i^2} \tag{15}$$

where O_i, S_i, n, and $\overline{O}$ refer to the measured value, simulated value, sample size, and average value, respectively.

The consistency between measured and simulated values was measured using RMSE to verify the model. The RMSE is a non-negative value and a low RMSE indicates good consistency between measured and simulated values [61]. R^2 is the coefficient of determination of the linear regression equation (y = x) between measured and simulated values and a large R^2 indicates good consistency between measured and simulated values [62]. Re is the relative error between measured and simulated values and a low Re indicates good consistency between measured and simulated values [63].

2.5. Orthogonal Tests

Orthogonal testing involves the use of orthogonal list-based multi-factor tests and result analysis. Representative points with a homogeneous distribution and good comparability were selected for testing based on the principle of orthogonality. The orthogonal test is used to assess result trends with reduced testing cycles. The sensitivity of each factor on the evaluation index was determined by range analysis of the results of orthogonal tests.

Suppose M and N refer to different influencing factors, t refers to the factor level, M_i refers to the ith ($i = 1, 2, \dots , t$) level of factor M, and X_{ij} refers to the ith ($I = 1, 2, \dots , n$) level of factor j ($j = M, N$). F_s ($s = 1, 2, \dots , n$) refers to the testing result at X_{ij}. The statistical parameters were calculated by:

$$K_{ij} = \frac{1}{n}\sum_{s=1}^{n}F_s - \overline{F} \tag{16}$$

where K_{ij}, n, F_s, and $\overline{F}$ refer to the average of factor j at the ith level, the testing cycles of factor j at the ith value, the value of the evaluation index in the sth test, and the average of the evaluation index, respectively.

The range (R_j) is the evaluation parameter for range analysis of factor sensitivity and can be calculated by:

$$R_j = \text{Max}\{K_{1j}, K_{2j}, \cdots\} - \text{Min}\{K_{1j}, K_{2j}, \cdots\} \tag{17}$$

A large R_j suggests that variation of a specific factor has a significant effect on the evaluation parameter, indicating high sensitivity of hyporheic exchange to this factor and vice versa.

Based on the 2D dune-shaped surface−subsurface coupling mathematical model, u, H, h, L, κ, θ, and D_m were selected as the factors affecting hyporheic exchange. Three levels (−20%, average, +20%) were designed for each factor. Table 2 summarizes the calculation parameters and average level of each factor.

Table 2. Orthogonal tests of sensitivities of affecting factors and their levels in the model.

Factor Level	u (m s^{-1})	H (m)	h (m)	L (m)	K (m^2)	Θ (%)	D_m (m^2 s^{-1})
1	0.056	0.080	0.016	0.160	1.2×10^{-11}	32	4.0×10^{-11}
2	0.070	0.100	0.020	0.200	1.5×10^{-11}	40	5.0×10^{-11}
3	0.084	0.120	0.024	0.240	1.8×10^{-11}	48	6.0×10^{-11}

3. Results and Discussion

3.1. Model Validation

The model was calibrated using data obtained at monitoring points c–m on the 6th dune under low flow conditions (2.1 L s^{-1}) and the calibrated model was further validated by using data obtained at monitoring points a, b, n, and o on the 5th and the 7th dunes, as shown in Figure 2. The average flow velocity was 0.07 m s^{-1}. Vertical simulated and measured values of velocity vectors at the monitoring points were obtained by the particle image velocimetry (PIV) technique by unifying model size and boundary conditions, as shown in Figure 3. Table 3 shows the results of the evaluation of the model simulation accuracy. A close correlation between observed and measured values was found, with RMSEs at monitoring points c–m on the 6th dune ranging between 0.0025–0.0055 m s^{-1}, significantly lower than the average flow velocity; R^2 at monitoring point k was 0.8935, whereas those at the other monitoring points exceeded 0.9; Re was 3.26–7.13%. Monitoring points a and b on the 5th dune and monitoring points n and o on the 7th dune were then investigated. The RMSEs obtained were 0.0033–0.0063 m s^{-1}, significantly lower than the average flow velocity; the R^2 values of b, n, and o exceeded 0.95, whereas that of a was 0.88; Re was in a reasonable range (4.7–8.53%). In summary, the proposed model exhibited excellent simulation accuracy and can precisely describe the dynamic migration of bed solutes.

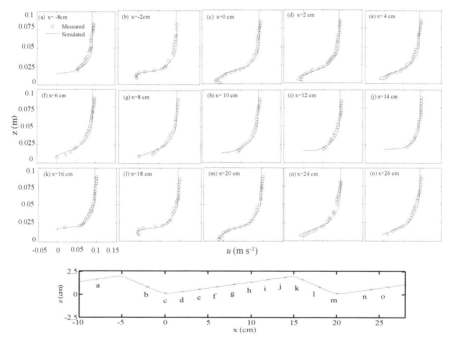

Figure 3. Comparison of vertical simulation results and experimental velocity vector results at different monitoring points (adapted from Janssen et al. [55]).

Table 3. Accuracy of vertical simulation of velocity vectors at different monitoring points.

Monitoring Points	RMSE (m s^{-1})	R^2	Re %
a	0.0046	0.8831	5.61
b	0.0033	0.9918	4.70
c	0.0044	0.9890	5.81
d	0.0047	0.9808	6.48
e	0.0054	0.9653	7.13
f	0.0055	0.9242	6.76
g	0.0050	0.9606	6.53
h	0.0053	0.9038	6.59
i	0.0044	0.8935	5.13
j	0.0034	0.9009	3.79
k	0.0042	0.9480	4.93
l	0.0041	0.9856	5.25
m	0.0025	0.9957	3.26
n	0.0063	0.9553	8.53
o	0.0039	0.976	5.05

3.2. Effects of Flow Velocity on Hyporheic Exchange

As flow velocity is a key factor affecting hyporheic exchange on a bed downstream of a dam, hyporheic exchanges at $u = 0.056$ m s^{-1}, 0.070 m s^{-1}, and 0.084 m s^{-1} were investigated. Figure 4 shows the cloud chart of concentrations. Depth and range of hyporheic exchange of solutes per unit time showed a positive relationship with u owing to interactions of surface water and subsurface water; seepage depth was 3.8 cm, 4.5 cm, and 5.5 cm and at $u = 0.056$ m s^{-1}, 0.070 m s^{-1}, and 0.084 m s^{-1}, respectively. In other words, the flow velocity was positively related to hyporheic exchange.

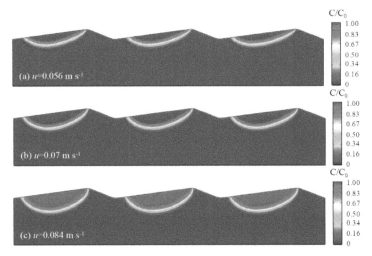

Figure 4. Distribution of concentration.

Table 4 shows the velocity effect on fluxes in hyporheic exchange, with flow velocity showing a positive relationship with water and solute fluxes on the interface of water and sand. In the vertical direction, opposite directions in water flux were observed between the upstream and downstream faces, demonstrating the hyporheic exchange process.

Table 4. Hyporheic exchange flux at different flow velocities.

	Vertical Water Flux on the Upstream Face $(m^2\ s^{-1})$	Vertical Water Flux on the Downstream Face $(m^2\ s^{-1})$	Overall Water Flux at the Interface $(m^2\ s^{-1})$	Overall Solute Flux at the Interface $(mol\ m^{-1}\ s^{-1})$
$u = 0.056$ m s^{-1}	-1.42×10^{-8}	5.20×10^{-9}	8.12×10^{-8}	2.49×10^{-7}
$u = 0.070$ m s^{-1}	-2.25×10^{-8}	8.28×10^{-9}	1.29×10^{-7}	4.26×10^{-7}
$u = 0.084$ m s^{-1}	-3.28×10^{-8}	1.21×10^{-8}	1.88×10^{-7}	6.51×10^{-7}

Figure 5 shows stress distributions at $u = 0.056$ m s^{-1}, 0.070 m s^{-1}, and 0.084 m s^{-1}. Stress distributions were consistent across different flow velocities. The maximum, minimum, and negative pressure zones were observed at midstream of the upstream face, at the crest, and the downstream face, respectively. In addition, the pressure increased continuously with u.

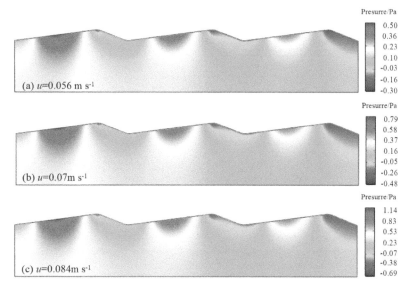

Figure 5. Distribution of the pressure field.

Figure 6 shows concentrations at the observation points over time. The curve gradient reflects the hyporheic exchange rate. As shown in Figure 6, u had negative and positive relationships with the time to equilibrium for concentrations at observation points (t) and the curve gradient, respectively, indicating an increasing hyporheic exchange rate. A minimum t of observation point B was evident at the upstream slope, indicating a maximum hyporheic exchange rate at this point. This can be attributed to the fact that maximum pressure was observed at midstream of the upstream face, which has a greater impact on observation point B.

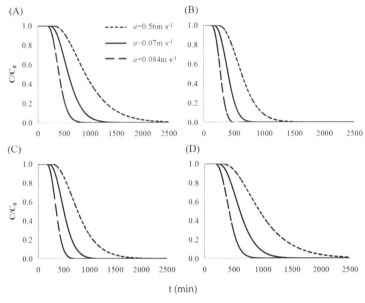

t (min)

Figure 6. Concentrations at different monitoring points over time: (**A**) Observation point A, (**B**) Observation point B, (**C**) Observation point C, (**D**) observation point D.

3.3. Effects of Water Depth on Hyporheic Exchange

Hyporheic exchanges at $H = 0.08$ m, 0.1 m, and 0.12 m were investigated as H has a significant effect on hyporheic exchange. Figure 7 shows the cloud chart of concentrations. H evidently showed negative relationships with hyporheic exchange and the migration depth of solute per unit time, resulting in a reduced affected area. More specifically, the seepage depth was 5.1 cm, 4.5 cm, and 4.3 cm at $H = 0.08$ m, 0.1 m, and 0.12 m, respectively. In other words, H was negatively related to hyporheic exchange.

Table 5 summarizes hyporheic exchange fluxes as a function of H. As observed, hyporheic exchange flux at the interface of water and sand showed a negative relationship with H. In the vertical direction, fluxes on the upstream and downstream faces were different and in opposite directions, indicating the presence of hyporheic exchange.

Table 5. Hyporheic exchange flux at different water depths.

	Vertical Water Flux on the Upstream Face $(m^2\,s^{-1})$	Vertical Water Flux on the Downstream Face $(m^2\,s^{-1})$	Overall Water Flux at the Interface $(m^2\,s^{-1})$	Overall Solute Flux at the Interface $(mol\,m^{-1}\,s^{-1})$
$H = 0.08$ m	-2.86×10^{-8}	1.70×10^{-8}	1.62×10^{-7}	5.50×10^{-7}
$H = 0.10$ m	-2.25×10^{-8}	8.28×10^{-9}	1.29×10^{-7}	4.26×10^{-7}
$H = 0.12$ m	-2.05×10^{-8}	7.45×10^{-9}	1.19×10^{-7}	3.86×10^{-7}

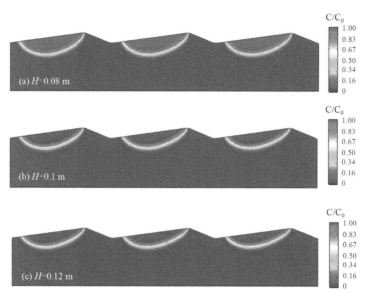

Figure 7. Distribution of concentrations.

Figure 8 shows that the stress distributions as a function of *H* were highly consistent with those as a function of *u*. In other words, the maximum and minimum pressures were observed at midstream of the upstream face and the crest, whereas the pressure values were different. In addition, the pressure decreased as *H* increased.

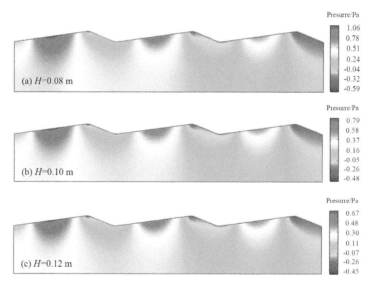

Figure 8. Distribution of the pressure field.

Figure 9 shows the solute concentrations at observation points A, B, C, and D over time as a function of *H*. A positive relationship between *H* and *t* was evident, indicating that hyporheic exchange was negatively related to *H*. The curve gradient, and thus the hyporheic exchange rate, was maximum at observation point B, followed by that at observation point C, whereas the hyporheic exchange rates of observation points A and D at the trough were similar and relatively low.

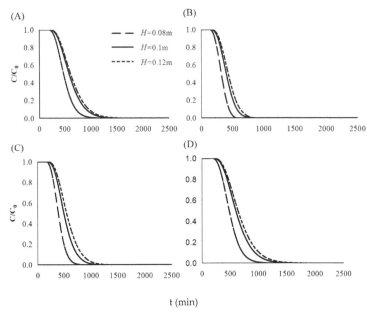

Figure 9. Concentrations at different monitoring points over time: (**A**) Observation point A, (**B**) Observation point B, (**C**) Observation point C, (**D**) observation point D.

3.4. Effects of Dune Wave Height on Hyporheic Exchange

As irregular bedform morphology has significant effects on hyporheic exchange, hyporheic exchanges at $h = 0.016$ m, 0.02 m, and 0.024 m were investigated. Figure 10 shows the cloud chart of concentrations. h showed positive relationships with surface roughness, depth, and range of solutes. In other words, hyporheic exchange increased with h; seepage depth was 4.1 cm, 4.5 cm, and 5.2 m at $h = 0.016$ m, 0.02 m, and 0.024 cm, respectively.

Table 6 shows hyporheic exchange flux as a function of h, where it is evident that bedform morphology roughness showed a positive relationship with the overall water and solute fluxes in hyporheic exchange at the sediment−water interfaces.

Table 6. Hyporheic exchange flux at different dune wave heights.

	Vertical Water Flux on the Upstream Face $(m^2\ s^{-1})$	Vertical Water Flux on the Downstream Face $(m^2\ s^{-1})$	Overall Water Flux at the Interface $(m^2\ s^{-1})$	Overall Solute Flux at the Interface $(mol\ m^{-1}\ s^{-1})$
$h = 0.16$ m	-1.62×10^{-8}	8.18×10^{-9}	1.03×10^{-7}	3.20×10^{-7}
$h = 0.20$ m	-2.25×10^{-8}	8.28×10^{-9}	1.29×10^{-7}	4.26×10^{-7}
$h = 0.24$ m	-2.93×10^{-8}	8.63×10^{-9}	1.58×10^{-7}	5.53×10^{-7}

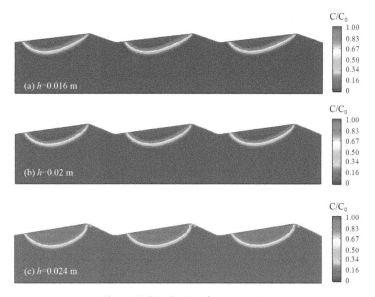

Figure 10. Distribution of concentration.

Figure 11 shows that the stress distributions as a function of *h* were highly consistent with those as a function of *u* and *H*, although the stress values were different. Indeed, the maximum pressure at *h* = 0.024 m was twice that at *h* = 0.016 m, whereas pressures in the negative pressure zone were homogeneous.

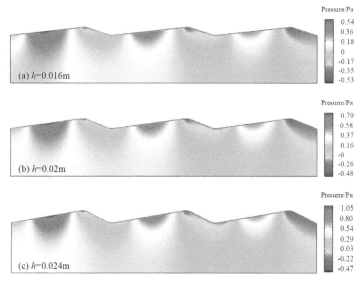

Figure 11. Distribution of pressure field.

Figure 12 shows the solute concentrations at observation points A, B, C, and D over time as a function of *h*. *h* showed a negative relationship with *t* at observation points A, B, C, and D, resulting in enhanced and accelerated hyporheic exchange. Hence, *h* was positively related to hyporheic exchange.

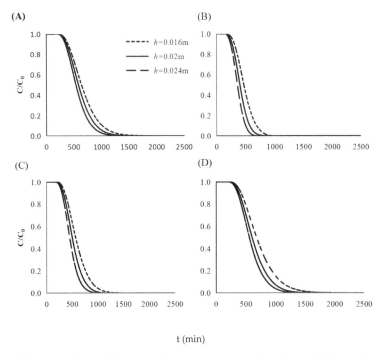

t (min)

Figure 12. Concentrations at different monitoring points as a function of time: (**A**) Observation point A, (**B**) Observation point B, (**C**) Observation point C, (**D**) observation point D.

3.5. Effects of Bed Substrate Permeability on Hyporheic Exchange

As the structure of bed sediments has a significant effect on hyporheic exchange and the permeability of bed substrate has a determining effect on the hyporheic exchange depth, hyporheic exchanges at $\kappa = 1.2 \times 10^{-11}$ m^2, 1.5×10^{-11} m^2, and 1.8×10^{-11} m^2 were investigated. Figure 13 shows the cloud chart of concentrations in these cases. Seepage depth was 4.2 cm, 4.5 cm, and 5 cm at $\kappa = 1.2 \times 10^{-11}$ m^2, 1.5×10^{-11} m^2, and 1.8×10^{-11} m^2, respectively. In other words, the depth and range of solute exchange increased with κ, indicating that hyporheic exchange was positively related to κ.

Table 7 shows hyporheic exchange flux as a function of κ. Bed substrates with high permeabilities accelerated hyporheic exchange by accelerating the penetration of fluids and solutes through the interface of water and sand.

Table 7. Hyporheic exchange flux at different bedform permeabilities.

	Vertical Water Flux on the Upstream Face (m^2 s^{-1})	Vertical Water Flux on the Downstream Face (m^2 s^{-1})	Overall Water Flux at the Interface (m^2 s^{-1})	Overall Solute Flux at the Interface (mol m^{-1} s^{-1})
$\kappa = 1.2 \times 10^{-11}$ m^2	-1.80×10^{-8}	6.63×10^{-9}	1.03×10^{-7}	3.30×10^{-7}
$\kappa = 1.5 \times 10^{-11}$ m^2	-2.25×10^{-8}	8.28×10^{-9}	1.29×10^{-7}	4.26×10^{-7}
$\kappa = 1.8 \times 10^{-11}$ m^2	-2.70×10^{-8}	9.94×10^{-9}	1.55×10^{-7}	5.23×10^{-7}

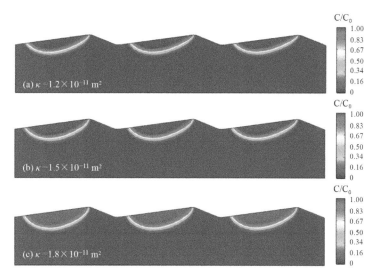

Figure 13. Distribution of concentration.

Figure 14 shows stress distributions as a function of κ, with no variations in stress as u, H, and h remained constant. Figure 15 shows the solute concentrations at observation points A, B, C, and D over time. A negative relationship was evident between κ and t, demonstrating that increasing κ accelerates hyporheic exchange. In addition, stress distributions as a function of κ were highly consistent with those as a function of u, H, and h. More specifically, the order of the curve gradient from largest to smallest was that at observation point B (maximum hyporheic exchange rate), followed by that at observation point C at the crest, and then that at observation points A and D at the trough.

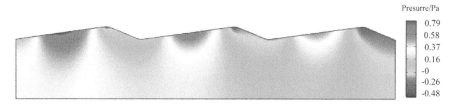

Figure 14. Distribution of pressure fields on bedforms with different permeabilities.

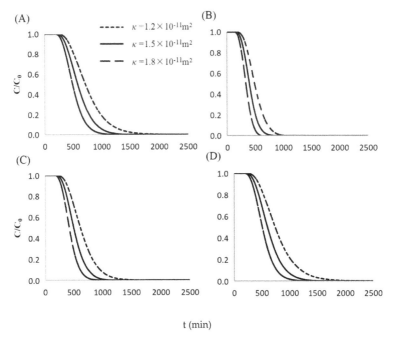

Figure 15. Concentrations at different monitoring points as a function of time: (**A**) Observation point A, (**B**) Observation point B, (**C**) Observation point C, (**D**) observation point D.

3.6. Sensitivity Analysis

To investigate the process and intensity of hyporheic exchange, 18 orthogonal tests were designed. Observation points I and III at the trough and II at the crest on the 6th dune (see Figure 16) were selected and injections of dye ($CaCl_2$ solution in this case) into the bed substrate were simulated by circles with observation points as centers and radii of 0.5 cm to monitor the hyporheic exchange routes. t at observation points I, II, and III (t_I, t_{II}, and t_{III}, respectively) were employed as the evaluation indices. As shown in Figure 16, the minimum, negative, and maximum pressure zones were observed at the crest, the downstream face, and trough and the upstream face, respectively.

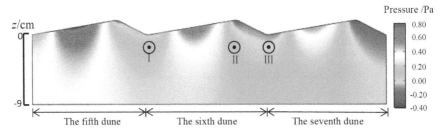

Figure 16. Cloud charts of pressure distributions at the 5th, 6th and 7th dunes of observation points I, II, and III.

Table 8 summarizes the schemes and results of the L_{18} (2×3^7) orthogonal tests of factors affecting hyporheic exchange. Tables 9–11 show the results of range analysis of factors affecting hyporheic exchange at observation points I, II, and III, respectively, based on t_I, t_{II}, and t_{III}, respectively.

Table 8. Schemes and results of the L_{18} (2×3^7) orthogonal tests of factors affecting hyporheic exchange.

Test	Empty	u (m s^{-1})	H (m)	h (m)	L (m)	κ (m^2)	θ (%)	D_m (m^2 s^{-1})	t_I (min)	t_{II} (min)	t_{III} (min)
1	1	0.056	0.08	0.016	0.16	1.20×10^{-11}	32	4.00×10^{-11}	3183	2155	3411
2	1	0.056	0.1	0.02	0.2	1.50×10^{-11}	40	5.00×10^{-11}	3800	2776	4027
3	1	0.056	0.12	0.024	0.24	1.80×10^{-11}	48	6.00×10^{-11}	4537	3485	4949
4	1	0.07	0.08	0.016	0.2	1.50×10^{-11}	48	6.00×10^{-11}	2390	1852	2621
5	1	0.07	0.1	0.02	0.24	1.80×10^{-11}	32	4.00×10^{-11}	884	714	952
6	1	0.07	0.12	0.024	0.16	1.20×10^{-11}	40	5.00×10^{-11}	3493	2590	3718
7	1	0.084	0.08	0.02	0.16	1.80×10^{-11}	40	6.00×10^{-11}	758	557	791
8	1	0.084	0.1	0.024	0.2	1.20×10^{-11}	48	4.00×10^{-11}	1661	1132	1727
9	1	0.084	0.12	0.016	0.24	1.50×10^{-11}	32	5.00×10^{-11}	885	852	986
10	2	0.056	0.08	0.024	0.24	1.50×10^{-11}	40	4.00×10^{-11}	2013	1347	2080
11	2	0.056	0.1	0.016	0.16	1.80×10^{-11}	48	5.00×10^{-11}	3760	2819	3995
12	2	0.056	0.12	0.02	0.2	1.20×10^{-11}	32	6.00×10^{-11}	4702	3585	5169
13	2	0.07	0.08	0.02	0.24	1.20×10^{-11}	48	5.00×10^{-11}	2463	1774	2463
14	2	0.07	0.1	0.024	0.16	1.50×10^{-11}	32	6.00×10^{-11}	1553	1083	1620
15	2	0.07	0.12	0.016	0.2	1.80×10^{-11}	40	4.00×10^{-11}	1624	1350	1761
16	2	0.084	0.08	0.024	0.2	1.80×10^{-11}	32	5.00×10^{-11}	449	321	449
17	2	0.084	0.1	0.016	0.24	1.20×10^{-11}	40	6.00×10^{-11}	1646	1510	1782
18	2	0.084	0.12	0.02	0.16	1.50×10^{-11}	48	4.00×10^{-11}	1680	1271	1816

Table 9. Results of range analysis of factors affecting hyporheic exchange at observation point I.

Factors	u	H	h	L	κ	θ	D_m
K_1	1361.3	−428.5	−56.5	100.0	553.5	−361.8	−463.7
K_2	−236.7	−87.2	76.7	133.2	−251.0	−82.2	170.5
K_3	−1124.7	515.7	−20.2	−233.2	−302.5	444.0	293.2
R_j	2486.0	944.2	133.2	366.3	856.0	805.8	756.8
Susceptibility			$u > H > \kappa > \theta > D_m > L > h$				

Table 10. Results of range analysis of factors affecting hyporheic exchange at observation point II.

Factors	u	H	h	L	κ	θ	D_m
K_1	962.7	−397.5	24.5	14.0	392.5	−280.2	−403.7
K_2	−171.3	−59.5	47.7	104.2	−201.7	−43.5	123.5
K_3	−791.3	457.0	−72.2	−118.2	−190.8	323.7	280.2
R_j	1754.0	854.5	119.8	222.3	594.2	603.8	683.8
Susceptibility			$u > H > D_m > \theta > \kappa > L > h$				

Table 11. Results of range analysis of factors affecting hyporheic exchange at observation point III.

Factors	u	H	h	L	κ	θ	D_m
K_1	1476.4	−492.9	−36.1	96.4	582.9	−364.2	−504.2
K_2	−272.9	−111.6	74.3	163.6	−270.4	−102.2	144.3
K_3	−1203.6	604.4	−38.2	−260.1	−312.6	466.4	359.9
R_j	2680.0	1097.3	112.5	423.7	895.5	830.7	864.2
Susceptibility			$u > H > D_m > \theta > \kappa > L > h$				

Figure 5 shows a histogram of range analysis of t for CaCl$_2$ concentration at the observation points. As shown in Figure 17, u and H (especially κ) have dominant effects on hyporheic exchange; k, D_m, and θ also have significant effects on hyporheic exchange, whereas L and h ($L > h$) have relatively low effects. These results are consistent with those of previous studies [46,64]. For instance, Wörman et al. [64] reported a dominant effect of surface water flow velocity on hyporheic exchange and a significant effect of H.

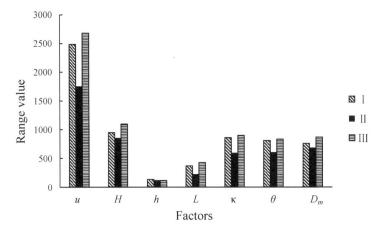

Figure 17. Ranges of factors affecting hyporheic exchange at observation points I, II, and III.

3.7. Migration Routes of Solutes

Cloud charts of stress distribution at the 6th dune and solute migration routes at different moments in different schemes were obtained based on the sensitivities of factors affecting the hyporheic exchange. Figure 18 shows the cloud chart of stress distribution at the 6th dune in Test 16 (red arrows denote the pore seepage field) and Figure 19 shows solute migration routes at 0 min, 60 min, 180 min, 360 min, 540 min, and 720 min. The maximum pressure was observed at half of the upstream face, whereas the downstream face and the trough were negative pressure zones. As the pressure difference causes exchange of surface water and subsurface water, the pore seepage field was divided into two parts. On the upstream face, reverse flows were observed at locations with heights lower than that corresponding to the maximum pressure, whereas normal flows were observed at other locations. Meanwhile, solute fields at the three observation points approached the surface water until disappearance, as shown in Figure 19.

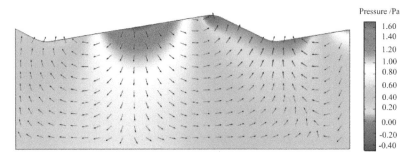

Figure 18. Cloud chart of pressures and distribution of seepage field at the 6th dune in Test 16.

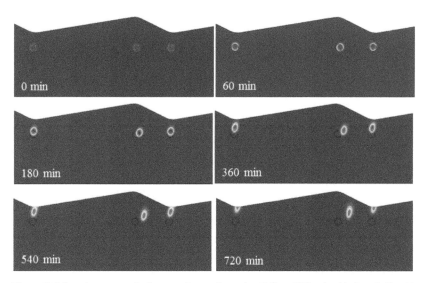

Figure 19. Migration routes of solutes at observation points I, II, and III at the 6th dune in Test 16.

4. Conclusions

A 2D dune-shaped surface−subsurface coupling mathematical model based on the RANS equation, the $k-\omega$ turbulence model, and a steady state groundwater flow model was proposed. Janseen et al. [55] used exactly the same combination of the multiphysics computational fluid dynamic (CFD) approach and COMSOL Multiphysics to solve these equations. The coupling mathematical model is verified by previous flume experiments [55], and numerical model simulations were verified against observations using the RMSE, R^2, and Re as evaluation indices. The maximum RMSE obtained was 0.0055 m s^{-1}, significantly lower than the average flow velocity; $R^2 > 0.9$ in all cases, indicating good fitting effectiveness; Re was 3.26–8.53%. In summary, simulation results were highly consistent with experimental results; therefore, it is argued that the proposed calculation model is reliable.

The single-factor effect of u, H, h, and κ on hyporheic exchange were investigated by numerical simulations. The results indicated that u, h, and κ were positively related to hyporheic exchange, whereas H was negatively related to hyporheic exchange.

Sensitivity analysis of parameters in the surface−subsurface coupling model and the solute migration model was conducted using orthogonal tests based on the simulation of dye injection into the dune with t used as the evaluation parameter. The results indicated that the sensitivities of u and H were the highest, followed by those of κ, D_m, θ, L, and h ($L > h$).

Solute migration routes at different moments in different schemes were then obtained. Owing to the exchange of surface water and subsurface water, solute fields at monitoring points A, C, and D approached the surface water until disappearance, whereas the solute field at monitoring point B approached the bed substrate.

The hyporheic exchange was shown to be affected by interactions of multiple factors and u and H exhibited the largest effects on hyporheic exchange. The present study provides knowledge vital to the protection and recovery of riverine ecology.

Author Contributions: J.R. and X.W. jointly analyzed the data and wrote the paper. Y.Z., B.C., L.M. provided critical feedback on the manuscript.

Funding: This work was supported by the National Natural Science Foundation of China (Grant No. 51679194, 51579014).

Conflicts of Interest: The authors declare no conflict of interest.

References

1. Angermann, L.; Lewandowski, J.; Fleckenstein, J.H.; Nutzmann, G. A 3D analysis algorithm to improve interpretation of heat pulse sensor results for the determination of small−scale flow directions and velocities in the hyporheic zone. *J. Hydrol.* **2012**, *475*, 1–11. [CrossRef]
2. Boano, F.; Harvey, J.W.; Marion, A.; Packman, A.I.; Revelli, R.; Ridolfi, L.; Worman, A. Hyporheic flow and transport processes: Mechanisms, models, and biogeochemical implications. *Rev. Geophys.* **2014**, *52*, 603–679. [CrossRef]
3. Bencala, K.E. Hyporheic zone hydrological processes. *Hydrol. Process.* **2000**, *14*, 2797–2798. [CrossRef]
4. Boano, F.; Poggi, D.; Revelli, R.; Ridolfi, L. Gravity-driven water exchange between streams and hyporheic zones. *Geophys. Res.* **2009**, *36*, 146–158. [CrossRef]
5. Zheng, L.; Cardenas, M.B.; Wang, L. Temperature effects on nitrogen cycling and nitrate removal−production efficiency in bed form−induced hyporheic zones. *J. Geophys. Res.* **2016**, *121*, 1086–1103. [CrossRef]
6. Packman, A.I.; Salehin, M.; Zaramella, M. Hyporheic Exchange with Gravel Beds: Basic Hydrodynamic Interactions and Bedform−Induced Advective Flows. *J. Hydraul. Eng.* **2004**, *130*, 647–656. [CrossRef]
7. Pinay, G.; Okeefe, T.C.; Edwards, R.T.; Naiman, R.J. Nitrate removal in the hyporheic zone of a salmon river in Alaska. *River Res. Appl.* **2009**, *25*, 367–375. [CrossRef]
8. Jones, J.B.; Mulholland, P.J. *Streams and Ground Waters*; Academic: San Diego, CA, USA, 2000.
9. Boulton, A.J.; Hancock, P.J. Rivers as groundwater−dependent ecosystems: A review of degrees of dependency, riverine processes and management implications. *Aust. J. Bot.* **2006**, *54*, 133–144. [CrossRef]
10. Fleckenstein, J.H.; Krause, S.; Hannah, D.M.; Boano, F. Groundwater−surface water interactions: New methods and models to improve understanding of processes and dynamics. *Adv. Water Resour.* **2010**, *33*, 1291–1295. [CrossRef]
11. Swanson, T.E.; Cardenas, M.B. Diel heat transport within the hyporheic zone of a pool−riffle−pool sequence of a losing stream and evaluation of models for fluid flux estimation using heat. *Limnol. Oceanogr.* **2010**, *55*, 1741–1754. [CrossRef]
12. Norman, F.A.; Cardenas, M.B. Heat transport in hyporheic zones due to bedforms: An experimental study. *Water Resour. Res.* **2014**, *50*, 3568–3582. [CrossRef]
13. Marion, A.; Zaramella, M. Diffusive behavior of bedform−induced hyporheic exchange in rivers. *J. Environ. Eng.* **2005**, *131*, 1260–1266. [CrossRef]
14. Lautz, L.K.; Siegel, D.I.; Bauer, R.L. Impact of debris dams on hyporheic interaction along a semi-arid stream. *Hydrol. Process.* **2006**, *20*, 183–196. [CrossRef]
15. Bardini, L.; Boano, F.; Cardenas, M.B.; Revelli, R.; Ridolfi, L. Nutrient cycling in bedform induced hyporheic zones. *Geochim. Cosmochim. Acta* **2012**, *84*, 47–61. [CrossRef]
16. Zarnetske, J.P.; Haggerty, R.; Wondzell, S.M.; Baker, M.A. Dynamics of nitrate production and removal as a function of residence time in the hyporheic zone. *J. Geophys. Res.* **2011**, *116*, 1–12. [CrossRef]
17. Battin, T.J.; Kaplan, L.A.; Findlay, S.E.; Hopkinson, C.S.; Marti, E.; Packman, A.I.; Newbold, J.D.; Sabater, F. Biophysical controls on organic carbon fluxes in fluvial networks. *Nat. Geosci.* **2008**, *1*, 95–100. [CrossRef]
18. Boano, F.; Revelli, R.; Ridolfi, L. Effect of streamflow stochasticity on bedform−driven hyporheic exchange. *Adv. Water Resour.* **2010**, *33*, 1367–1374. [CrossRef]
19. Karwan, D.L.; Saiers, J.E. Hyporheic exchange and streambed filtration of suspended particles. *Water Resour. Res.* **2012**, *48*. [CrossRef]
20. Liu, D.; Zhao, J.; Chen, X.; Li, Y.; Weiyan, S.; Feng, M. Dynamic processes of hyporheic exchange and temperature distribution in the riparian zone in response to dam−induced water fluctuations. *Geosci. J.* **2018**, *22*, 465–475. [CrossRef]
21. Hester, E.T.; Doyle, M.W. In-stream geomorphic structures as drivers of hyporheic exchange. *Water Resour. Res.* **2008**, *44*. [CrossRef]
22. Menichino, G.T.; Hester, E.T. Hydraulic and thermal effects of in−stream structure−induced hyporheic exchange across a range of hydraulic conductivities. *Water Resour. Res.* **2014**, *50*, 4643–4661. [CrossRef]
23. Rana, S.M.M.; Scott, D.T.; Hester, E.T. Effects of in−stream structures and channel flow rate variation on transient storage. *J. Hydrol.* **2017**, *548*, 157–169. [CrossRef]

24. Worman, A.; Packman, A.I.; Marklund, L.; Harvey, J.W.; Stone, S.H. Exact three-dimensional spectral solution to surface-groundwater interactions with arbitrary surface topography. *Geophys. Res. Lett.* **2006**, *33*. [CrossRef]

25. Cardenas, M.B.; Wilson, J.L. Hydrodynamics of coupled flow above and below a sediment−water interface with triangular bedforms. *Adv. Water Resour.* **2007**, *30*, 301–313. [CrossRef]

26. Caruso, A.; Ridolfi, L.; Boano, F. Impact of watershed topography on hyporheic exchange. *Adv. Water Resour.* **2016**, *94*, 400–411. [CrossRef]

27. Fox, A.; Laube, G.; Schmidt, C.; Fleckenstein, J.H.; Arnon, S. The effect of losing and gaining flow conditions on hyporheic exchange in heterogeneous streambeds. *Water Resour. Res.* **2016**, *52*, 7460–7477. [CrossRef]

28. Pryshlak, T.T.; Sawyer, A.H.; Stonedahl, S.H.; Soltanian, M.R. Multiscale hyporheic exchange through strongly heterogeneous sediments. *Water Resour. Res.* **2015**, *51*, 9127–9140. [CrossRef]

29. Su, X.; Shu, L.; Lu, C. Impact of a low-permeability lens on dune−induced hyporheic exchange. *Hydrolog. Sci. J.* **2018**, *63*, 818–835. [CrossRef]

30. Bhaskar, A.S.; Harvey, J.W.; Henry, E.J. Resolving hyporheic and groundwater components of streambed water flux using heat as a tracer. *Water Resour. Res.* **2012**, *48*. [CrossRef]

31. Chen, X.; Cardenas, M.B.; Chen, L. Three−dimensional versus two dimensional bed form−induced hyporheic exchange. *Water Resour. Res.* **2015**, *51*, 2923–2936. [CrossRef]

32. Endreny, T.; Lautz, L.; Siegel, D. Hyporheic flow path response to hydraulic jumps at river steps: Hydrostatic model simulations. *Water Resour. Res.* **2011**, *47*, 1198–1204. [CrossRef]

33. Laattoe, T.; Werner, A.D.; Post, V.E.A. Spatial periodicity in bed form−scale solute and thermal transport models of the hyporheic zone. *Water Resour. Res.* **2014**, *50*, 7886–7899. [CrossRef]

34. Langston, G.; Hayashi, M.; Roy, J.W. Quantifying groundwater−surface water interactions in a proglacial moraine using heat and solute tracers. *Water Resour. Res.* **2013**, *49*, 5411–5426. [CrossRef]

35. Stonedahl, S.H.; Harvey, J.W.; Wörman, A.; Salehin, M.; Packman, A.I. A multiscale model for integrating hyporheic exchange from ripples to meanders. *Water Resour. Res.* **2010**, *46*, 308–316. [CrossRef]

36. Trauth, N.; Schmidt, C.; Maier, U.; Vieweg, M.; Fleckenstein, J.H. Coupled 3−D stream flow and hyporheic flow model under varying stream and ambient groundwater flow conditions in a pool−riffle system. *Water Resour. Res.* **2013**, *49*, 5834–5850. [CrossRef]

37. Harvey, J.W.; Wagner, B.J. Quantifying hydrologic interactions between streams and their subsurface hyporheic zones. In *Streams Ground Waters*; Academic Press: Cambridge, MA, USA, 2000; pp. 3–44.

38. Cardenas, M.B.; Wilson, J.L. Dunes, turbulent eddies, and interfacial exchange with permeable sediments. *Water Resour. Res.* **2007**, *430*, 199–212. [CrossRef]

39. Cardenas, M.B.; Wilson, J.L. Exchange across a sediment-water interface with ambient groundwater discharge. *J. Hydrol.* **2007**, *346*, 69–80. [CrossRef]

40. Sawyer, A.H.; Cardenas, M.B.; Buttles, J. Hyporheic exchange due to channel−spanning logs. *Water Resour. Res.* **2011**, *47*, W08502. [CrossRef]

41. Schmadel, N.M.; Ward, A.S.; Lowry, C.S.; Malzone, J.M. Hyporheic exchange controlled by dynamic hydrologic boundary conditions. *Geophys. Res. Lett.* **2016**, *43*, 4408–4417. [CrossRef]

42. Peng, Y.; Zhou, J.G.; Burrows, R. Modelling the free surface flow in rectangular shallow basins by lattice Boltzmann method. *J. Hydraul. Eng.* **2011**, *137*, 1680–1685. [CrossRef]

43. Peng, Y.; Zhou, J.G.; Burrows, R. Modelling solute transport in shallow water with the lattice boltzmann method. *Comput. Fluids* **2011**, *50*, 181–188. [CrossRef]

44. Peng, Y.; Zhang, J.M.; Zhou, J.G. Lattice Boltzmann Model Using Two−Relaxation−Time for Shallow Water Equations. *J. Hydraul. Eng.* **2016**, *142*, 06015017. [CrossRef]

45. Peng, Y.; Zhang, J.M.; Meng, J.P. Second order force scheme for lattice Boltzmann model of shallow water flows. *J. Hydraul. Res.* **2017**, *55*, 592–597. [CrossRef]

46. Elliott, A.H.; Brooks, N.H. Transfer of nonsorbing solutes to a streambed with bed forms: Laboratory experiments. *Water Resour. Res.* **1997**, *33*, 137–151. [CrossRef]

47. Ju, L.; Zhang, J.J.; Chen, C.; Wu, L.S.; Zeng, L.Z. Water flux characterization through hydraulic head and temperature data assimilation: Numerical modeling and sandbox experiments. *J. Hydrol.* **2018**, *558*, 104–114. [CrossRef]

48. Stonedahl, S.H.; Roche, K.R.; Stonedahl, F.; Packman, A.I. Visualizing hyporheic flow through bedforms using dye experiments and simulation. *J. Vis. Exp.* **2015**, *105*. [CrossRef]

49. Zhou, T.; Endreny, T.A. Reshaping of the hyporheic zone beneath river restoration structures: Flume and hydrodynamic experiments. *Water Resour. Res.* **2013**, *49*, 5009–5020. [CrossRef]

50. Tonina, D.; Buffington, J.M. Hyporheic exchange in gravel bed rivers with pool−riffle morphology: Laboratory experiments and three−dimensional modeling. *Water Resour. Res.* **2007**, *43*, 208–214. [CrossRef]

51. Wu, Y.X.; Hunkeler, D. Hyporheic exchange in a karst conduit and sediment system—A laboratory analog study. *J. Hydrol.* **2013**, *501*, 125–132. [CrossRef]

52. Sawyer, A.H.; Cardenas, M.B.; Buttles, J. Hyporheic temperature dynamics and heat exchange near channel−spanning logs. *Water Resour. Res.* **2012**, *48*. [CrossRef]

53. Fox, A.; Boano, F.; Arnon, S. Impact of losing and gaining streamflow conditions on hyporheic exchange fluxes induced by dune−shaped bed forms. *Water Resour. Res.* **2014**, *50*, 1895–1907. [CrossRef]

54. Lu, C.; Zhuang, W.; Wang, S.; Zhu, X.; Li, H. Experimental study on hyporheic flow varied by the clay lens and stream flow. *Environ. Earth Sci.* **2018**, *77*, 482. [CrossRef]

55. Janssen, F.; Cardenas, M.B.; Sawyer, A.H.; Dammrich, T.; Krietsch, J.; Beer, D. A comparative experimental and multiphysics computational fluid dynamics study of coupled surface-subsurface flow in bed forms. *Water Resour. Res.* **2012**, *48*, 8514. [CrossRef]

56. Wilcox, D.C. *Turbulence Modeling for CFD*; DCW Industries, Inc.: La Canada, CA, USA, 1998; 540p.

57. Yoon, J.Y.; Patel, V.C. Numerical model of turbulent flow oversand dune. *J. Hydraul. Eng.* **1996**, *122*, 10–18. [CrossRef]

58. Cardenas, M.B.; Wilson, J.L. Comment on "Flow resistanceand bed form geometry in a wide alluvial channel" by Shu-Qing Yang, Soon-KeatTan, and Siow-Yong Lim. *Water Resour. Res.* **2006**, *42*. [CrossRef]

59. Cardenas, M.B.; Wilson, J.L.; Haggerty, R. Residence time of bedform−driven hyporheic exchange. *Adv. Water Resour.* **2008**, *31*, 1382–1386. [CrossRef]

60. de Marsily, G. *Quantitative Hydrogeology: Groundwater Hydrology for Engineers*; Academic Press: Orlando, FL, USA, 1986.

61. Mentaschi, L.; Besio, G.; Cassola, F.; Mazzino, A. Problems in RMSE−based wave model validations. *Ocean Model.* **2013**, *72*, 53–58. [CrossRef]

62. Quinino, R.C.; Reis, E.A.; Bessegato, L.F. Using the coefficient of determination R^2 to test the significance of multiple linear regression. *Teach. Stat.* **2013**, *35*, 84–88. [CrossRef]

63. Suñé, V.; Carrasco, J.A. Efficient implementations of the randomization method with control of the relative error. *Comput. Oper. Res.* **2005**, *32*, 1089–1114. [CrossRef]

64. Wörman, A.; Packman, A.I.; Johansson, H.; Jonsson, K. Effect of flow-induced exchange in hyporheic zones on longitudinal transport of solutes in streams and rivers. *Water Resour. Res.* **2002**, *38*, 2-1–2-15. [CrossRef]

Article

An Experimental Study on Mechanisms for Sediment Transformation Due to Riverbank Collapse

Anping Shu [1,*], Guosheng Duan [1], Matteo Rubinato [2], Lu Tian [1], Mengyao Wang [1] and Shu Wang [1]

[1] School of Environment, Key Laboratory of Water and Sediment Sciences of MOE, Beijing Normal University, Beijing 100875, China; duanguosheng@mail.bnu.edu.cn (G.D.); tianlu2011@mail.bnu.edu.cn (L.T.); miawang@mail.bnu.edu.cn (M.W.); wangshu861217@gmail.com (S.W.)

[2] Department of Civil and Structural Engineering, The University of Sheffield, Sir Frederick Mappin Building, Mappin Street, Sheffield S1 3JD, UK; m.rubinato@sheffield.ac.uk

* Correspondence: shuap@bnu.edu.cn; Tel.: +86-10-5880-2928

Received: 16 December 2018; Accepted: 4 March 2019; Published: 14 March 2019

Abstract: Riverbank erosion is a natural process in rivers that can become exacerbated by direct and indirect human impacts. Unfortunately, riverbank degradation can cause societal impacts such as property loss and sedimentation of in-stream structures, as well as environmental impacts such as water quality impact. The frequency, magnitude, and impact of riverbank collapse events in China and worldwide are forecasted to increase under climate change. To understand and mitigate the risk of riverbank collapse, experimental/field data in real conditions are required to provide robust calibration and validation of hydraulic and mathematical models. This paper presents an experimental set of tests conducted to characterize riverbank erosion and sediment transport for banks with slopes of 45°, 60°, 75°, and 90° and quantify the amount of sediments transported by the river, deposited within the bank toe or settled in the riverbed after having been removed due to erosion. The results showed interesting comprehension about the characterization of riverbank erosion and sediment transport along the river. These insights can be used for calibration and validation of new and existing numerical models.

Keywords: sediment; transforming mechanism; riverbank collapse

1. Introduction

Rivers and streams are dynamic systems and are continuously changing their structure due to different flow conditions. Riverbank erosion [1] commonly refers to the removal of bank material by flowing water or carried sediment. This is a phenomenon that derives from two categories of factors [2]: natural (e.g., climate parameters such as precipitation type, intensity, and variability, or soil properties like water content and shear strength and type of vegetation) and human actions [3] (e.g., construction of dams, logging, and intensive grazing).

The phenomenon of riverbank collapse incorporates a variety of bank and riverbed deformations in the affected sections (e.g., the longitudinal erosion and deposition of material on the riverbed or the transverse deformation of riverbed channels) [4]. These deformations assume different shapes and typically alter the cross-sectional morphology of the river affected. This phenomenon plays a very important role in the evolution of rivers, and despite the fact that some stable rivers have a healthy amount of erosion from which they benefit, unstable rivers and the erosion taking place on their banks are a cause for economic, environmental, and social concern: for example, (i) people are forced to migrate due to land erosion; (ii) riverbank collapse causes the loss of large areas of farmland; (iii) riverbank collapse can change the original boundary conditions of the river as well as the water and sediment conditions when large amounts of sediments enter the river channel and siltation occurs; (iv) the sediments eroded due to riverbank collapse are one of the main sources of river sediments and make the river muddy, originating environmental and ecological problems. Therefore, this phenomenon is a great concern for the society and researchers should accurately characterize and assess the several causes that typically accelerate this phenomenon that leads to major impacts, and they should identify feasible solutions for mitigation and adaptation strategies to be implemented.

For non-cohesive riverbanks, the sediment particles on the bank slope are mainly affected by the thrust on the bank, the uplift force, and the gravity effect generated by the water flowing in the river [5]. When the slope of the riverbank is larger than the underwater angle formed by the deposition of eroded sediment, the soil within the upper layer typically collapses along a sliding surface, usually in the form of "shallow collapse" [6]. Particle entrainment can be quantified using the magnitude of the shear stress and the particle size [7,8] for each soil type. On the other hand, cohesive riverbanks, more commonly found worldwide within river streams, are not only subject to these forces but also to the inter-particle cohesion magnitude. Lohnes et al. [9] completed a study to analyze the stability of cohesive riverbanks by calculating the ratio of the driving forces to provide resistance on the collapse surface, and proposed a hypothetical collapse model to be associated with cohesive riverbanks. However, the model that Lohnes et al. [9] developed does not take into consideration the effects of tensile cracks [10], pore water pressure [11], and hydrostatic pressure, and only assumes that the collapse surface can pass through the slope foot and therefore this method can only be applied to relatively steep cohesive riverbanks [12,13]. On this basis, Osman et al. [10] established an additional collapse model for cohesive riverbanks, which takes into account the effects of tensile cracks and assumes that the collapse surface passes through the bank slope foot. The study conducted by Darby et al. [14] also considered the actual topography of the riverbank, plus the pore water pressure and the hydrostatic pressure. Few years later, Rinaldi and Casagli [15] introduced the suction component of saturated and unsaturated parts of the riverbank into the identified collapse mode. Over the last decade, the bank stability and toe erosion model (BSTEM) proposed by the US National Sediment Laboratory has been widely used to numerically quantify the erosion within riverbanks. In this model, the process of riverbank collapse is divided into two parts [16]: (a) the slope foot erosion and (b) the riverbank stability analysis. However, other studies have been completed to provide alternative numerical and mathematical solutions. Huang et al. [17] also established a mathematical model for the collapse, quantifying the factors affecting the riverbank stability. Wang and Kuang [18] derived an equation for calculating the critical height of the initial and secondary riverbank collapse while Xia et al. [19] established a secondary collapse model to be used for cohesive riverbanks that analyzes the forces applied on the bank using the soil typical of the lower Yellow River.

Furthermore, the composite riverbank of an alluvial river generally exhibits a dual structure of two single layers (with different thickness and same distribution) or a tiered structure (with different thickness and distributions). For this last configuration, cantilever collapse [20] and piping collapse [21] are more likely to occur. Xia et al. [22] studied the dual structure experienced within the lower Jingjiang River and specified the three stages of dual structure cantilever collapse, quantifying the influencing factors and analyzing the collapse process. Once the riverbank collapse has taken place, the eroded material is transported within the river and this process is composed of three steps: (i) riverbank collapse and movement of eroded particles; (ii) sediment deposition and its interaction within the riverbed; and (iii) sediment transport along the river. Nagata et al. [23] combined the riverbank collapse model with a two-dimensional mathematical model to simulate the deformation process of the river channel. Darby and Delbono [24] combined a two-dimensional model with the collapse mode of cohesive riverbanks to calculate the deformation process of curved channels. Simon et al. [25] comprehensively considered the impact of sediment accumulation from eroded banks and collapsed banks on the riverbed, using the BSTEM model to estimate the riverbank collapse and material sedimentation under different flood conditions. Darby et al. [26] predicted the evolution of river channels composed of fine sand by coupling an infiltration model and the collapse model. Jia et al. [27] combined the dual structure collapse mode with the three-dimensional model to simulate the evolution of Shishou Bay in the lower Jingjiang River. Nardi et al. [28] simulated the evolution of rivers where bed is composed by medium grained sand. Xiao et al. [29] simulated the river evolution process under the influence of vegetation by combining non-cohesive riverbank collapse models with the shallow water equation. Xia et al. [30] established a mixed model of two-dimensional riverbed deformation in the orthogonal curvilinear coordinate system, and simulated the evolution process of the wandering sections of the lower Yellow River. Jia et al. [31], based on the Osman model, built a three-dimensional water sediment model with bank deformation considered and effectively simulated the horizontal oscillations of river channels caused by cohesive riverbank collapse. These studies demonstrate that the riverbed erosion is a phenomenon clearly observed but they do not consider the characterization of the motion of the sediment eroded and the load in the riverbed. Without considering these aspects, it is very difficult to replicate the riverbed erosion phenomenon with numerical models due to the paucity of existing datasets useful for calibration and validation of numerical tools. Mathematical models developed to date can be used to obtain riverbed and riverbank deformations but are applicable only under limited boundary conditions. Furthermore, it is still very challenging to numerically replicate multiple conditions associated with various slopes and flow conditions because of the paucity of high resolution localized data for this type of phenomenon. Hence, more field or experimental studies are needed to calibrate and validate the dynamic features associated with riverbank erosion under numerous conditions.

Focusing on previous studies based on physical models, Yu et al. [32] used a flume test to qualitatively analyze the interactions between material eroded due to hydraulic effect and its deposition in the river. Yu and Guo [33] studied the coupling relationship between material eroded and its transportation and deposition in the riverbed using a curved channel flume. Wu and Yu [34] revealed new insights on this natural phenomenon for cohesive riverbank collapse, non-cohesive riverbank collapse, and riverbed deposition via experimental tests. Deng et al. [35] simulated the collapse process of the upper Jingjiang Riverbank by combining the longitudinal deposition of the riverbed surface with the secondary collapse mode of cohesive riverbanks. Yu et al. [36] studied the interactions between bank collapse and riverbed deposition for different near-shore riverbed compositions by using a curved channel flume to complete the experimental tests.

Physical experiments can qualitatively measure the amount of material eroded and the amount of material deposited within the close riverbed sections to fulfill the gaps previously described. This paper presents the results obtained with an experimental flume constructed at the Key Laboratory of Water and Sediment located within the School of Environment of Beijing Normal University, investigated the phenomenon of riverbank collapse utilizing a variety of slopes (45°, 60°, 75°, and 90°) for cohesive banks, and characterized the transport of riverbank eroded material within rivers under dissimilar flow conditions (45 and 60 L/s). The datasets provided led to new insights for riverbank erosion under novel specific physical and hydraulic conditions and can be used by numerical modelers to validate relationships between variables associated with this natural phenomenon.

2. Methodology

2.1. Experiments

The experiments were conducted in the multi-function flume (0.8 m wide, 0.8 m deep, and total length of 25 m), located at the School of Environment, Key Laboratory of Water and Sediment of Beijing Normal University, China. By using a set of rigorous procedures authors guaranteed the continuous accurate re-construction of the bank slopes tested (four configurations) within this facility, ensuring the regularity of the experiments.

2.1.1. Experimental Setup

Gravel material was located upstream and downstream of the bank re-created to enable constant boundary conditions. Authors constructed in the laboratory a container with the dimensions of the banks to be replicated and fit it (once filled with material) within the flume for each experiment, guaranteeing the same position each time (±2 mm) by controlling reference points identified at the edge of the investigation area. Six sections (1#, 2#, 3#, 4#, 5#, and 6# as displayed in Figure 1) were identified within the bank to monitor the relevant parameters for this study. Four water level gauges were fitted in the flume to monitor the water level, and a pore water pressure gauge was embedded to monitor changes in the pore water pressure inside the bank as shown in Figure 2b. An example of constructed bank slope fitted within the experimental flume is shown in Figure 2a. For each experiment, velocities associated to each flow rate were measured at multiple locations within sections by using a propeller (accuracy 0.01 m/s), as represented in Figure 3, for the six sections illustrated in Figure 4. Water samples containing material were taken every three minutes at three monitoring section (1#, 3#, 5#) plus at the tailgate to measure the correspondent sediment concentration. A camera was set up on the side of the flume to record the whole process of bank slope collapse.

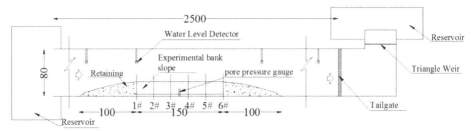

Figure 1. Schematic diagram of test flume and its equipment (not to scale) (Unit: cm).

Figure 2. Scheme of the experimental model. (**a**) Constructed bank slope fitted within the experimental flume; (**b**) Position of the pore water pressure gauge embedded inside the bank.

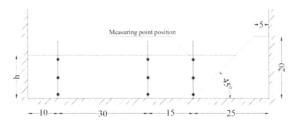

Figure 3. Example of one cross section and correspondent monitoring point for the velocity (Unit: cm).

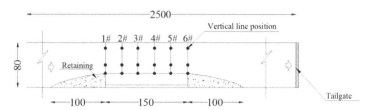

Figure 4. Side-view of the experimental setup and location of the sections for data collection (Unit: cm).

2.1.2. Experimental Testing Conditions

Based on field observations [37], four bank slope configurations (No.1, No.2, No.3, and No.4 as shown in Table 1) were considered and replicated with the experimental facility. Material utilized for this study was collected in the natural riverbank located in the upstream section of the Yellow River, from Dengkou County of the Ningmeng reach (Figure 5a) and the particle size distribution is shown in Figure 5b. All the physical properties of the tested materials are listed in Table 2.

Table 1. Configurations and bank morphology details.

Group	Slope Degree (°)	Bank Morphology			Flux (L/s)	Water Discharge Time (min)	Water Level (cm)
		Bank Top Width (cm)	Bank Height (cm)	Bank Toe Width (cm)			
No.1	45	5	20	25	45	40	11.2
					60	30	13.2
No.2	60	13.45	20	25	45	40	11.5
					60	30	12.8
No.3	75	19.64	20	25	45	40	11.8
					60	30	13.9
No.4	90	25	20	25	45	40	11.25
					60	30	13

(a)

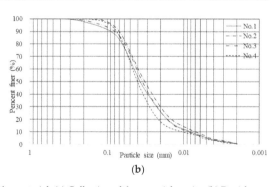

(b)

Figure 5. Real site and characteristics of the material. (**a**) Collection of the material on site; (**b**) Particle size distribution of the material collected.

Table 2. Physical properties of the material tested for each configuration.

Group	Soil Position	Moisture Content (%)	Unit Weight (g/cm^3)	Cohesion (kPa)	Friction Angle (°)
No.1	Bank toe	15.50	1.83	14.77	19.46
	Bank top	15.33	1.82	14.45	19.19
No.2	Bank toe	15.50	2.01	14.77	19.46
	Bank top	15.33	1.76	14.45	19.19
No.3	Bank toe	14.92	1.66	13.66	18.54
	Bank top	14.54	1.58	12.93	17.92
No.4	Bank toe	14.98	1.80	13.77	18.63
	Bank top	15.80	1.61	15.37	19.96

2.1.3. Experimental Procedure

Each experiment was conducted following these steps:

(1) Before each group of experiments, soil samples were taken from the monitored sections on the bank slope to obtain parameters such as unit weight, water content, and size distribution associated to initial conditions and hence parameter of comparison with data collected successively.

(2) In order to avoid bank toe erosion caused by a drastic change of water level during the initial water discharge released, the tailgate was kept close and water was entering the flume very slowly. When the water reached the designed level, the release was interrupted to let the part of the bank under slope to be soaked as in the natural scenario. When the soil was saturated, water was then discharged according to the designed flow, and the tailgate was opened as designed. This moment was considered the start of the experiment.

(3) After each experiment began, the monitoring devices in the flume were turned on to monitor the water level in real time, as well as the camera to record the collapse process.

(4) Samples were taken regularly at the monitoring section (1#, 3#, 5#) plus at the tailgate to measure the sediment concentration. A ladle was used to take samples of sandy water from the flow, then the samples collected were put into a beaker. The weight of the beaker and sandy water could be obtained by using a scale. Successively, the beaker was inserted in the oven, and once it was dry, the weight of sediment remained by the evaporation of the water could be obtained by using the weight scale. Finally, the concentration could be obtained dividing the weight of sediment by the weight of sandy water. The accuracy of scale is 0.01 g.

(5) For each configuration group, 45 L/s flowed over a pre-constructed bank for 40 min and 60 L/s flowed over a water worked bank for additional 30 min.

(6) When there was no more observation of riverbank material being removed or eroded, the erosion process was considered completed, the water was stopped and the remaining terrain (and its shape) of the riverbank was measured by regularly removing 20 cm of section each time and create profiles at each removal step.

2.2. Data Processing

2.2.1. Estimation of Riverbank Collapse Volume

After each experiment, the amount of material collapsed was obtained by combining the volume associated with the topography of the riverbank before and after the entire process and calculating the difference. As previously stated, multiple measuring sections were selected every 20 cm in the direction of water flow to characterize the final riverbank shape, and for each step the full topography profile was measured by using a system of gridlines as shown in Figure 6. Finally, the coordinates obtained were entered into AutoCAD© [38] to recreate the profiles acquired and an example (first configuration, 45° slope) is shown in Figure 7.

Figure 6. Schematic diagram of a topographical measurement conducted after the collapse section using a system of gridlines.

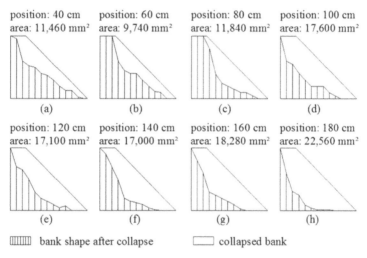

Figure 7. Example of riverbank profiles reconstructed via AutoCAD for the 45° riverbank configuration. (**a**) Position 40 cm; (**b**) Position 60 cm; (**c**) Position 80 cm; (**d**) Position 100 cm; (**e**) Position 120 cm; (**f**) Position 140 cm; (**g**) Position 160 cm; (**h**) Position 180 cm.

2.2.2. Estimation of Velocity for Incipient Sediment Motion

When the sediments enter the channel due to riverbank collapse, part of them was carried downstream by the water flowing, and the remaining part deposited at the foot of the existing bank. The deposited sediment, under the action of water flow, was then further activated and transformed into bed load and suspended load, which was transported towards the downstream section of the channel [39]. Therefore, to study the sediment transport, it was necessary to analyze the characteristics of the sediment deposited at the bank toe. The particle size of the experimental soil tested was between 0.0016 to 0.352 mm, and these values are typical of cohesive particles. It is generally believed that the newly deposited cohesive soil is compacted sediment that deposited rapidly. It can still be treated as single-particle sediment, but there is an additional bonding force between particles applied to it. Based on this view, the particle size was calculated by using the equation provided by Qian and Wan [40]. This equation enabled to compute D (sediment particle size) under the experimental conditions as showed in Table 3.

$$\frac{U_i}{\sqrt{gD}} = \sqrt{\frac{\gamma_s - \gamma}{\gamma}\left(6.25 + 41.6\frac{h}{Ha}\right) + \left(111 + 740\frac{h}{Ha}\right)\frac{Ha\delta_0}{D^2}} \tag{1}$$

where Ha is the atmospheric pressure expressed in terms of water column height, and $Ha = 10$ m; δ_0 is thickness of a water molecule, and $\delta_0 = 3.0 \times 10^{-8}$ cm; γ_s is unit weight of sediment, and $\gamma_s = 17{,}542$ N/m^3; γ is unit weight of water, and $\gamma = 9800$ N/m^3; g is the gravitational acceleration, and $g = 9.8$ m/s^2, D is sediment particle size, m; U_i is velocity for incipient sediment motion and U is the velocity, m/s (for this study $U_i = U$); h is water depth, m.

Table 3. Characterization of percentage of transported sediment under different experimental conditions.

Group	Slope Gradient (°)	Flow Rate (L/s)	Average Water Level (cm)	Average Flow Rate (m/s)	Incipience Motion Particle Size (Lower Limits) (μm)	Incipience Motion Particle Size (Upper Limits) (mm)	Incipience Motion Percentage (%)
No.1	45	45	12.40	0.62	9.25	7.29	89.85
		60	13.25	0.70	7.23	7.84	91.46
No.2	60	45	11.50	0.69	7.46	8.95	91.06
		60	12.80	0.74	6.38	10.42	92.92
No.3	75	45	11.80	0.65	8.48	8.88	90.56
		60	13.90	0.72	6.88	10.63	92.06
No.4	90	45	11.25	0.65	8.20	9.52	90.81
		60	13.00	0.77	5.98	11.10	93.60

2.2.3. Estimation of Sediment Load Ratio

In a sediment gravity flow, the coarse sand is mainly bed load movement, and the energy is primarily provided by rapid whirlpools revolving the current in the river [41]. The fine sand is mainly suspended load movement, and its energy is provided by smooth change in water direction. In spite of the mutual transformation of the two in the process of movement, under specific water flow conditions, the amount of bed load and suspended load carried by the water flow would barely change in a certain period of time. Therefore, from the point of view of quantity, it is reasonable to divide the sediment in water into coarse particles and fine particles moving with the water flow. Then, it is particularly important to introduce a boundary particle size between the bed load and the suspended load. The following procedure was applied to estimate the amount of collapsed bank sediment converted to bed load and suspended sediment after entering the river channel.

Different from the clear water flow, the movement of sediment gravity flow needs to overcome the internal cohesive resistance of water flow carrying suspended load and the friction of the bed surface during the movement of the bed load, in addition to the boundary resistance. Therefore, in the presence of the sediment gravity flow, total energy slope loss (J) is equal to the sum of the energy slope loss of water flow carrying suspended sand (J_l) and the energy slope loss of bed load motion (J_s). The automatic adjustment of alluvial riverbed is accompanied by energy loss, transformation and modification, that is, various elements (such as bed sand, section, and longitudinal slope) tend to regulate toward minimum energy consumption of the water flow. In the process of sediment transport, the adjustment tends to minimize the total energy consumption of the suspension movement and the layer movement, which can be expressed in the form of energy slope [42].

$$J_l + J_s = J_{\min} \tag{2}$$

The water flow with suspended sediment is composed of water and fine particles in suspension motion and its energy slope loss can be expressed by the Darcy–Weisbach equation. After considering the unit weight correction of the suspended sediment flow, its energy slope loss can be calculated as follows:

$$J_l = \frac{f}{8} \times \frac{U^2}{gR} \times \frac{\gamma_f}{\gamma_m} \tag{3}$$

where γ_f and γ_m are the unit weight of the suspended sediment flow and the sediment gravity flow, respectively, N/m^3; R is the hydraulic radius, m. The resistance coefficient of the suspended sediment flow f is expressed as [43,44] provided:

$$f = 0.11a\left(\frac{d_0}{4R} + \frac{68}{Re}\right)^{0.25} \tag{4}$$

$$a = 1 - 0.41g(\mu_r) + 0.08(\lg\mu_r)^2\left(\frac{4R}{d_{90}}\right)^{\frac{1}{6}} \tag{5}$$

where d_0 is the boundary particle size of the suspended sediment and bed load, m; a is resistance-reducing coefficient; Re is the Reynolds number of the sediment gravity flow.

$$Re = \frac{4RU_c\gamma_m}{g\mu} \tag{6}$$

where μ is dynamic viscosity of the sediment gravity flow (water flowing with both coarse sand and fine sand), Pa·s; U_c is the non-silting critical velocity of the suspended sediment flow (water flowing with fine sand), m/s; μ_r is the relative viscosity of the sediment gravity flow, it can be obtained based on Equation (7).

$$\mu_r = \mu/\mu_0 = 1 + 2.5S_v \tag{7}$$

where μ_0 is dynamic viscosity of water in the same temperature as the sediment gravity flow, Pa·s; S_v is the sand content of the sediment gravity flow.

According to the movement characteristics of sediment gravity flows, suspended sediment and bed load are continuously exchanged in the process of moving with the water flow, and the mechanism is very complicated. Based on previous research results, it is believed that the fine particles are carried away by the water flow in the form of suspended sediment, while the coarse particles move in the form of bed load. Then, the non-silting critical flow rate of the fine particles in sediment gravity flow can be obtained.

$$U_c = 27.8\sqrt{\frac{8}{f}}\omega_0 \cdot S_v^{\frac{2}{3}}\left(\frac{4R}{d_{90}}\right)^{\frac{1}{9}} \tag{8}$$

where ω_0 stands for the corresponding sedimentation rate of d_0. Within the range of Stokes, ω_0 can be calculated using Equation (9):

$$\omega_0 = (\gamma_m - \gamma_w)gd_0^2/25.6\gamma_w\nu \tag{9}$$

where γ_w is the unit weight of water, N/m^3; ν is the kinematic viscosity of water, m^2/s.

If the upper particle size is replaced by the boundary particle size d_0, then the weight percentage of bed load in all solids is X. The suspended sediment concentration and the bed load concentration are S_{vf} and S_{vc}, which can be expressed as follows:

$$S_{vc} = X \cdot S_v \tag{10}$$

$$S_{vf} = \frac{S_v(1 - X)}{1 - XS_v} \tag{11}$$

If the sediment concentration (S_v) in Equation (8) is expressed by the suspended sediment concentration (S_{vf}), the non-silting critical flow rate of the fine particles in sediment gravity flow can be obtained.

$$U_c = 27.8 \sqrt{\frac{8}{f}} \omega_0 \cdot S_{vf}^{\frac{2}{3}} \left(\frac{4R}{d_0} \right)^{\frac{1}{9}} \tag{12}$$

For this study, $U = U_c$, and by including Equation (12) into Equation (3), it is possible to calculate the energy slope loss of the water flowing with the suspended sediment.

$$J_l = \frac{1}{gR} \left[27.8 \omega_0 \cdot S_{vf}^{\frac{2}{3}} \left(\frac{4R}{d_0} \right)^{\frac{1}{9}} \right]^2 \frac{\gamma_f}{\gamma_m} \tag{13}$$

As can be seen, the energy slope loss J_l of the suspended sediment water flow is closely related to the boundary particle size d_0.

According to Bagnold's research, the energy slope consumed by the movement of solid-phase particles in solid–liquid flow to overcome internal resistance is related to the coarse particle concentration (S_{vc}) and the macroscopic coefficient of friction between particle interactions (tanα). After considering the buoyancy of suspended sediment water unit weight on bed load particles, their movement resistance can be expressed in the form of energy slope.

$$J_s = XS_{vc} \left(\frac{\gamma_s - \gamma_f}{\gamma_m} \right) \tan \alpha \tag{14}$$

where $\gamma_f = \gamma + S_{vf} (\gamma_s - \gamma)$ is the unit weight of suspended sediment flow in the sediment gravity flow, and tan α is the macroscopic coefficient of friction between bed load particles, α is friction angle between bed load particles, $^\circ$. For this study, the Begnold parameter tan α was selected to be 0.63.

Based on the principle of Equation (2), the critical particle size of different groups of simulation experiments is obtained by using the sediment gradation in the flume experiment. According to the experiment soil gradation of each group in Figure 5b and characterization of incipience motion particle size under different experimental conditions, the amount of collapsed bank sediment converted to bed load and suspended sediment after entering the river channel can be obtained.

The calculating process can be summarized as follows:

- Firstly, it is essential to assume a series of d_0 values (The value of d_0 is from 0.050 mm to 0.070 mm, and d_0 should be taken every 0.002 mm, for example, 0.050 mm, 0.052 mm, ..., 0.070 mm). For each value of d_0, the corresponding X can be obtained through Figure 5b;
- Consequently, because the parameters (S_v, γ_f, γ_m, γ_s, γ_w, g, R, and ν) are the basic data and can be obtained through experimental conditions, S_{vc} and S_{vf} can be obtained through Equations (10) and (11);
- Then ω_0 can be obtained by applying Equation (9);
- J_l can be obtained from Equation (13), J_s can be obtained from Equation (14);
- $J = J_l + J_s$ can be obtained from Equation (2);

A series of J values can then be obtained: $J_{0.050}$, $J_{0.052}$, $\cdots$, $J_{0.070}$. It is important to choose the smallest value each time for each range selected, which corresponds to the critical particle size.

3. Results

3.1. Dynamic Characteristics of the Bank Collapse Process

3.1.1. Collapse Process of Cohesive Riverbanks

The various collapse processes of cohesive riverbank associated with different hydraulic conditions were recorded (some examples are shown in Figure 8) and we have characterized the collapse process via exhaustive and reiterated observations of the experiments completed. After releasing the water, the bank toe was initially scoured by the water flowing in the channel.

Consequently, small fragments of material begun to fall down along the bank slope, occasionally, and a groove was gradually formed on the bank toe, as shown in Figure 8a.

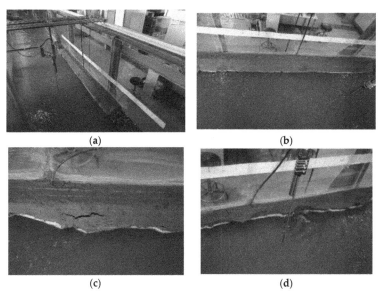

Figure 8. Example of collapse processes recorded for cohesive riverbank. (**a**) Upstream bank toe erosion; (**b**) Lateral view of the bank; (**c**) Tension crack; (**d**) Complete bank collapse.

As the water continued to flow washing the bank, the groove became deeper and deeper but due to the strong bonding effect typical of the cohesive soil, the upper part of the bank affected by the groove was maintained suspended. When the gravity of the suspended soil was then greater than its anti-sliding force and exceeded its stable slope, multiple cracks appeared on the bank, as shown in Figure 8b. The magnitude of these cracks deepened with the time and eventually the collapse occurred on the bank causing the removal of partial blocks as shown in Figure 8c. Finally, in the case in which the bank slope was steep, its collapse surface was recorded to be almost vertical, as shown in Figure 8d, which is in agreement with the natural phenomenon observed in the Dengkou river section [37].

3.1.2. Velocity Distribution

Flow velocity is a major factor influencing the phenomenon of riverbank collapse, hence it was important to obtain the localized velocity estimation for the experiments conducted under different flow conditions. Figure 9 shows an example of velocity distribution recorded for section #3 (displayed in Figure 4). These velocity measurements were taken before the bank collapse and among the collapse process. The main reason is due to the fact that the exact location and timing of the collapse were difficult to predict. The greater riverbank slope, the higher the nearshore velocity measured under the same hydraulic conditions. Typically, a larger slope of the bank corresponds to a larger obstructed area since the bottom width of the bank is imposed, and as a consequence, velocities are larger when the contraction ratio is larger.

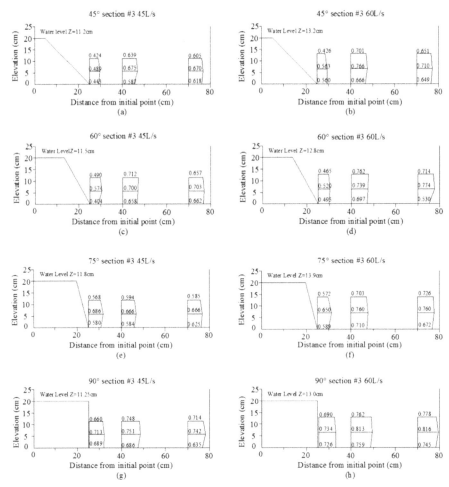

Figure 9. Velocity distribution of section #3 under different flow conditions (45 L/s and 60 L/s) and slope angles (45°, 60°, 75°, and 90°) (Unit: m/s). (**a**) 45° section #3 45 L/s; (**b**) 45° section #3 60 L/s; (**c**) 60° section #3 45 L/s; (**d**) 60° section #3 60 L/s; (**e**) 75° section #3 45 L/s; (**f**) 75° section #3 60 L/s; (**g**) 90° section #3 45 L/s; (**h**) 90° section #3 60 L/s.

3.1.3. Pore Water Pressure

Water content within the riverbank is another important factor to consider when replicating riverbank collapse within experimental facilities. An example showing the changes of pore water pressure for each slope tested (45°, 60°, 75°, and 90°) and the flow rate 45 L/s is shown in Figure 10. By an analysis completed by the authors assessing all the tests conducted including scenarios with all the slopes tested and flow rate 60 L/s, it can be detected that the steeper is the slope of the riverbank, faster is the initial collapse. For gentle slopes, the internal pore water pressure of the soil fluctuates for a longer time, particles seem to interact more frequently, and the shear resistance is stronger than the one measured for steep slopes.

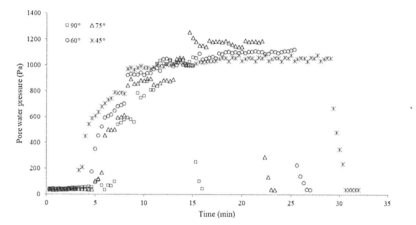

Figure 10. Pore water pressure measured under the flow rate of 45 L/s for each slope angle (45°, 60°, 75°, and 90°).

As shown in Figure 10, pore water pressure is very small at the beginning of the experiment because there is a low quantity of water in the riverbank. Gradually the pore water pressure increases as the water penetrates the bank, reaching an almost constant value prior the bank collapse and consequently drastically decreases because due to the removal of the bank material, the pore water pressure gauge used to measure this parameter is directly exposed to the water in the flume.

3.2. Quantitative Analysis of the Sediments Transformation

3.2.1. Sediment Concentration Variation

In order to analyze the variation in time of sediment concentration during the collapse process, this parameter was monitored in real time for the sections selected for this study during each experiment (see Methodology section) and an example is provided in Figure 3 where results collected at Section #3 (Figure 11a) and the tailgate (Figure 11b) are displayed.

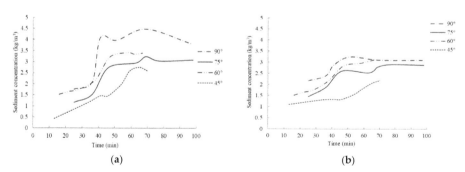

Figure 11. The diagram of sediment concentration change in typical sections. (**a**) Sediment concentration measured in Section #3; (**b**) Sediment concentration measured at the tailgate section.

By comparing the sediment concentration recorded in Section #3 for each slope configuration of 45°, 60°, 75°, and 90°, it is possible to notice a consistent trend regardless of the different initial slope (Figure 11a). At the beginning, for each configuration, the sediment concentration increases slowly, and then rises dramatically. The rapid increase in sediment concentration corresponds with the moment when the bank slope collapses. When this phenomenon happens, a large amount of debris enters the river, mixes with the water, increasing its sediment concentration that reaches its maximum level within the stream. When the water continues to flow towards the downstream section, the sediment concentration recorded gradually decreases and becomes stable (Figure 11b). For the tests recorded, the time measured to reach the collapse and hence the highest sediment concentration in the river is between 35 and 45 min. Furthermore, it has been noticed that highest is the slope and faster is this process.

3.2.2. Sediment Transformation from Bank Collapsed Materials to River Sediment

The duration to achieve the collapse phenomenon (min), the amount of material collapsed from the riverbank (kg) and the average collapse magnitude (g/s) for each configuration tested (45°, 60°, 75°, and 90°) is displayed in Table 4. The steeper the slope, the less time it takes the collapse to occur and higher the average collapse magnitude. The larger the near bank velocity, the larger is the bank collapse magnitude (Figure 9 and Table 4).

Table 4. Experimental parameters measured within the experimental tests.

Group	Slope Gradient (°)	Collapse Time (min)	Collapse Amount (kg)	Average Collapse Magnitude (g/s)
No.1	45	29	38.87	9.25
No.2	60	27	41.58	9.90
No.3	75	23	62.45	14.87
No.4	90	16	82.43	19.63

As displayed in Figure 5b the sediment particle sizes used in the experiment range 2 μm–0.32 mm. As shown in Table 3, all the incipience motion particle sizes have lower and upper limits and these can be calculated using Equation (1). For example, considering 45° and 45 L/s, when the sediment particle distribution is between 9.25 μm and 7.29 mm, material will start to move as observed under the experimental conditions. When the size of the sediment is smaller than 9.25 μm, the force between particles is greater than the flow shear force, hence the sediments tend to deposit at the bank toe. When the size of the sediment particle is bigger than 7.29 mm, it requires a greater flow shear force and this means that only the sediments smaller than 9.25 μm (10.15%) deposit at the bank toe, while the remaining part (89.85%) moves. For all the experimental groups, about 89.85–93.6% of the collapse amount typically moves, and 6.4–10.15% deposits at the bank toe.

3.2.3. Sediment Transformation from Bed Load to Suspended Load

The critical particle size distribution associated with different experiments was obtained by using the sediment gradation in the flume experiment, as shown in Table 5. It can be seen that the critical particle size of the bed load and suspended sediment obtained by this method is about 0.06–0.068 mm, which is close to the critical particle size (0.05 mm) between coarse and fine sands in the upper reaches of the Yellow River.

Table 5. The critical particle size of collapsed riverbank.

Group	Slope Degree (°)	Flow Rate (L/s)	Critical Particle Size (mm)
No.1	45	45	0.06
		60	0.062
No.2	60	45	0.064
		60	0.062
No.3	75	45	0.068
		60	0.066
No.4	90	45	0.068
		60	0.068

It can be seen from Table 6 that after the collapsed bank sediment enters the river, it is mainly converted into three parts. About 7–10% of the sediment typically deposits locally, and the rest continues to move downstream with the water flowing, part of which (70–77%) forms suspended sediment and another part (15–20%) forms bed load.

Table 6. Mass percentage of sediment fractions.

Group	Bank Slope	Flow Rate (L/s)	Deposited Sediment (%)	Suspended Sediment Percentage (%)	Bed Load Percentage (%)
No.1	45°	45	10.15	70.35	19.50
		60	8.54	71.61	19.85
No.2	60°	45	8.94	75.76	15.30
		60	7.08	77.31	15.61
No.3	75°	45	9.44	70.91	19.65
		60	7.94	72.08	19.98
No.4	90°	45	8.39	73.38	18.23
		60	9.06	74.16	16.78

4. Conclusions and Discussion

The conclusions can be summarized as follows:

(1) Experimental tests conducted have enabled the characterization of the riverbank collapse process within four different slopes simulated. This phenomenon can be subdivided into multiple steps: (i) the foot of the bank is frequently affected by the washing effect of the water flowing within the river; (ii) small fragments start to fall down along the bank and a groove is usually gradually formed on the bank toe. The groove becomes deeper and deeper as the water flow continues to wash the bank. Due to the strong bonding effect, the cohesive soil in the upper part of the groove is suspended. When the gravity of the suspended soil is greater than its anti-sliding force and exceeds its stable slope, cracks form along the bank slope. The cracks deepen until the collapse occurs and part of the bank falls in blocks inside the river. When the bank slope is steep, its collapse surface is almost flat, which is consistent with the collapse observed at Dengkou River section [37].

(2) Topographic surveys were completed to characterize the riverbank collapse under different slope scenarios and the bank collapse magnitude is directly correlated with the water flow rate; in fact, the greater is the flow rate, the faster the collapse occurs and faster is the time for the collapse to initiate. Moreover, the pore water pressure monitored during the occurrence of the riverbank collapse, is an indicator of how quickly this phenomenon arises (e.g., for gentle slopes, the internal pore water pressure of soil fluctuates for a longer duration, particles interact more frequently, and the shear resistance seems to be stronger than the shear resistance of steep slopes).

(3) The process of sediment transformation from bank collapsed materials to river sediment was analyzed through monitoring the sediment concentration at typical sections and the tailgate. It was found that the change of sediment concentration with time is basically consistent regardless of the initial bank slope. At the beginning, the sediment concentration increases slowly, and then rises dramatically. The rapid rise of the curve corresponds with the occurrence of the collapse along the riverbank. When the high amount of debris enters the river, it mixes with the water, increasing the sediment concentration along the stream and reaching the maximum value recorded. Consequently, it gradually decreases and becomes stable. The rapid increase of sediment concentration due to collapse is between 35 and 45 min after the start of the experiment, but for the tailgate section, the maximum sediment concentration is between 40 and 50 min. The maximum sediment concentration of the tailgate section arrives later than that of section #3. This is due to the fact that when collapsed soil enters into the water, the water requires time to

carry it downstream. Additionally, it was observed that the maximum sediment concentration of the tailgate is smaller than that of section #3. That is due to the typical sediments transport process within river.

(4) When the sediment enters the river, part of it is carried downstream by the current, and the remaining part deposits at the foot of the bank. The deposited sediment, under the action of water flowing, is further activated and then transformed into bed load and suspended load, which is transported downstream as well within the water. In terms of quantity, about 7–10% of the sediment typically deposits locally, and the rest continues to move downstream with the water flowing, part of which (70–77%) forms suspended sediment and another part (15–20%) forms bed load.

Despite the insights provided with this study, it is fundamental to state its limitations considering the complexity of the riverbank collapse phenomenon. To study the mechanisms of riverbank collapse, it is necessary to consider not only the longitudinal erosion and the siltation effect of the riverbed due to water and sand, but also the transverse deformation of the river channel (or the collapse of riverbank). In natural rivers, the longitudinal erosion and siltation of the riverbed, riverbank collapse, and sediment transportation are simultaneous, hence it has been very complex to simulate all these phases. Additionally, since the mechanism of bank collapse and the interaction between sediments and silt bed are not completely explored yet, it is difficult to accurately simulate the process of sediment transportation in the bank collapse section by using either a mathematical or a physical model. However, we were able to provide new insights regarding the longitudinal deformation and the transverse deformation of the river, involving the degree of deformation in a certain period of time under various flow conditions, furnishing details about the amount of riverbed erosion, the amount of siltation, and the amount of riverbank collapse. Hence, this study can be a considered as a preliminary phase to investigate bank erosion and sediment transportation mechanisms due bank collapse. Future research should also address other limitations. Firstly, bank shapes in natural rivers are irregular, so more bank angles should be considered. Secondly, the bank is typically eroded by sediment gravity flow in natural rivers, so the effect of sediment concentration on bank collapse should be added when simulating the water flowing, which is never completely clear. However, results obtained with this study should be used to calibrate existing and new numerical models and such recommendations can support current best practices for river management and environmentally sustainable restoration, preventing riverbank destabilization.

Author Contributions: All the authors jointly contributed to this research. A.S. was responsible for the proposition and design of experiment and the calculation method; G.D. and M.R. analyzed the experimental datasets and wrote the paper; L.T., M.W., and S.W. participated in the experiments.

Funding: This research was supported by the National Basic Research Program of China (Grant No. 2011CB403304) and National Natural Science Foundation of China (Grant No. 11372048).

Conflicts of Interest: The authors declare no conflict of interest.

References

1. Yao, Z.Y.; Ta, W.Q.; Jia, X.P.; Xiao, J.H. Bank erosion and accretion along the Ningxia-Inner Mongolia reaches of the Yellow River from 1958 to 2008. *Geomorphology* **2011**, *1–2*, 99–106. [CrossRef]
2. Shu, A.; Gao, J.; Duan, G.S.; Zhang, X. Cluster analysis for factor classification and riverbank collapse along the desert wide valley reach of the upper Yellow River. *J. Tsinghua Univ. (Sci. Tech.)* **2014**, *8*, 1044–1048. (In Chinese) [CrossRef]
3. Ta, W.Q.; Xiao, H.L.; Dong, Z.B. Long-term morphodynamic changes of a desert reach of the Yellow River following upstream large reservoirs' operation. *Geomorphology* **2008**, *97*, 249–259. [CrossRef]
4. Yu, M.H.; Wei, H.Y.; Liang, Y.J.; Hu, C.W. Study on the stability of non-cohesive riverbank. *Int. J. Sediment. Res.* **2010**, *4*, 391–398. [CrossRef]
5. Xia, J.Q.; Wang, G.Q.; Zhang, H.W.; Fang, H.W. A review of the research on lateral widenning mechanisms of alluvial rivers and their simulation approaches. *J. Sediment. Res.* **2001**, *6*, 71–78. (In Chinese) [CrossRef]

6. Thorne, C.R. ASCE task committee on hydraulic, bank mechanics, and modeling of river width adjustment. River width adjustment. I: Processes and mechanisms. *J. Hydraul. Eng.* **1998**, *9*, 881–902. [CrossRef]

7. Buffington, J.M.; Montgomery, D.R. A systematic analysis of eight decades of incipient motion studies, with special reference to gravel-bedded rivers. *J. Water Resour. Res.* **1997**, *8*, 1993–2029. [CrossRef]

8. Hooke, J.M. An analysis of the processes of riverbank erosion. *J. Hydrol.* **1979**, *1–2*, 39–62. [CrossRef]

9. Lohnes, R.A.; Handy, R.L. Slope Angles in Friable Loess. *J. Geol.* **1968**, *3*, 247–258. [CrossRef]

10. Osman, A.M.; Thorne, C.R. Riverbank Stability Analysis. I: Theory. *J. Hydraul. Eng.* **1988**, *2*, 134–150. [CrossRef]

11. Simon, A.; Wolfe, W.J.; Molinas, A. Mass wasting algorithms in an alluvial channel model. In Proceedings of the 5th Federal Interagency Sedimentation Conference, Las Vegas, NV, USA, 18–21 March 1991; pp. 22–29.

12. Taylor, D.W. Fundamentals of Soil Mechanics. *Soil Sci.* **1948**, *2*, 161. [CrossRef]

13. Millar, R.G.; Quick, M.C. Discussion: Development and Testing of Riverbank-Stability Analysis. *J. Hydraul. Eng.* **1997**, *11*, 1051–1053. [CrossRef]

14. Darby, S.E.; Thorne, C.R. Development and Testing of Riverbank-Stability Analysis. *J. Hydraul. Eng.* **1996**, *8*, 443–454. [CrossRef]

15. Rinaldi, M.; Casagli, N. Stability of streambanks formed in partially saturated soils and effects of negative pore water pressures: The Sieve River (Italy). *Geomorphology* **1999**, *4*, 253–277. [CrossRef]

16. Simon, A.; Curini, A.; Darby, S.E.; Langendoen, E.J. Bank and near-bank processes in an incised channel. *Geomorphology* **2000**, *3–4*, 193–217. [CrossRef]

17. Huang, B.S.; Bai, Y.C.; Wan, Y.C. Model for dilapidation mechanism of riverbank. *J. Hydraul. Eng.* **2002**, *9*, 49–54. (In Chinese) [CrossRef]

18. Wang, Y.G.; Kuang, S.F. Critical height of bank collapse. *J. Hydraul. Eng.* **2007**, *10*, 1158–1165. (In Chinese) [CrossRef]

19. Xia, J.Q.; Wu, B.S.; Wang, Y.P.; Zhao, S.G. An analysis of soil composition and mechanical properties of riverbanks in a braided reach of the Lower Yellow River. *Chin. Sci. Bull.* **2008**, *15*, 2400–2409. [CrossRef]

20. Thorne, C.R.; Tovey, N.K. Stability of composite riverbanks. *Earth Surf. Proc. Land.* **1981**, *5*, 469–484. [CrossRef]

21. Hagerty, D.J. Piping/Sapping Erosion. I: Basic Considerations. *J. Hydraul. Eng.* **1991**, *8*, 991–1008. [CrossRef]

22. Xia, J.Q.; Zong, Q.L.; Xu, Q.X.; Deng, C.Y. Soil properties and erosion mechanisms of composite riverbanks in Lower Jingjiang Reach. *Adv. Water Sci.* **2013**, *6*, 810–820. (In Chinese) [CrossRef]

23. Nagata, N.; Hosoda, T.; Muramoto, Y. Numerical Analysis of River Channel Processes with Bank Erosion. *J. Hydraul. Eng.* **2000**, *4*, 243–252. [CrossRef]

24. Darby, S.E.; Delbono, I. A model of equilibrium bed topography for meander bends with erodible banks. *Earth Surf. Proc. Landf.* **2002**, *10*, 1057–1085. [CrossRef]

25. Simon, A.; Pollenbankhead, N.; Mahacek, V.; Langendoen, E. Quantifying reductions of mass-failure frequency and sediment loadings from streambanks using toe protection and other means: Lake Tahoe, United States. *J. Am. Water Resour. Assoc.* **2009**, *1*, 170–186. [CrossRef]

26. Darby, S.E.; Trieu, H.Q.; Carling, P.A.; Sarkkula, J.; Koponen, J.; Kummu, M.; Conlan, I.; Leyland, J. A physically based model to predict hydraulic erosion of fine-grained riverbanks: The role of form roughness in limiting erosion. *J. Geophys. Res. Earth Surf.* **2010**, *115*, F04003. [CrossRef]

27. Jia, D.D.; Shao, X.J.; Wang, H.; Zhou, G. Three-dimensional modeling of bank erosion and morphological changes in the Shishou bend of the middle Yangtze River. *Adv. Water Res.* **2010**, *3*, 348–360. [CrossRef]

28. Nardi, L.; Rinaldi, M.; Solari, L. An experimental investigation on mass failures occurring in a riverbank composed of sandy gravel. *Geomorphology* **2012**, *9*, 56–59. [CrossRef]

29. Xiao, Y.; Shao, X.J.; Zhou, G.; Zhou, J.Y. 2D mathematical modeling of fluvial processes considering influences of vegetation and bank erosion. *J. Hydraul. Eng.* **2012**, *6*, 149–153. (In Chinese)

30. Xia, J.Q.; Wang, G.Q.; Wu, B.S. *Evolution of Wandering Rivers and Its Numerical Simulation*; Deng, Q., Zhang, J., Li, L., Eds.; China Water Conservancy and Hydropower Press: Beijing, China, 2005; pp. 182–188, ISBN 9787508426259.

31. Jia, D.D.; Shao, X.J.; Wang, H.; Zhou, G. 3D mathematical modeling for fluvial process considering bank erosion. *Adv. Water Sci.* **2009**, *3*, 311–317. (In Chinese) [CrossRef]

32. Yu, M.H.; Shen, K.; Wu, S.B.; Wei, H.Y. An experimental study of interaction between bank collapse and riverbed evolution. *Adv. Water Sci.* **2013**, *5*, 675–682. (In Chinese) [CrossRef]

33. Yu, M.H.; Guo, X. Experimental study on the interaction between the hydraulic transport of failed bank soil and near-bank bed evolution. *Adv. Water Sci.* **2014**, *5*, 677–683. (In Chinese) [CrossRef]

34. Wu, S.B.; Yu, M.H. Experimental study on bank failure process and interaction with riverbed deformation due to fluvial hydraulic force. *J. Hydraul. Eng.* **2014**, *6*, 649–657. (In Chinese) [CrossRef]

35. Deng, S.S.; Xia, J.Q.; Zhou, M.R.; Li, J. Conceptual model of bank retreat processes in the Upper Jingjiang Reach. *Chin. Sci. Bull.* **2016**, *33*, 3606–3615. (In Chinese) [CrossRef]

36. Yu, M.H.; Chen, X.; Wei, H.Y.; Hu, C.W.; Wu, S.B. Experiment of the influence of different near-bank riverbed compositions on bank failure. *Adv. Water Sci.* **2016**, *2*, 176–185. (In Chinese) [CrossRef]

37. Duan, G.S.; Shu, A.; Matteo, R.; Wang, S.; Zhu, F.Y. Collapsing mechanisms of the typical cohesive riverbank along the Ningxia-Inner Mongolia catchment. *Water* **2018**, *9*, 1272. [CrossRef]

38. Xue, Y. *Basic Course of AutoCAD 2010*; Tsinghua University Press: Beijing, China, 2009; pp. 113–135, ISBN 9787302207504. (In Chinese)

39. Dong, Y.H. Review on 10 groups of demarcating grain sizes of river sediment. *J. Yangtze Sci. Res. Inst.* **2018**, *2*, 1–7. (In Chinese) [CrossRef]

40. Qian, N.; Wan, Z.H. *Sediment Motion Mechanics*; Yang, J., Ed.; Science Press: Beijing, China, 1983; pp. 265–268, ISBN 9787030112606.

41. Qian, N.; Zhang, R.J.; Zhou, Z.D. *Riverbed Evolution*; Zhu, S., Ed.; Science Press: Beijing, China, 1987; pp. 258–269, ISBN 13031-3498.

42. Shu, A.; Wang, L.; Yang, K.; Fei, X.J. Investigation on movement characteristics for non-homogeneous and solid-liquid two-phase debris flow. *Chin. Sci. Bull.* **2010**, *31*, 3006–3012. (In Chinese) [CrossRef]

43. Shu, A.; Zhang, Z.D.; Wang, L.; Fei, X.J. Method for determining the critical grain size of viscous debris flow based on energy dissipation principle. *J. Hydraul. Eng.* **2008**, *3*, 257–263. (In Chinese) [CrossRef]

44. Fei, X.J. A model for calculating viscosity of sediment carrying flow in the middle and lower Yellow River. *J. Sediment. Res.* **1991**, *2*, 1–13. (In Chinese) [CrossRef]

MDPI

St. Alban-Anlage 66

4052 Basel

Switzerland

Tel. +41 61 683 77 34

Fax +41 61 302 89 18

www.mdpi.com

Water Editorial Office

E-mail: water@mdpi.com

www.mdpi.com/journal/water